CW00645837

British Railways

LOCOMOTIVES &

COACHING STOCK

1998

The Complete Guide to all
Locomotives & Coaching Stock
Vehicles which run on
Britain's Mainline Railways

Peter Fox

ISBN 1 902336 00 3

© 1998. Platform 5 Publishing Ltd., 3 Wyvern House, Sark Road, Sheffield,
S2 4HG, England.

CONTENTS

SECTION 1 - LOCOMOTIVES

SECTION 2 - LOCO-HAULED PASSENGER STOCK

SECTION 3 - DIESEL MULTIPLE UNITS

SECTION 4 - ELECTRIC MULTIPLE UNITS

CONTENTS

ACQUISITION OF INFORMATION

This book has been published with great difficulty. Privatisation of the railways and the splitting up of BR into different companies has been used as an excuse to deny the railway press access to official rolling stock library information, breaking a tradition of freely-supplied information which has existed for around half a century. We hope that readers will find the information accurate, but cannot be responsible for any inaccuracies.

We would like to thank the companies and individuals which have been co-operative in supplying information and would ask other companies which find this book useful to help us in future to make the book as accurate as possible.

This book is updated to 1st January 1998.

Cover Photographs:

Front: Great North Eastern Railway liveried Class 91 No. 91009 'The Samaritans' passes north through Burn with an express from King's Cross on 4th June 1997.   **Ian A. Lyall**

Back: West Yorkshire PTE liveried Class 321 No. 321 901 passes Ardsley on 21st October 1996 with a Leeds–Doncaster service.   **Les Nixon**

ORGANISATION OF BRITAIN'S RAILWAY SYSTEM

INFRASTRUCTURE

Britains national railway infrastructure, i.e. the track, signalling, stations and overhead line equipment is now owned by a private company called 'Railtrack'. Many stations and maintenance depots are leased to train operating companies.

DOMESTIC PASSENGER TRAIN OPERATIONS

Passenger trains are operated by train operating companies (TOCs). These TOCs operate on fixed term franchises. A list of these is appended below:

TOC	Operator	New Name
Anglia Railways	GB Trains	
Inter City East Coast	Sea Containers Ltd.	Great North Eastern Railway
Inter City West Coast	Virgin Group	Virgin Trains
Cross Country Trains	Virgin Group	Virgin Trains
Great Western Trains	Great Western Holdings	
North West Regional Railways	Great Western Holdings	North Western Trains
Midland Main Line	National Express	
Gatwick Express	National Express	
North London Railways	National Express	Silverlink
Central Trains	National Express	
Scotrail	National Express	
Merseyrail Electrics	MTL Holdings	
Regional Railways North East	MTL Holdings	Northern Spirit
LTS Rail	Prism Rail	
South Wales & West Railway	Prism Rail	Wales & West Passenger Trains
Cardiff Railway Co.	Prism Rail	
West Anglia Great Northern	Prism Rail	
South West Trains	Stagecoach	
Island Line	Stagecoach	
Network South Central	Connex	Connex South Central
South East Trains	Connex	Connex South Eastern
Great Eastern	FirstBus	
Thameslink	GOVIA	
Chiltern Railways	M40 Trains	
Thames Trains	Victory Rail	

NOTES ON TRAIN OPERATING COMPANY OWNERS

Connex

This is a French company owned by Société Générale des Entreprises Automobiles, a subsidiary of Compagnie Générale des Eaux.

FirstBus

This is a large bus company which was originally formed by the amalgamation of Badgerline and GRT bus group.

GB Trains

This is a company set up for rail privatisation.

GOVIA

A joint venture between the Go-Ahead bus company and VIA, a French public transport operating company.

Great Western Holdings

This is a jointly owned by the former Great Western Trains management. 3i plc and FirstBus. _FirstBus now majority shareholder_

National Express

This is an operator which runs express coach services, mainly by sub-contracting them to various bus companies. It also owns East Midlands Airport.

M40 Trains

This is owned by the former management of Chiltern Railways.

MTL Holdings

This is the former municipal bus operator Merseyside PTE which operates buses in Merseyside and London.

Prism

This is a company whose shares are owned by individuals and financial institutions. Its chairman is joint managing director of EYMS, a bus company.

Sea Containers

This is a Bermuda-based shipping company which also owns the Venice-Simplon-Orient Express.

Stagecoach

Tha largest private bus operator in the UK.

Victory Railway Holdings

This is a joint venture between the Go Ahead group and Thames Trains managment.

Virgin Group

This is the well-known company headed by Richard Branson which has interests in travel, leisure and retailing.

CHANNEL TUNNEL PASSENGER TRAIN OPERATIONS

Eurostar trains are operated by Eurostar (UK) Ltd. jointly with French Railways (SNCF) and Belgian Railways (NMBS/SNCB). Eurostar (UK) will also operate the Night Service trains jointly with SNCF, Netherlands Railways (NS) and German Railways (DB). Eurostar (UK) is now owned by a private company, London & Continental.

FREIGHT TRAIN OPERATIONS

The three trainload freight companies Loadhaul, Mainline and Transrail which were set up on the goverment's orders in readiness for privatisation and Railfreight Distribution have been sold to the North & South Railway Company whose main shareholder is Wisconsin Central Transportation Corporation of the USA. Rail Express Systems, which operates mail and charter trains has also been sold to this company. The five concerns have been combined and now known as the English. Welsh and Scottish Railway Ltd. (EWS).

The container train operation known as Freightliner has been sold to a managament buyout known as Freightliner (1995) Ltd.

Certain other companies e.g. Direct Rail Services and National Power operate freight trains with their own locomotives.

OWNERSHIP OF LOCOMOTIVES AND ROLLING STOCK

The locomotives of EWS and those of Eurostar are owned by those companies. Most locomotives, hauled coaching stock and multiple unit vehicles used by the passenger train operating companies are owned by three leasing companies which were originally set up by British Railways as subsidiaries and then privatised. These are:

- Forward Trust Rail (formerly Eversholt Leasing) owned by Hong Kong Shanghai Bank.
- Angel Train Contracts (owned by the Royal Bank of Scotland).
- Porterbrook Leasing Company Ltd (owned by Stagecoach Holdings).

Other vehicles are owned or operated on behalf of private owners by various private companies such as Fragonset Railways, West Coast Railway CompanyLtd., Riviera Trains Ltd. and the Venice-Simplon-Orient Express Ltd.

Further details of these companies will be found in the section on abbreviations and codes. Thus for each vehicle it is generally necessary to specify both the owner and the company which currently operates the vehicle.

A number of 'service' type vehicles are owned by Railtrack (e.g. Sandite vehicles) and others are owned by former BR Headquarters organisations which have now been privatised e.g. Serco Railtest or by railway vehicle manufacturing and repair companies. Royal Train vehicles are owned by Railtrack.

LOCOMOTIVES - INTRODUCTION

The following notes are applicable to locomotives:

DETAILS & DIMENSIONS

Principal details and dimensions are given for each class in metric units. Imperial equivalents are also given for power. Maximum speeds are still quoted in miles per hour since imperial units are still used in day to day railway operations in Britain. Since the present maximum permissible speed of certain classes of locomotives is different from the design speed, these are now shown separately in class details. In some cases certain low speed limits are arbitrary and may occasionally be raised when necessary if a locomotive has to be pressed into passenger service.

LOCOMOTIVE DETAIL DIFFERENCES

Detail differences which affect the areas and types of train which locos work are shown. Where detail differences occur within a class or part class of locomotives., these are shown against the individual locomotive number. Except where shown, diesel locomotives have no train heating equipment and have train air brakes only. Electric or electro-diesel locomotives are assumed to have train heating unless shown otherwise. Standard abbreviations used are:

c	Fitted with Scharfenberg couplers for Eurostar working.
e	Fitted with electric heating apparatus (ETH).
j	Fitted with RCH jumper cables for operating with PCVs (propelling control vehicles)
r	Fitted with radio electronic token block equipment.
s	Slow speed control fitted (and operable).
t	Fitted with automatic vehicle identification transponders.
v	Train vacuum brakes only.
x	Dual train brakes (air & vacuum).
y	ETH equipped but equipment isolated.
+	Extended range locos with Additional fuel tank capacity compared with others in class.

After the locomotive number notes are shown regarding braking, heating etc., the livery code, the owner code, the operation code where applicable, the depot code and name if any. Locomotives which have been renumbered in recent years show the last number in parentheses after the current number. For previous numbers of other locos, please refer to the Platform 5 Book 'Diesel & Electric Loco Register'.

NAMES

All official names are shown as they appear on the locomotive i.e. all upper case or upper & lower case lettering. Where only a few locomotives in a class are named, these are shown in a separate table at the end of the class or sub-class.

DEPOT ALLOCATIONS

The depot at which a locomotive is allocated is the one at which it receives its main examinations. This depot may be a long way away from where it normally performs its duties. Pool codes were allocated under the TOPS system and are still used, but have now become superfluous as, now that English Welsh & Scottish Railway have changed to a common user system for locos, locos of the same type at the same depot have the same pool code. A list of pool codes will be found on page 92. (S) denotes stored serviceable and (U) stored unserviceable. For stored locos the last known storage location is shown. Thus the layout is as follows:

No.	Old No.	Notes	Livery	Owner	Operation	Depot	Name
20301	(20047)	**DR**	D	*DR*	SD	FURNESS RAILWAY 150

GENERAL INFORMATION ON BRITISH RAILWAYS' LOCOMOTIVES

CLASSIFICATION & NUMBERING

Initially BR diesel locomotives were allocated numbers in the 1xxxx series, with electrics allotted numbers in the 2xxxx series. Around 1957 diesel locomotives were allocated new numbers with between one and four digits with 'D' prefixes. Diesel electric shunters in the 13xxx series had the '1' replaced by a 'D', but diesel mechanical shunters were completely renumbered. Electric locomotives retained their previous numbers but with an 'E' prefix.

When all standard gauge steam locomotives had been withdrawn, the prefix letter was removed. In 1972, the present TOPS numbering system was introduced whereby the loco number consisted of a two-digit class number followed by a serial number. In some cases the last two digits of the former number were generally retained (Classes 20, 37, 50), but in other classes this is not the case.

Diesel locomotives are classified as 'types' depending on their engine horsepower as follows:

Type	Engine hp.	Old Number Range	Current Classes
1	800–1000	D 8000–D 8999	20.
2	1001–1499	D 5000–D 6499/D 7500–D 7999	31.
3	1500–1999	D 6500–D 7499	33, 37.
4	2000–2999	D 1–D 1999	46, 47, 50.
5	3000+	D 9000–D 9499	55, 56, 58, 59, 60.
Shunter	300–799	D 3000–D 4999	08, 09.

Class 14 (650 hp diesel hydraulics) were numbered in the D95xx series.

Electric and electro-diesel locomotives are classified according to their supply system. Locomotives operating on a d.c. system are allocated classes 71-

80, whilst a.c. or dual voltage locomotives start at Class 81.

WHEEL ARRANGEMENT

For main line diesel and electric locomotives the system whereby the number of driven axles on a bogie or frame is denoted by a letter (A=1, B=2, C=3 etc.) and the number of undriven axles is noted by a number is used. The letter 'o' after a letter indicates that each axle is individually powered and a + sign indicates that the bogies are intercoupled. For shunters the Whyte notation is used. In this notation, generally used in Britain for steam locomotives, the number of leading wheels are given, followed by the number of driving wheels and then the trailing wheels.

HAULING CAPABILITY OF DIESEL LOCOS

The hauling capability of a diesel locomotive depends basically upon three factors:

1. Its adhesive weight. The greater the weight on its driving wheels, the greater the adhesion and thus more tractive power can be applied before wheel slip occurs.

2. The characteristics of its transmission. In order to start a train the locomotive has to exert a pull at standstill. A direct drive diesel engine cannot do this, hence the need for transmission. This may be mechanical, hydraulic or electric. The current British standard for locomotives is electric transmission. Here the diesel engine drives a generator or alternator and the current produced is fed to the traction motors. The force produced by each driven wheel depends on the current in its traction motor. In other words the larger the current, the harder it pulls.

As the locomotive speed increases, the current in the traction motors falls hence the *Maximum Tractive Effort* is the maximum force at its wheels that the locomotive can exert at a standstill. The electrical equipment cannot take such high currents for long without overheating. Hence the *Continuous Tractive Effort* is quoted which represents the current which the equipment can take continuously.

3. The power of its engine. Not all of this power reaches the rail as electrical machines are approximately 90% efficient. As the electrical energy passes through two such machines (the generator/alternator and the traction motors), the *Power At Rail* is about 81% (90% of 90%) of the engine power, less a further amount used for auxiliary equipment such as radiator fans, traction motor cooling fans, air compressors, battery charging, cab heating, ETH, etc. The power of the locomotive is proportional to the tractive effort times the speed. Hence when on full power there is a speed corresponding to the continuous tractive effort.

HAULING CAPABILITY OF ELECTRIC LOCOS

Unlike a diesel locomotive, an electric locomotive does not develop its power on board and its performance is determined only by two factors, namely its weight and the characteristics of its electrical equipment. Whereas a diesel locomotive tends to be a constant power machine, the power of an electric loco-motive varies considerably. Up to a certain speed it can produce virtually a constant tractive effort. Hence power rises with speed according to the formula given in section 3 above, until a maximum speed is reached at which tractive effort falls, such that the power also falls. Hence the power at the speed corresponding to the maximum tractive effort is lower than the maximum.

BRAKE FORCE

The brake force is a measure of the braking power of a locomotive. This is shown on the locomotive data panels so that railway staff can ensure that sufficient brake power is available on freight trains.

TRAIN HEATING AND POWER EQUIPMENT

The standard system in use in Britain for heating loco hauled trains is by means of electricity and is now known as ETS (Electric train supply). Locomotives which were equipped to provide steam heating have had this equipment removed or rendered inoperable (isolated). Electric heat is provided from the locomotive by means of a separate alternator on the loco, except in the case of Class 33 which have a d.c. generator. The ETH Index is a measure of the electrical power available for train heating. All electrically heated coaches have an ETH index and the total of these in a train must not exceed the ETH power of a locomotive.

ROUTE AVAILABILITY

This is a measure of a railway vehicle's axle load. The higher the axle load of a vehicle, the higher the RA number on a scale 1 to 10. Each route on BR has an RA number and in theory no vehicle with a higher RA number may travel on that route without special clearance. Exceptions are made, however.

MULTIPLE AND PUSH-PULL WORKING

Multiple working between diesel locomotives in Britain has usually been provided by means of an electro-pneumatic system, with special jumper cables connecting the locos. A coloured symbol is painted on the end of the locomotive to denote which system is in use. Class 47s nos. 47701-17 used a time-division multiplex (t.d.m.) system which utilised the existing RCH (an abbreviation for the former railway clearing house, a pre-nationalisation standards organisation) jumper cables for push-pull working. These had in the past only been used for train lighting control, and more recently for public address (pa) and driver-guard communication. A new standard t.d.m. system is now fitted to all a.c. electric locomotives and other vehicles, enabling them to work in both push-pull and multiple working modes.

COMMUNICATION

Virtually all main line locomotives are now fitted with cab to shore radio communication. Where locomotives are fitted with train communication this is stated in the class headings.

1.1. DIESEL LOCOMOTIVES

CLASS 08 BR SHUNTER 0–6–0

Built: 1953–62 by BR at Crewe, Darlington, Derby, Doncaster or Horwich Works.
Engine: English Electric 6KT of 298 kW (400 hp) at 680 rpm.
Main Generator: English Electric 801.
Traction Motors: Two English Electric 506.
Max. Tractive Effort: 156 kN (35000 lbf).
Cont. Tractive Effort: 49 kN (11100 lbf) at 8.8 m.p.h.
Power At Rail: 194 kW (260 hp). **Length over Buffers:** 8.92 m.
Brake Force: 19 t. **Wheel Diameter:** 1372 mm.
Design Speed: 20 m.p.h. **Weight:** 50 t.
Max. Speed: 15 or 20* m.p.h. **RA:** 5.

Non-standard liveries:
08077 & 08785 are RFS grey with blue and yellow bodyside stripes.
08296, 08602, 08846 & 08943 are grey and carry numbers 001, D 3769, D 4144 and 002 respectively.
08414 is **D** with RfD brandings and also carries its former number D 3529.
08460 is light grey with a dark grey roof, black cab doors and window surrounds and 'TLF South East' branding.
08500 is red lined out in black & white.
08519 is BR black.
08527 is light grey with a black roof, blue bodyside stripe and 'Ilford Level 5' branding.
08593 is Great Eastern blue lined out in red and also carries its former number D 3760.
08601 is London Midland & Scottish Railway black.
08629 is Royal purple.
08642 is London & South Western Railway black and also carries its former number D 3809.
08649 is grey with blue, red and white bodyside stripes and also carries its former number D 3816.
08689 is **D** with Railfreight general markings.
08715 is in experimental dayglo orange livery.
08721 is blue with a red & yellow stripe ('Red Star' livery).
08730 is BR black.
08743 & 09903 are Trafalgar blue.
08805 is LMS maroon and also carries its former number 3973.
08867 is BR black.
08879 is turquoise with full yellow ends, black cab doors, black numbers on a yellow background and RfD brandings.
08883 is Caledonian blue.
08907 is London & North Western Railway black.
08938 is grey and red.

08616 carries its former number D 3783.
08830 is on long-term lease to the East Somerset Railway and carries its former number D 3998.

n Waterproofed for working at Oxley Carriage Depot.
z Fitted with buckeye adaptor at nose end for HST depot shunting.

Formerly numbered in series 3000–4192.

Class 08/0. Standard Design.

					Depot	Location/Sub-depot
08077	a	**0**	FL	*FL*	CD	Millbrook FLT
08296	x	**0**	AD	*AD*	ZC	
08331	x	**GN**	RF		ZB (U)	
08388	a	**F**	E		IM (U)	
08389	a		E	*EW*	AN	Wembley
08393	a	**D**	E	*EW*	AN	Wembley
08397	a	**F**	E	*EW*	CD	
08401	a	**D**	E	*EW*	IM	
08402	a	**D**	E	*EW*	CD	
08405	a	**D**	E	*EW*	IM	
08410	a	**D**	GW	*GW*	PM	
08411	a		E	*EW*	ML	
08413	a	**D**	E		TI (U)	
08414	a*	**0**	E	*EW*	OC	
08417	a		SO		DY (U)	
08418	a	**F**	E	*EW*	DR	
08428	a		E		BS (U)	
08441	a		E	*EW*	TO	
08442	a	**F**	E	*EW*	KY	
08445	a		E		IM (U)	
08448	a		E		BS (U)	
08449	a		E		TO (U)	
08451	x		VT	*VW*	WN	
08454	x		VT	*VW*	WN	
08460	a	**0**	E	*EW*	CD	
08466	a	**F0**	E		IM (U)	
08472	a		GN	*GN*	EC	
08480	a	**G**	E	*EW*	EH	
08481	x		E	*EW*	CF	MG
08482	a	**D**	E	*EW*	AN	Wembley
08483	a	**D**	GW	*GW*	PM	
08484	a	**D**	RC	*RC*	ZN	
08485	a		E	*EW*	CD	
08489	a	**F**	E		WA (U)	
08492	a		E	*EW*	TO	
08493	a		E	*EW*	CF	MG
08495	x		E	*EW*	TO	
08499	a	**F**	E	*EW*	KY	
08500	x	**0**	E	*EW*	DR	
08506	a		E		ZB (U)	
08509	a	**F**	E	*EW*	DR	
08510	a		E	*EW*	DR	
08511	a		E	*EW*	TO	
08512	a	**F**	E	*EW*	DR	

08514	a		E	*EW*	DR	
08516	a	D	E	*EW*	TO	PB
08517	a		E		SF (U)	
08519	a	0	E		BS (U)	
08523	x	ML	E	*EW*	OC	
08525	x	F	MM	*ML*	NL	
08526	x		E	*EW*	OC	
08527	x	0	AD	*AD*	ZI	
08528	x	D	E	*EW*	TO	
08529	x		E	*EW*	TO	
08530	x	D	FL	*FL*	CD	Tilbury FLT
08531	x	D	FL	*FL*	CD	Tilbury FLT
08534	x	D	E	*EW*	ML	
08535	x	D	E	*EW*	TI	SY
08536	x		MM		DY (U)	
08538	x	D	E	*EW*	TO	PB
08540	x	D	E		ZB (U)	
08541	x	D	E	*EW*	SF	
08542	x	F	E	*EW*	BS	
08543	x	D	E	*EW*	BS	
08561	x		E	*EW*	ML	
08567	x		E	*EW*	BS	
08568	x		RC	*RC*	ZH	
08569	x		E	*EW*	AN	
08571	xz		GN	*GN*	EC	
08573	x		AD	*AD*	ZI	
08575	x	BS	FL	*FL*	CD	Millbrook FLT
08576	x		E	*EW*	CF	BZ
08577	x		E	*EW*	TE	TY
08578	x	R	E	*EW*	OC	
08580	x		E	*EW*	TO	PB
08581	x	BS	E		ZB (U)	
08582	a	D	E	*EW*	TE	
08585	x		FL	*FL*	CD	Basford Hall Yard
08586	a	F	E		AY (U)	
08587	x		E	*EW*	DR	
08588	xz	BS	MM	*ML*	NL	
08593	x	0	E	*EW*	SF	IP
08594	x		E		TO (U)	
08597	x		E	*EW*	KY	
08599	x		E		SP (U)	
08601	x	0	E	*EW*	BS	
08602	x	0	AD	*AD*	ZD	
08605	x		E	*EW*	KY	
08607	x		E		TO (U)	
08610	x		E		BS (U)	
08611	x	V	VT	*VW*	LO	
08616	x	G	CT	*CT*	TS	
08617	x		VT	*VW*	WN	
08619	x		E		LO (U)	
08622	x		E		ML (U)	

08623	x		E	EW	BS	
08624	x		FL	FL	CD	Coatbridge FLT
08625	x		E		CF (U)	
08628	x		E	EW	BS	
08629	x	0	RC	RC	ZN	
08630	x		E	EW	ML	
08632	x		E	EW	IM	
08633	x	RX	E	EW	HT	
08635	x		E	EW	SF	
08641	xz	D	GW	GW	LA	PZ
08642	x*	0	FL	FL	CD	Felixstowe North FLT
08643	xz	D	GW	GW	PM	
08644	xz	M	GW	GW	LA	
08645	xz	D	GW	GW	LA	
08646	x	F	E	EW	OC	
08648	x*	D	GW	GW	LA	
08649	x	0	WT	WT	ZG	
08651	x	D	E	EW	CF	
08653	x*	RD	E	EW	AN	
08655	x*	F	E		AN (U)	
08661	a	RD	E		AN (U)	
08662	x		E	EW	KY	
08663	a	D	GW	GW	LA	
08664	x		E	EW	EH	
08665	x		E	EW	IM	
08670	a		E	EW	ML	
08675	x	F	E	EW	ML	
08676	x		E		ZB (U)	
08682	x		AD	AD	ZF	
08683	x		E		CF (U)	
08685	x		E	EW	ML	
08689	a	0	E		OC (U)	
08690	x		MM	ML	DY	
08691	x	G	FL	FL	CD	Trafford Park FLT
08692	x		AD		ML (U)	
08693	x		E			
08694	x		E	EW	AN	Wembley
08695	x		E	EW	CD	
08696	a	D	VT	VW	WN	
08697	x		MM	ML	DY	
08698	a		E		ZB (U)	
08699	x	D	AD	AD	ZC	
08700	a		E		SF (U)	
08701	x	RX	E	EW	CD	
08702	x		E	EW	OC	
08703	a		E	EW	AN	
08706	x		E	EW	KY	
08709	x		E		ZB (U)	
08711	x	RX	E	EW	OC	
08713	a		E		ZB (U)	
08714	x	RX	E	EW	TO	PB

08715	v	**0**	E		SF (U)		
08718	x		E		AY (U)		
08720	a	**D**	E	*EW*	ML	MH	
08721	x	**0**	VT	*VW*	LO		
08723	x		E		TO (U)		
08724	x		GN		ZB (U)		
08730	x	**0**	RC	*RC*	ZH		
08731	x		E		ML (U)		
08734	x		E		CF (U)		
08735	x		E	*EW*	ML	FW	
08737	x	**RD**	E		AN (U)		
08738	x	**D**	E	*EW*	CD		
08739	x		E		AN (U)		
08740	x	**F**	E		SF (U)		
08742	x	**RX**	E	*EW*	CD		
08743	x	**0**	I	*IC*	BH		
08745	x	**RD**	FL	*FL*	CD	Ipswich Upper Yard	
08746	x	**D**	E	*EW*	ML	AY	
08750	x		E		SF (U)		
08751	x	**RD**	E		TI (U)		
08752	x	**C**	E	*EW*	TO	PB	
08754	x		SR	*SR*	IS		
08756	x	**D**	E	*EW*	CF		
08757	x	**RX**	E	*EW*	HT		
08758	x		E		SF (U)		
08762	x		SR	*SR*	IS		
08765	xn	**D**	E	*EW*	BS		
08768	x		E	*EW*	ML		
08770	a	**D**	E	*EW*	CF	MG	
08773	x		E		TO (U)		
08775	x		E	*EW*	OC		
08776	a	**D**	E	*EW*	KY		
08780	x		GW	*GW*	LE		
08782	a		E	*EW*	KY		
08783	x		E	*EW*	KY		
08784	x		E		AN (U)		
08785	x		FL		ZB (U)		
08786	a	**D**	E	*EW*	CF	BZ	
08790	x		VT	*VW*	LO		
08792	x		E	*EW*	CF	EX	
08795	x	**M**	GW	*GW*	LE		
08798	x		E	*EW*	CF	Tavistock Junction	
08799	x		E	*EW*	AN		
08801	x		E	*EW*	CF	MG	
08802	x	**RX**	E	*EW*	CD		
08804	x		E	*EW*	BK		
08805	x	**0**	CT	*CT*	TS		
08806	a	**F**	E	*EW*	TE	TY	
08807	x	**BS**	E	*EW*	CD		
08810	a		AR	*AR*	NC		
08813	a	**D**	E	*EW*	TE		

08815	x		E		SP (U)
08817	x	BS	E		SP (U)
08818	x		MR		CP (U)
08819	x	D	E	*EW*	CF
08822	x	M	GW	*GW*	LE
08823	a		AD		ZF (U)
08824	a	F	E	*EW*	IM
08825	a		E	*EW*	AN	Wembley
08826	a		E		ML (U)
08827	a		E	*EW*	ML
08828	a	E	E	*EW*	CF	MG
08830	x*	G	CA	*SS*	CO
08834	x	F	GN		ZB (U)
08836	x	I	GW	*GW*	OO
08837	x*	D	E	*EW*	AN
08842	x		E	*EW*	AN
08844	x		E	*EW*	AN	Wembley
08846	x	0	AD	*AD*	ZC
08847	x*		WT	*WT*	ZG
08853	xr		GN	*GN*	BN
08854	x*		E	*EW*	OC
08856	x		E	*EW*	AN
08865	x		E	*EW*	SF	IP
08866	x		E	*EW*	CD
08867	x	0	E	*EW*	CD
08868	x		MR	*JF*	ZB	Connington Tip
08869	x	G	AR	*AR*	NC
08872	x	D	E	*EW*	AN	Wembley
08873	x	RX	E	*EW*	CD
08877	x	D	E	*EW*	KY
08879	x	0	E	*EW*	TI
08881	x	D	E	*EW*	ML	AY
08882	x		E	*EW*	ML	AB
08883	x	0	E	*EW*	ML	PH
08884	x		E	*EW*	BS
08886	x	E	E	*EW*	TO
08887	x	0	VT	*VW*	WN
08888	x	E	E		ZB (U)
08890	x	D	E	*EW*	OC
08891	x		FL	*FL*	CD	Garston FLT
08892	x*	GN	RF	*GN*	BN
08893	x	D	E		BS (U)
08894	x		E		SP (U)
08896	x	E	E	*EW*	BK
08897	x	D	E	*EW*	CD
08899	x	MM	MM	*ML*	DY
08900	x	D	E	*EW*	CF
08901	xn		E		BS (U)
08902	x		E		AN (U)
08903	x	0	I	*IC*	BH
08904	x		E	*EW*	OC

Number							
08905	x	**RD**	E	*EW*	TI	SY
08906	x		E	*EW*	ML	
08907	x	**0**	E	*EW*	AN	
08908	xz		MM	*ML*	NL	
08909	x		E	*EW*	BS	
08910	x		E		ZB (U)	
08911	x	**D**	E	*EW*	CD	
08912	x	**BS**	E	*EW*	ML	Carlisle
08913	x	**D**	E	*EW*	AN	Wembley
08914	x		E		BS (U)	
08915	x	**F**	E	*EW*	CD	
08918	x	**D**	E		SP (U)	
08919	x	**RX**	E	*EW*	OC	
08920	x	**F**	E	*EW*	BS	
08921	x	**E**	E	*EW*	CD	LM
08922	x	**D**	E	*EW*	ML	Carlisle
08924	x	**D**	E	*EW*	OC	
08925	x		E		ZB (U)	
08926	x		E		AN (U)	
08927	x		E	*EW*	BS	
08928	x	**FR**	AR	*AR*	NC	
08931	x		E		TE (U)	
08932	x		E	*EW*	CF	
08933	x*	**E**	E	*EW*	EH	
08934	x		VT	*VW*	WN	
08938	xr	**0**	E		ML (U)	
08939	x		E	*EW*	AN	
08940	x		E		EH (U)	
08941	x		E	*EW*	CF	BZ
08942	x		E	*EW*	CF	
08943	x	**0**	AD	*AD*	ZT	
08944	x	**D**	E	*EW*	OC	
08946	x	**RD**	E	*EW*	TI	SY
08947	x		E	*EW*	OC	
08948	xc	**EP**	LC	*ES*	OC	PI
08950	x	**I**	MM	*ML*	NL	
08951	x	**D**	E	*EW*	TI	SY
08952	x		E		ML (U)	
08953	x	**D**	E	*EW*	CF	BZ
08954	x	**FT**	E	*EW*	CF	
08955	x		E	*EW*	CF	
08956	x		SO	*SO*	DY	ZA
08957	x	**E**	E	*EW*	CF	
08958	x		E		SF (U)	

Names:

08578	Libert Dickinson	08714	Cambridge
08649	G.H. Stratton	08743	ANGIE
08661	Europa	08790	M.A. SMITH
08682	Lionheart	08869	The Canary
08701	The Sorter	08879	Sheffield Children's Hospital

08711	EAGLE C.U.R.C.	08888	Postman's Pride
08896	Stephen Dent	08950	Neville Hill 1st
08919	Steep Holm		

Class 08/9. Fitted with cut-down cab and headlight for Cwmmawr branch.

				Depot	Location/Sub-depot
08993	x	**FT**	E	EW	CF
08994	a	**D**	E		ZB (U)
08995	a	**FT**	E		ZB (U)

CLASS 09   BR SHUNTER   0–6–0

Built: 1959–62 by BR at Darlington or Horwich Works.
Engine: English Electric 6KT of 298 kW (400 hp) at 680 rpm.
Main Generator: English Electric 801.
Traction Motors: English Electric 506.
Max. Tractive Effort: 111 kN (25000 lbf).
Cont. Tractive Effort: 39 kN (8800 lbf) at 11.6 m.p.h.
Power At Rail: 201 kW (269 hp).
Brake Force: 19 t.
Weight: 50 t.
Max. Speed: 27 m.p.h.
Train Brakes: Air & Vacuum.

Length over Buffers: 8.92 m.
Wheel Diameter: 1372 mm.
RA: 5.

Non-standard livery:
09017 is BR blue with a grey cab and is numbered 97806.

Class 09/0 were originally numbered 3665–71, 3719–21, 4099–4114.

Class 09/0. Built as Class 09.

				Depot	Location/Sub-depot
09001		E	EW	CF	MG
09003		E	EW	CF	
09004		SC	SC	SU	
09005	**D**	E		TE (U)	
09006	**ML**	E	EW	HG	
09007	**ML**	E	EW	HG	
09008	**D**	E	EW	CF	Tavistock Junction
09009	**E**	E	EW	HG	
09010	**D**	E	EW	SF	IP
09011	**D**	E	EW	TI	SY
09012	**D**	E	EW	CD	
09013	**D**	E	EW	CF	Tavistock Junction
09014	**D**	E	EW	KY	
09015	**D**	E	EW	CF	
09016	**D**	E	EW	OC	
09017	**0**	E	EW	CF	Sudbrook
09018	**ML**	E		ZB (U)	
09019	**ML**	E	EW	HG	
09020		E		OC (U)	
09021	**RD**	E	EW	TI	SY

09022		E		AN (U)
09023		E		ZB (U)
09024	**ML**	E	*EW*	HG
09025		SC	*SC*	SU
09026	**D**	SC	*SC*	SU

Names:

09009	Three Bridges C.E.D.	09026	William Pearson
09012	Dick Hardy		

Class 09/1. Converted from Class 08. 110 V electrical equipment.

				Depot	*Location/Sub-depot*
09101	**D**	E	*EW*	OC	
09102	**D**	E	*EW*	OC	
09103	**D**	E	*EW*	ML	AB
09104	**D**	E	*EW*	BS	
09105	**D**	E	*EW*	CF	
09106	**D**	E	*EW*	TE	
09107	**D**	E	*EW*	CF	

Class 09/2. Converted from Class 08. 90 V electrical equipment.

				Depot	*Location/Sub-depot*
09201	**D**	E	*EW*	DR	
09202	**D**	E	*EW*	ML	
09203	**D**	E	*EW*	CF	
09204	**D**	E	*EW*	TE	
09205	**D**	E	*EW*	ML	MH

CLASS 20   ENGLISH ELECTRIC TYPE 1   Bo–Bo

Built: 1957–68 by English Electric Company at Vulcan Foundry, Newton le Willows or Robert Stephenson & Hawthorn, Darlington.
20007–128/301–5/901–4 were originally built with disc indicators whilst 20131–215/905/6 were built with four character headcode panels.
Engine: English Electric 8SVT Mk. II of 746 kW (1000 hp) at 850 rpm.
Main Generator: English Electric 819/3C.
Traction Motors: English Electric 526/5D (20007–42/301/904) or 526/8D (others).
Max. Tractive Effort: 187 kN (42000 lbf).
Cont. Tractive Effort: 111 kN (25000 lbf) at 11 m.p.h.

Power At Rail: 574 kW (770 hp).	**Length over Buffers:** 14.25 m.
Brake Force: 35 t.	**Wheel Diameter:** 1092 mm.
Design Speed: 75 m.p.h.	**Weight:** 73.5 t.
Max. Speed: 60 m.p.h.	**RA:** 5.

Train Brakes: Air & Vacuum.
Multiple Working: Blue Star Coupling Code.

Non-standard liveries:
20042 is Waterman Railways black wih cream and red lining.
20075, 20128, 20131 & 20187 are Racal-BRT two-tone grey.

20088, 20102, 20105, 20108, 20133, 20145, 20159 & 20194 are RFS grey with blue and yellow bodyside stripes and carry the following numbers:

20088: 2017	20105: 2016	20133: 2005	20159: 2010
20102: 2008	20108: 2001	20145: 2019	20194: 2006

Originally numbered in series 8007–8190, 8315–8325.

Class 20/0. Standard Design.

20007	st		D	ZB (U)
20016	st		E	BS (U)
20032	s		D	ZB (U)
20042		**0**	D	ZB (U)
20057	st		E	BS (U)
20059	st	**FR**	E	MG (U)
20066			E	BS (U)
20072	st		D	ZB (U)
20075	st	**0**	D	ZB (U)
20081	st		E	BS (U)
20087	st	**BS**	E	BS (U)
20088		**0**	D	ZB (U)
20092		**CS**	E	BS (U)
20102		**0**	D	ZB (U)
20104	st	**FR**	D	ZB (U)
20105		**0**	D	ZB (U)
20108		**0**	D	BL (U)
20117	st		D	ZB (U)
20118		**FR**	E	BS (U)
20121	st		D	ZB (U)
20128	st	**0**	D	ZB (U)
20131	st	**0**	D	ZB (U)
20132	st	**FR**	E	BS (U)
20133		**0**	D	ZB (U)
20138		**FR**	E	BS (U)
20145		**0**	D	ZB (U)
20159		**0**	D	ZB (U)
20165		**FR**	E	BS (U)
20168	st		E	MG (U)
20169	st	**CS**	E	BS (U)
20187	st	**0**	D	ZB (U)
20190	st		D	ZB (U)
20194		**0**	D	ZB (U)
20209			H	ZK (U)
20215	st	**FR**	D	ZB (U)

Class 20/3. Refurbished locos for Direct Rail Services.
All have train air brakes only, twin fuel tanks and non-standard multiple working jumpers. Disc indicators or headcode panels removed.

20301	(20047)	**DR**	D	*DR*	SD	FURNESS RAILWAY 150
20302	(20084)	**DR**	D	*DR*	SD	
20303	(20127)	**DR**	D	*DR*	SD	
20304	(20120)	**DR**	D	*DR*	SD	

20305	(20095)	**DR**	D	*DR*	SD
20306	(		)		
20307	(		)		
20308	(		)		
20309	(		)		
20310	(		)		
20311	(		)		

Class 20/9. Hunslet–Barclay Ltd.
All have train air brakes only.

20901	t	**HB**	H	*HB*	ZK	NANCY
20902		**HB**	H	*HB*	ZK	LORNA
20903		**HB**	H	*HB*	ZK	ALISON
20904		**HB**	H	*HB*	ZK	JANIS
20905	t	**HB**	H	*HB*	ZK	IONA
20906		**HB**	H	*HB*	ZK	Kilmarnock 400

CLASS 31   BRUSH TYPE 2   A1A–A1A

Built: 1957–62 by Brush Traction at Loughborough.
31102/6/7/10/25/34/44/444/50/61 retain two headcode lights. Others have roof-mounted headcode boxes.
Engine: English Electric 12SVT of 1100 kW (1470 hp) at 850 rpm.
Main Generator: Brush TG160-48.
Traction Motors: Brush TM73-68.
Max. Tractive Effort: 160 kN (35900 lbf) (190 kN (42800 lbf)*).
Cont. Tractive Effort: 83 kN (18700 lbf) at 23.5 m.p.h. (99 kN (22250 lbf) at 19.7 m.p.h. *.)

Power At Rail: 872 kW (1170 hp).	**Length over Buffers:** 17.30 m.
Brake Force: 49 t.	**Driving Wheel Diameter:** 1092 mm.
Design Speed: 90 (80*) m.p.h.	**Centre Wheel Diameter:** 1003 mm.
Weight: 107–111 t.	**Train Brakes:** Air & Vacuum.
RA: 5 or 6.	**ETH Index (Class 31/4):** 66.

Max. Speed: 60 m.p.h. (90 m.p.h. Class 31/4).
Multiple Working: Blue Star Coupling Code.

Non-standard livery:
31116 is red, yellow, red and grey with 'Infrastructure' branding.

Originally numbered 5520–5699, 5800–5862 (not in order).

Class 31/1. Standard Design. RA5.

31102		**C**	E		CD (U)
31106	*	**C**	E		BS (U)
31107		**C**	E		BS (U)
31110		**C**	E	*EW*	BS
31113		**C**	E	*EW*	BS
31116		**O**	E		TO (U)
31119		**C**	E		CL (U)
31125		**C**	F		BS (U)
31126		**C**	E		SP (U)
31128		**FO**	E		BS (U)

31130		F	E		BS (U)
31132		F0	E		BS (U)
31134		C	E		SP (U)
31135		C	E		TO (U)
31142		C	E	EW	BS
31144		C	E		CL (U)
31145		C	E		SP (U)
31146	r	C	E	EW	BS Brush Veteran
31147	r	C	E		BS (U)
31149		FR	E		TO (U)
31154		C	E	EW	BS
31155		F	E		BS (U)
31158		C	E		BS (U)
31160		F	E		SP (U)
31163		C	E	EW	BS
31164		F0	E		BS (U)
31165		G	E		TO (U)
31166	r	C	E	EW	BS
31171		F0	E		BS (U)
31174		C	E		BS (U)
31178		C	E		BS (U)
31181		C	E		TO (U)
31185		C	E		BS (U)
31186		C	E		TO (U)
31187		C	E		TO (U)
31188		C	E	EW	BS
31190		C	E		CL (U)
31191		C	E		TO (U)
31199		F	E		TO (U)
31200		F	E		CD (U)
31201		F	E	EW	BS
31203		C	E	EW	BS
31205		FR	E		TO (U)
31206		C	E		BS (U)
31207		C	E	EW	BS
31219		C	E		TO (U)
31224		C	E		CL (U)
31229		C	E	EW	BS
31230	*	F0	E		TO (U)
31232		C	E		BS (U)
31233		C	E	EW	BS Severn Valley Railway
31235		C	E		CL (U)
31237		C	E		BS (U)
31238		C	E		SP (U)
31242		C	E		CL (U)
31247		FR	E		TO (U)
31248		F0	E		BS (U)
31250		C	E		TO (U)
31252		F0	E		PB (U)
31255		C	E	EW	BS
31263		C	E		BS (U)

31268	**C**	E		TO (U)
31270	**F**	E		CL (U)
31271	**F**	E		TO (U)
31273	**C**	E	*EW*	BS
31275	**F**	E		CN (U)
31276	**F**	E		TO (U)
31285	**C**	E		CL (U)
31294	**F**	E		TO (U)
31301	**FR**	E		BS (U)
31302	**F**	E		SP (U)
31304	**F**	E		SP (U)
31306	**C**	E	*EW*	BS
31308	**C**	E	*EW*	BS
31312	**F**	E		SP (U)
31317	**FO**	E		BS (U)
31319	**F**	E		CD (U)
31327	**FR**	E		CL (U)

Class 31/4. Equipped with Train Heating. RA6.
Class 31/5. Train Heating Equipment isolated. RA6.

31405	**M**	E		TO (U)
31407	**ML**	E	*EW*	BS
31408		E		SP (U)
31410	**RR**	E		CN (U)
31411	**D**	E		BS (U)
31512	**C**	E	*EW*	BS
31514	**C**	E		BS (U)
31415		E		BS (U)
31516	**C**	E		BS (U)
31417	**D**	E		BS (U)
31519	**C**	E		SP (U)
31420	**M**	E	*EW*	BS
31421	**RR**	E		CD (U)
31422	**M**	E		BS (U)
31423	**M**	E		BS (U)
31524	**C**	E		BS (U)
31526	**C**	E		BS (U)
31427		E		CL (U)
31530	**C**	E	*EW*	BS
31531	**C**	E		TO (U)
31432		E		SP (U)
31533	**C**	E		BS (U)
31434		E	*EW*	BS
31435	**C**	E		BS (U)
31537	**C**	E		BS (U)
31538		E		CL (U)
31439	**RR**	E		BS (U) North Yorkshire Moors Railway
31541	**C**	E		HG (U)
31444	**C**	E		SP (U)
31545		E		BS (U)
31546	**C**	E		BS (U)

31548	C	E		BS (U)
31549	C	E		TO (U)
31450		E	EW	BS
31551	C	E		TO (U)
31552	C	E		TO (U)
31554	C	E	EW	BS
31455	RR	E		SP (U)
31556	C	E		CL (U)
31558	C	E		TO (U)
31459		E		TO (U)
31461	D	E		TO (U)
31462	D	E		BS (U)
31563	C	E		TO (U)
31465	RR	E	EW	BS
31466	C	E	EW	BS
31467		E	EW	BS
31468	C	E		TO (U)

CLASS 33   BRCW TYPE 3   Bo–Bo

Built: 1960–62 by the Birmingham Railway Carriage & Wagon Company, Smethwick.
Engine: Sulzer 8LDA28 of 1160 kW (1550 hp) at 750 rpm.
Main Generator: Crompton Parkinson CG391B1.
Traction Motors: Crompton Parkinson C171C2.
Max. Tractive Effort: 200 kN (45000 lbf).
Cont. Tractive Effort: 116 kN (26000 lbf) at 17.5 m.p.h.
Power At Rail: 906 kW (1215 hp).   **Length over Buffers:** 15.47 m.
Brake Force: 35 t.   **Wheel Diameter:** 1092 mm.
Design Speed: 85 m.p.h.   **Weight:** 77.5 t (78.5 t Class 33/1).
Max. Speed: 60 m.p.h.   **RA:** 6.
Train Heating: Electric (y isolated). **ETH Index:** 48.
Train Brakes: Air & Vacuum (Class 33/1 also has electro-pneumatic).
Multiple Working: Blue Star Coupling Code.

Non-standard livery:
33021 is Post Office red.

33021 is owned by Alan & Tracey Lear and managed by Fragonset Railways.
33116 carries its original number D 6535.
33208 carries its original number D 6593.

Originally numbered in series 6500–97 but not in order.

Class 33/0. Standard Design.

33019		C	E	EW	EH	
33021		0	FG		TM (U)	Eastleigh
33025		C	E	EW	EH	
33026		C	E	EW	EH	
33030		C	E	EW	EH	
33046	y	C	E	EW	EH	
33051			E	EW	EH	Shakespeare Cliff

Class 33/1. Fitted with Buckeye Couplings & SR Multiple Working Equipment for use with SR EMUs, TC stock & Class 73.
Also fitted with flashing light adaptor for use on Weymouth Quay line.

33116		E	*EW*	EH	Hertfordshire Rail Tours

Class 33/2. Built to Former Loading Gauge of Tonbridge–Battle Line.
All equipped with slow speed control.

33202	y	**C**	E	*EW*	EH
33208		**G**	MH	*CA*	RL

CLASS 37   ENGLISH ELECTRIC TYPE 3   Co–Co

Built: 1960–5 by English Electric Company at Vulcan Foundry, Newton le Willows or Robert Stephenson & Hawthorn, Darlington.
37003–115/340/1/3/350/1/9 with the exception of 37019*/047/065*/072*/073/074/075*/100* (* one end only) retain box-type route indicators, the remainder having central headcode panels/marker lamps.
Engine: English Electric 12CSVT of 1300 kW (1750 hp) at 850 rpm.
Main Generator: English Electric 822/10G.
Traction Motors: English Electric 538/A.
Max. Tractive Effort: 245 kN (55500 lbf).
Cont. Tractive Effort: 156 kN (35000 lbf) at 13.6 m.p.h.

Power At Rail: 932 kW (1250 hp).	**Length over Buffers:** 18.75 m.
Brake Force: 50 t.	**Wheel Diameter:** 1092 mm.
Design Speed: 90 m.p.h.	**Weight:** 103–108 t.
Max. Speed: 80 m.p.h.	**RA:** 5 or 7.

Train Heating: Electric (Class 37/4 only). **ETH Index:** 30.
Train Brakes: Air & Vacuum.
Multiple Working: Blue Star Coupling Code.

Non-standard livery:
37116 is BR blue with Transrail markings.

a Vacuum brake isolated.

Originally numbered 6600–8, 6700–6999 (not in order). 37274 is the second loco to carry that number. It was renumbered to avoid confusion with Class 37/3 locos.

Class 37/0. Unrefurbished Locos. Technical details as above. RA5.

37003	+	**C**	E		IM (U)	
37010		**C**	E	*EW*	TO	
37012		**C**	E	*EW*	TO	
37013	+	**ML**	E	*EW*	TO	
37019	+	**F**	E		HM (U)	
37023		**ML**	E	*EW*	TO	Stratford TMD Quality Approved
37025		**BR**	E	*EW*	EH	Inverness TMD Quality Assured
37026	+	**F**	E		SP (U)	
37035		**C**	E		TO (U)	
37037		**F**	E	*EW*	EH	
37038		**C**	E	*EW*	TO	
37040		**E**	E	*EW*	EH	

37042	+	**E**	E	*EW*	TO
37043		**CT**	E	*EW*	ML
37045	+	**F**	E		TO (U)
37046		**C**	E	*EW*	IM
37047	+	**ML**	E	*EW*	EH
37048		**FM**	E		TO (U)
37051		**E**	E	*EW*	TO	Merehead
37054		**C**	E	*EW*	EH
37055	+	**ML**	E	*EW*	TO	RAIL Celebrity
37057	+	**E**	E	*EW*	TO	Viking
37058	+	**C**	E	*EW*	IM
37059	+	**F**	E		IM (U)
37063	+	**F**	E		TE (U)
37065	+	**ML**	E	*EW*	EH
37068	+	**F**	E		IM (U)
37069	+	**C**	E	*EW*	ML
37071	+	**C**	E	*EW*	IM
37072	+	**D**	E		IM (U)
37073	+	**FT**	E	*EW*	TO	Fort William/An Gearasdan
37074	+	**ML**	E	*EW*	EH
37075	+	**F**	E		IM (U)
37077		**ML**	E	*EW*	EH
37078	+	**F**	E		ML (U)
37079	+	**F**	E	*EW*	TO
37083	+	**C**	E		IM (U)
37087		**C**	E		CD (U)
37088		**CT**	E		ML (U)	Clydesdale
37092		**C**	E		TO (U)
37095	+	**C**	E	*EW*	TO
37097		**C**	E	*EW*	EH
37098	+	**C**	E	*EW*	IM
37100	+	**FT**	E	*EW*	ML
37101	+	**F**	E		IM (U)
37104		**C**	E		IM (U)
37106	+	**C**	E	*EW*	EH
37107	+	**F**	E		SP (U)
37108	+	**F**	E		BS (U)
37109		**E**	E	*EW*	EH
37110	+	**F**	E		IM (U)
37111		**FT**	E		TO (U)
37114	+	**E**	E	*EW*	TO	City of Worcester
37116	+	**0**	E	*EW*	EH	Sister Dora
37131	+	**F**	E	*EW*	IM
37133		**C**	E	*EW*	TO
37137		**FM**	E		TO (U)
37139	+	**F**	E		TE (U)
37140		**C**	E	*EW*	EH
37141		**C**	E		IM (U)
37142		**C**	E		CD (U)
37144	r	**F**	E		IM (U)
37146		**C**	E	*EW*	TO

37152		I	E	EW	ML	
37153		CT	E	EW	ML	
37154	+	FT	E	EW	TO	
37156	r	FT	E	EW	IM	
37158		C	E	EW	TO	
37162	+	D	E	EW	TO	
37165	+	C	E	EW	ML	
37170	r	C	E	EW	ML	
37174		E	E	EW	EH	
37175		C	E	EW	ML	
37178	+	F	E	EW	CF	
37184		C	E		BS (U)	
37185	+	C	E	EW	TO	Lea & Perrins
37188		C	E		TO (U)	
37191		C	E	EW	TO	
37194	+	FM	E	EW	EH	British International Freight Association
37196		C	E	EW	TO	
37197	+	CT	E	EW	CF	
37198	+	ML	E	EW	EH	
37201		CT	E		BS (U)	
37203		ML	E	EW	EH	
37207		C	E		BS (U)	
37209		BR	E		DR (U)	
37211		C	E	EW	EH	
37212	+	FT	E	EW	IM	
37213	+	F	E		TO (U)	
37214	+	FT	E		BS (U)	
37216	r+	ML	E	EW	TO	Great Eastern
37217	+		E		IM (U)	
37218	+	F	E		IM (U)	
37219	r	ML	E	EW	EH	
37220	+	E	E	EW	TO	
37221		FT	E	EW	ML	
37222	+	FM	E		CF (U)	
37223	+	F	E		IM (U)	
37225	+	F	E	EW	IM	
37227	+	FM	E		SL (U)	
37229	+	F	E	EW	CF	
37230	+	CT	E	EW	CF	
37232	r	CT	E		ML (U)	The Institution of Railway Signal Engineers
37235	+	F	E		DR (U)	
37238	+	F	E	EW	TO	
37240	+	C	E		BS (U)	
37241		F	E		TO (U)	
37242	+	ML	E	EW	EH	
37244	+	F	E	EW	IM	
37245		C	E	EW	EH	
37248	I	ML	E	EW	TO	Midland Railway Centre
37250	+	FT	E	EW	ML	

37251	+	I	E		ML (U)
37254	+	C	E	EW	CF
37255	+	C	E	EW	TO
37258	+	C	E	EW	TO
37261	+	F	E	EW	ML	Caithness
37262	+	D	E	EW	EH
37263		C	E	EW	CF
37264		C	E	EW	TO
37274	+	ML	E	EW	EH
37275	+		E	EW	CF	Oor Wullie
37278	+	F	E		TO (U)
37293	+	ML	E	EW	EH
37294	+	C	E	EW	ML
37298	+	F	E		IM (U)

Class 37/3. Unrefurbished locos fitted with regeared (CP7) bogies. Details as Class 37/0 except:

Max. Tractive Effort: 250 kN (56180 lbf).
Cont. Tractive Effort: 184 kN (41250 lbf) at 11.4 m.p.h.

37330	(37128)	+	BR	E		TO (U)
37331	(37202)		F	E		DR (U)
37332	(37239)	+	F	E	EW	TO
37334	(37272)	+	F	E		IM (U)
37335	(37285)	+	F	E		IM (U)
37340	(37009)	+	F	E		IM (U)
37341	(37015)	+	F	E		TE (U)
37343	(37049)		C	E		TO (U)
37344	(37053)	+	F	E		IM (U)
37350	(37119)	+	F	E	EW	IM
37351	(37002)	+	CT	E	EW	ML
37358	(37091)		F	E		IM (U)
37359	(37118)		F	E		TE (U)
37370	(37127)		E	E	EW	EH
37371	(37147)	+	ML	E	EW	EH
37372	(37159)		ML	E	EW	EH
37375	(37193)	+	ML	E	EW	EH
37376	(37199)	+	F	E	EW	IM
37377	(37200)	+	C	E	EW	EH
37379	(37226)		ML	E	EW	TO	Ipswich WRD Quality Assured
37380	(37259)		FM	E	EW	EH
37381	(37284)	+	F	E		FH (U)
37382	(37145)		F	E		IM (U)
37383	(37167)	+	ML	E	EW	EH
37384	( )					

Class 37/4. Refurbished locos fitted with train heating. Main generator replaced by alternator. Regeared (CP7) bogies. Details as class 37/0 except:

Main Alternator: Brush BA1005A.
Max. Tractive Effort: 256 kN (57440 lbf).
Cont. Tractive Effort: 184 kN (41250 lbf) at 11.4 m.p.h.
Power At Rail: 935 kW (1254 hp).
All have twin fuel tanks.

37401	r	**FT**	E	*EW*	ML	Mary Queen of Scots
37402	r	**F**	E	*EW*	IM	Bont Y Bermo
37403	r	**G**	E	*EW*	TO	Ben Cruachan
37404	r	**FT**	E	*EW*	ML	Loch Long
37405	r	**E**	E	*EW*	ML	
37406	r	**FT**	E	*EW*	ML	The Saltire Society
37407	r	**FT**	E	*EW*	IM	Blackpool Tower
37408		**BR**	E	*EW*	IM	Loch Rannoch
37409	r	**FT**	E	*EW*	ML	Loch Awe
37410	r	**FT**	E	*EW*	ML	Aluminium 100
37411		**E**	E	*EW*	CF	Ty Hafan
37412		**FT**	E	*EW*	CF	Driver John Elliot
37413	r	**E**	E	*EW*	ML	The Scottish Railway Preservation Society
37414	r	**RR**	E	*EW*	CD	Cathays C&W Works 1846–1993
37415	r	**E**	E	*EW*	CD	
37416	r	**E**	E	*EW*	CF	
37417	r	**F**	E	*EW*	IM	Highland Region
37418	r	**E**	E	*EW*	CD	East Lancashire Railway
37419		**E**	E	*EW*	CD	
37420	r	**RR**	E	*EW*	CD	The Scottish Hosteller
37421	r	**E**	E	*EW*	CD	
37422	r	**RR**	E	*EW*	CD	Robert F. Fairlie Locomotive Engineer 1831–1885
37423	r	**FT**	E	*EW*	TO	Sir Murray Morrison 1873–1948 Pioneer of British Aluminium Industry
37424	r	**FT**	E	*EW*	ML	
37425	r	**RR**	E	*EW*	ML	Sir Robert McAlpine/ Concrete Bob (opposite sides)
37426	r	**E**	E	*EW*	CD	
37427	r	**E**	E	*EW*	CF	
37428	r	**F**	E	*EW*	ML	David Lloyd George
37429	r	**RR**	E	*EW*	IM	Eisteddfod Genedlaethol
37430	r	**FT**	E	*EW*	ML	Cwmbrân
37431	r	**M**	E	*EW*	TO	

Class 37/5. Refurbished locos. Main generator replaced by alternator. Regeared (CP7) bogies. Details as class 37/4 except:

Max. Tractive Effort: 248 kN (55590 lbf).
All have twin fuel tanks.

37503	**E**	E	*EW*	TO	
37505	**FT**	E	*EW*	IM	British Steel Workington

37509		**F**	E	*EW*	IM	
37510		**I**	E	*EW*	ML	
37513		**LH**	E	*EW*	IM	
37515	s	**E**	E	*EW*	IM	
37516	s	**LH**	E	*EW*	IM	
37517	ars	**LH**	E	*EW*	IM	
37518		**E**	E	*EW*	IM	
37519		**F**	E	*EW*	TO	
37520		**E**	E	*EW*	ML	
37521		**E**	E	*EW*	CF	English China Clays

Class 37/6. Refurbished locos modified for use with Nightstar stock. Main generator replaced by alternator. Class 50 bogies. Details as class 37/0 except:

Main Alternator: Brush BA1005A.
All have twin fuel tanks, train air brakes only, UIC brake and coaching stock jumpers, RCH jumpers, ETH through wires.

37601	(37501)	**EP**	LC	*ES*	OC
37602	(37502)	**EP**	LC	*ES*	OC
37603	(37504)	**EP**	LC	*ES*	OC
37604	(37506)	**EP**	LC	*ES*	OC
37605	(37507)	**EP**	LC	*ES*	OC
37606	(37508)	**EP**	LC	*ES*	OC
37607	(37511)	**EP**	D	*DR*	SD
37608	(37512)	**DR**	D	*DR*	SD
37609	(37514)	**DR**	D	*DR*	SD
37610	(37687)	**EP**	D	*DR*	SD
37611	(37690)	**DR**	D	*DR*	SD
37612	(37691)	**EP**	D	*DR*	SD

Class 37/5 continued.

37667	s	**E**	E	*EW*	ML	Meldon Quarry Centenary
37668	s	**E**	E	*EW*	CF	
37669		**E**	E	*EW*	CF	
37670		**E**	E	*EW*	CF	
37671		**FT**	E	*EW*	CF	Tre Pol and Pen
37672	s	**F**	E	*EW*	CF	
37673		**FT**	E	*EW*	CF	
37674		**FT**	E	*EW*	CF	Saint Blaise Church 1445–1995
37675	s	**FT**	E	*EW*	IM	
37676		**F**	E	*EW*	IM	
37677		**F**	E	*EW*	TO	
37678		**F**	E	*EW*	IM	
37679		**F**	E	*EW*	IM	
37680		**F**	E	*EW*	TO	
37682	r	**E**	E	*EW*	IM	Hartlepool Pipe Mill
37683		**FT**	E	*EW*	IM	
37684		**E**	E	*EW*	ML	Peak National Park
37685		**I**	E	*EW*	IM	
37686		**F**	E	*EW*	IM	

37688		E	E	EW	IM	
37689	s	F	E	EW	IM	
37692	s	F	E	EW	ML	The Lass O' Ballochmyle
37693	s	FT	E	EW	ML	
37694	s	E	E	EW	IM	
37695	s	E	E	EW	IM	
37696	s	FT	E	EW	CF	
37697	s	E	E	EW	IM	
37698	s	LH	E	EW	IM	

Class 37/7. Refurbished locos. Main generator replaced by alternator. Regeared (CP7) bogies. Ballast weights added. Details as class 37/4 except:

Main Alternator: GEC G564AZ (37796–803) Brush BA1005A (others).
Max. Tractive Effort: 276 kN (62000 lbf).
Weight: 120 t.   **RA:** 7.
All have twin fuel tanks.

37701	s	FT	E	EW	CF	
37702	s	FT	E	EW	ML	Taff Merthyr
37703	s	E	E	EW	EH	
37704	s	E	E	EW	CF	
37705		FM	E	EW	EH	
37706		E	E	EW	TO	
37707		E	E	EW	IM	
37708		F	E	EW	IM	
37709		FM	E	EW	EH	
37710		LH	E	EW	IM	
37711		E	E	EW	EH	
37712		E	E	EW	ML	
37713		LH	E	EW	IM	
37714		E	E	EW	ML	
37715		FM	E	EW	TO	British Petroleum
37716		E	E	EW	IM	
37717		E	E	EW	IM	St Margaret's Church of England
						Primary School City of Durham
						Railsafe Trophy Winners 1997
37718		E	E	EW	IM	
37719		F	E	EW	IM	
37796	s	E	E	EW	ML	
37797	s	E	E	EW	ML	
37798	s	ML	E	EW	TO	
37799	s	FT	E	EW	ML	Sir Dyfed/County of Dyfed
37800	s	E	E	EW	EH	
37801	s	E	E	EW	ML	
37802	s	FT	E	EW	ML	
37803	s	ML	E	EW	EH	
37883		E	E	EW	IM	
37884		LH	E	EW	IM	Gartcosh
37885		E	E	EW	IM	
37886		E	E	EW	IM	
37887	s	FT	E	EW	CF	

▲ Although it has been sold to Freightliner, Class 08 No. 08077 still carries the livery of its former owner, RFS. It is pictured here at Southampton Maritime Freightliner Terminal on 26th January 1997. **Brian Denton**

▼ Departmental liveried Class 09 No. 09107 passes through Newport with a Llanwern to Newport Alexandra Dock Jn transfer freight on 1st April 1997. **Bob Sweet**

Direct Rail Services liveried Class 20s Nos. 20305 & 20301 'FURNESS RAILWAY 150' head south through Lowgill on 14th July 1997, with a Penrith to Cricklewood milk train.

Kevin Conkey

A pair of Class 31s, Nos. 31467, in BR blue livery, and 31229, in Civil-link livery pass Slindon, Staffordshire on 25th July 1997 with the 11.10 Sheerness–Mossend Enterprise service.

Peter Fox

▲ One of only eight Class 33s still in traffic, No. 33046 descends from Whiteball tunnel with the 10.00 Westbury–Meldon Quarry empty ballast working on 7th April 1997. The loco carries Civil-link livery.   **Russell Ayre**

▼ The 13.34 Fawley–Tavistock Junction bogie tank train passes through St Denys on 22nd August 1997 with Loadhaul liveried Class 37 No. 37884 in charge.
David Brown

A pair of Transrail Class 37s Nos. 37897 & 37887 climb the bank at Stormy with eastbound steel working on 26th May 1997.
Russell Ayre

Midland Mainline liveried Class 43 No. 43085 leads a similarly liveried set forming the 11.33 Nottingham–London St Pancras at Harrowden, Wellingborough on 29th October 1997.

Michael J. Collins

Great Western Trains liveried Class 43 power cars Nos. 43132, leading, & 43185 'Great Western' provide the power for the 06.30 London Paddington–Plymouth as it passes Dawlish Warren on 20th June 1997.
C.J. Marsden

▲ The 08.11 Southampton–Ripple Lane Freightliner service enters Camden Road station behind Freightliner liveried Class 47 No. 47052 on 23rd September 1997.
Kevin Conkey

▼ Transrail class 56 No. 56070 passes through Swanley on 19th February 1996 with the 11.10 Sheerness–Willesden freight.　　**Rodney Lissenden**

EWS liveried Class 58 No. 58033 prepares to leave Bilsthorpe Colliery on 21st March 1997 with a trip to High Marnham Power Station. Bilsthorpe Colliery ceased mining less than a month later.
Nic Joynson

Class 59 No. 59204 'Vale of Glamorgan' approaches Gascoigne Wood with a train of coal hoppers on 2nd June 1997. Both the locomotive and wagons are in National Power livery.

John G. Teasdale

Mainline freight liveried Class 60 No. 60011 passes Spetchley on 9th May 1997 with the 07.31 Silverdale–Llanwern coal service.

Bob Sweet

▲ Class 73 No. 73131, in EWS livery, runs light from Stewarts Lane to Folkestone prior to working the Venice Simplon Orient Express on 1st May 1997.

Chris Wilson

▼ Class 86 No. 86101 'Sir William A Stanier FRS' passes Heastone Lane with the 16.00 London Euston–Manchester Piccadilly on 4th June 1997. The loco carries Intercity livery.

Chris Wilson

A full train of Virgin Trains liveried stock hauled by similarly liveried Class 87 No. 87006 is pictured on the southbound 'Royal Scot' service at Carlisle. The date is 20th September 1997.
Dave McAlone

▲ Great North Eastern Railway liveried Class 89 No. 89001 departs from London Kings Cross on 9th September 1997 with the 15.40 service to Bradford Foster Square. **Hugh Ballantyne**

▼ The 17.35 Leeds–London Kings Cross is pictured near Sandal on 31st May 1997 with Class 91 No. 91014 providing the power at the rear. **John G. Teasdale**

Rail express systems liveried Class 90 No. 90016 arrives at its destination, London Kings Cross, with the 14.03 service from Low Fell on 6th September 1996.
Russell Ayre

Eurostar (UK) locomotive livery is carried by all Class 92s. One of them, 92003 'Beethoven' is pictured with the 13.45 Dollands Moor–Wembley at Kemsing on 1st May 1997.
Rodney Lissenden

37888		F	E	*EW*	CF	
37889		FT	E	*EW*	CF	
37890	a	FM	E	*EW*	EH	The Railway Observer
37891		FM	E	*EW*	EH	
37892		FM	E	*EW*	EH	Ripple Lane
37893		E	E	*EW*	ML	
37894	s	E	E	*EW*	CF	
37895	s	E	E	*EW*	CF	
37896	s	FT	E	*EW*	CF	
37897	s	FT	E	*EW*	CF	
37898	s	FT	E	*EW*	CF	Cwmbargoed DP
37899	s	E	E	*EW*	TO	

Class 37/9. Refurbished Locos. Fitted with manufacturers prototype power units and ballast weights. Main generator replaced by alternator. Details as Class 37/0 except:

Engine: Mirrlees MB275T of 1340 kW (1800 hp) at 1000 rpm (37901–4), Ruston RK270T of 1340 kW (1800 hp) at 900 rpm (37905/6).
Main Alternator: Brush BA1005A (GEC G564, 37905/6).
Max. Tractive Effort: 279 kN (62680 lbf).
Cont. Tractive Effort: 184 kN (41250 lbf) at 11.4 m.p.h.
Weight: 120 t.   **RA:** 7.
All have twin fuel tanks.

37901		FT	E	*EW*	CF	Mirrlees Pioneer
37902		F	E	*EW*	CF	
37903		F	E	*EW*	CF	
37904		F	E		CF (U)	
37905	s	F	E	*EW*	CF	
37906	s	FT	E		CF (U)	

CLASS 43   HST POWER CAR   Bo–Bo

Built: 1976–82 by BREL Crewe Works. Formerly numbered as coaching stock but now classified as locomotives. Fitted with luggage compartment.
Engine: Paxman Valenta 12RP200L (Paxman VP185*) of 1680 kW (2250 hp) at 1500 rpm.
Main Alternator: Brush BA1001B.
Traction Motors: Brush TMH68–46 or GEC G417AZ (43124–151/180). Frame mounted.
Max. Tractive Effort: 80 kN (17980 lbf).
Cont. Tractive Effort: 46 kN (10340 lbf) at 64.5 m.p.h.
Power At Rail: 1320 kW (1770 hp).   **ETH:** Non standard 3-phase system.
Brake Force: 35 t.   **Length over Buffers:** 17.79 m.
Weight: 70 t.   **Wheel Diameter:** 1020 mm.
Max. Speed: 125 m.p.h.   **RA:** 5.
Train Brakes: Air.
Multiple Working: With one other similar vehicle.
Communication Equipment: All equipped with driver–guard telephone.

43002	**I**	A	*GW*	PM	
43003	**GW**	A	*GW*	PM	
43004	**GW**	A	*GW*	PM	Borough of Swindon
43005	**GW**	A	*GW*	PM	
43006	**I**	A	*GW*	LA	
43007	**I**	A	*GW*	LA	
43008	**GW**	A	*GW*	LA	
43009	**GW**	A	*GW*	PM	
43010	**GW**	A	*GW*	PM	
43011	**GW**	A	*GW*	PM	Reader 125
43012	**GW**	A	*GW*	PM	
43013	**I**	P	*VX*	EC	CROSSCOUNTRY VOYAGER
43014	**I**	P	*VX*	EC	
43015	**GW**	A	*GW*	PM	
43016	**GW**	A	*GW*	PM	
43017	**GW**	A	*GW*	LA	
43018	**GW**	A	*GW*	LA	The Red Cross
43019	**GW**	A	*GW*	LA	City of Swansea/Dinas Abertawe
43020	**GW**	A	*GW*	LA	John Grooms
43021	**I**	A	*GW*	LA	
43022	**GW**	A	*GW*	LA	
43023	**GW**	A	*GW*	LA	County of Cornwall
43024	**GW**	A	*GW*	LA	
43025	**I**	A	*GW*	LA	Exeter
43026	**GW**	A	*GW*	LA	City of Westminster
43027	**I**	A	*GW*	LA	Glorious Devon
43028	**I**	A	*VW*	LO	
43029	**I**	A	*VW*	LO	
43030	**GW**	A	*GW*	PM	
43031	**GW**	A	*GW*	PM	
43032	**GW**	A	*GW*	PM	The Royal Regiment of Wales
43033	**I**	A	*GW*	PM	
43034	**GW**	A	*GW*	PM	The Black Horse
43035	**I**	A	*GW*	PM	
43036	**GW**	A	*GW*	PM	
43037	**I**	A	*GW*	PM	
43038	**GN**	A	*GN*	NL	
43039	**GN**	A	*GN*	NL	
43040	**I**	A	*GW*	PM	
43041	**I**	A	*VW*	LO	City of Discovery
43042	**I**	A	*VW*	LO	
43043	**MM**	P	*ML*	NL	LEICESTERSHIRE COUNTY CRICKET CLUB
43044	**MM**	P	*ML*	NL	Borough of Kettering
43045	**MM**	P	*ML*	NL	
43046	**MM**	P	*ML*	NL	Royal Philharmonic
43047 *	**MM**	P	*ML*	NL	
43048	**I**	P	*ML*	NL	
43049	**MM**	P	*ML*	NL	Neville Hill
43050	**I**	P	*ML*	NL	
43051	**I**	P	*ML*	NL	The Duke and Duchess of York

43052	I	P	ML	NL	City of Peterborough
43053	I	P	ML	NL	Leeds United
43054	I	P	ML	NL	
43055	MM	P	ML	NL	Sheffield Star
43056	MM	P	ML	NL	
43057	I	P	ML	NL	Bounds Green
43058	MM	P	ML	NL	
43059 *	MM	P	ML	NL	MIDLAND PRIDE
43060	MM	P	ML	NL	County of Leicestershire
43061	MM	P	ML	NL	
43062	I	P	VX	EC	
43063	V	P	VX	EC	Maiden Voyager
43064	I	P	ML	NL	City of York
43065	I	P	VX	EC	City of Edinburgh
43066	MM	P	ML	NL	Nottingham Playhouse
43067	I	P	VX	EC	
43068	V	P	VX	EC	The Red Nose
43069	I	P	VX	EC	
43070	I	P	VX	EC	
43071	I	P	VX	EC	Forward Birmingham
43072	MM	P	ML	NL	Derby Etches Park
43073	I	P	ML	NL	
43074 *	MM	P	ML	NL	BBC EAST MIDLANDS TODAY
43075 *	I	P	ML	NL	
43076	MM	P	ML	NL	THE MASTER CUTLER 1947-1997
43077	MM	P	ML	NL	
43078	I	P	VX	EC	Golowan Festival Penzance
43079	I	P	VX	EC	
43080	I	P	VX	EC	
43081	MM	P	ML	NL	
43082	MM	P	ML	NL	DERBYSHIRE FIRST
43083	MM	P	ML	NL	
43084	V	P	VX	EC	County of Derbyshire
43085	MM	P	ML	NL	
43086	I	P	VX	EC	
43087	I	P	VX	LA	
43088	I	P	VX	LA	XIII Commonwealth Games Scotland 1986
43089	I	P	VX	LA	
43090	V	P	VX	LA	
43091	I	P	VX	LA	Edinburgh Military Tattoo
43092	V	P	VX	EC	Institution of Mechanical Engineers 150th Anniversary 1847-1997
43093	V	P	VX	EC	Lady in Red
43094	I	P	VX	EC	
43095	GN	A	GN	NL	
43096	GN	A	GN	NL	The Great Racer
43097	I	P	VX	EC	
43098	V	P	VX	EC	
43099	I	P	VX	EC	
43100	I	P	VX	EC	Craigentinny

43101	I	P	VX	LA	Edinburgh International Festival
43102	I	P	VX	LA	
43103	I	P	VX	LA	John Wesley
43104	I	A		LA (U)	County of Cleveland
43105	GN	A	GN	NL	
43106	GN	A	GN	NL	
43107	GN	A	GN	NL	
43108	GN	A	GN	NL	
43109	GN	A	GN	NL	
43110	GN	A	GN	EC	
43111	GN	A	GN	EC	
43112	GN	A	GN	EC	
43113	GN	A	GN	EC	
43114	GN	A	GN	EC	
43115	GN	A	GN	EC	
43116	GN	A	GN	EC	
43117	GN	A	GN	EC	
43118	GN	A	GN	EC	
43119	GN	A	GN	EC	
43120	GN	A	GN	EC	
43121	I	P	VX	LA	West Yorkshire Metropolitan County
43122	I	P	VX	LA	South Yorkshire Metropolitan County
43123	I	P	VX	EC	
43124	GW	A	GW	PM	
43125	GW	A	GW	PM	Merchant Venturer
43126	I	A	GW	PM	City of Bristol
43127	I	A	GW	PM	
43128	GW	A	GW	PM	
43129	GW	A	GW	PM	
43130	I	A	GW	PM	Sulis Minerva
43131	GW	A	GW	PM	Sir Felix Pole
43132	GW	A	GW	PM	
43133	I	A	GW	PM	
43134	I	A	GW	PM	County of Somerset
43135	GW	A	GW	PM	
43136	GW	A	GW	PM	
43137	GW	A	GW	PM	Newton Abbot 150
43138	GW	A	GW	PM	
43139	GW	A	GW	PM	
43140	GW	A	GW	PM	
43141	GW	A	GW	PM	
43142	GW	A	GW	PM	
43143	I	A	GW	PM	
43144	I	A	GW	PM	
43145	I	A	GW	PM	
43146	I	A	GW	PM	
43147	I	A	GW	PM	
43148	I	A	GW	PM	
43149	GW	A	GW	PM	BBC Wales Today
43150	I	A	GW	PM	Bristol Evening Post
43151	I	A	GW	PM	

43152	I	A	GW	PM	
43153	V	P	VX	LA	THE ENGLISH RIVIERA
					TORQUAY PAIGNTON BRIXHAM
43154	V	P	VX	LA	INTERCITY
43155	V	P	VX	LA	The Red Arrows
43156	I	P	VX	LA	
43157	I	P	VX	LA	Yorkshire Evening Post
43158	I	P	VX	LA	Dartmoor The Pony Express
43159	I	P	VX	LA	
43160	V	P	VX	LA	
43161	I	P	VX	LA	Reading Evening Post
43162	I	P	VX	LA	Borough of Stevenage
43163	I	A	GW	LA	
43164	I	A	VW	LO	
43165	I	A	VW	LO	
43166	I	A	VW	LO	
43167	*	GN	A	GN	NL
43168	*	GW	A	GW	LA
43169	*	GW	A	GW	LA	The National Trust
43170	*	GW	A	GW	LA	Edward Paxman
43171	I	A	GW	LA	
43172	I	A	GW	LA	
43173	*	GW	A		ZC (U)
43174	GW	A	GW	LA	Bristol - Bordeaux
43175	*	I	A	GW	LA
43176	I	A	GW	LA	
43177	*	GW	A	GW	LA	University of Exeter
43178	GW	A	GW	LA	
43179	*	GW	A	GW	LA	Pride of Laira
43180	I	P	GW	LA	
43181	I	A	GW	LA	Devonport Royal Dockyard 1693-1993
43182	GW	A	GW	LA	
43183	GW	A	GW	LA	
43184	I	A	GW	LA	
43185	GW	A	GW	LA	Great Western
43186	GW	A	GW	LA	Sir Francis Drake
43187	GW	A	GW	LA	
43188	GW	A	GW	LA	City of Plymouth
43189	GW	A	GW	LA	RAILWAY HERITAGE TRUST
43190	GW	A	GW	LA	
43191	*	GW	A	GW	LA	Seahawk
43192	GW	A	GW	LA	City of Truro
43193	I	P	VX	LA	Plymouth SPIRIT OF DISCOVERY
43194	I	P	VX	LA	
43195	I	P	VX	LA	British Red Cross 125th Birthday 1995
43196	I	P	VX	LA	The Newspaper Society Founded 1836
43197	I	P	VX	LA	Railway Magazine
					1897 Centenary 1997
43198	I	P	VX	LA	

CLASS 46   BR TYPE 4   1Co–Co1

Built: 1962 by BR Derby Locomotive Works.
Engine: Sulzer 12LDA28B of 1860 kW (2500 hp) at 750 rpm.
Main Generator: Brush TG160-60.
Traction Motors: Brush TM73-68 Mk3 (axle hung).
Max. Tractive Effort: 245 kN (55000 lbf).
Cont. Tractive Effort: 141 kN (31600 lbf) at 22.3 m.p.h.

Power At Rail: 1460 kW (1960 hp).	**Length over Buffers:** 20.70 m.
Brake Force: 63 t.	**Wheel Diameter:** 914/1143 mm.
Design Speed: 90 m.p.h.	**Weight:** 141 t.
Max. Speed: 75 m.p.h.	**RA:** 7.
Train Brakes: Air & Vacuum.	**Multiple Working:** Not equipped.

Carries original number D 172.

46035	**G**	LN	*SS*	CQ	Ixion

CLASS 47   BRUSH TYPE 4   Co–Co

Built: 1963–67 by Brush Traction, Loughborough or BR Crewe Works.
Engine: Sulzer 12LDA28C of 1920 kW (2580 hp) at 750 rpm.
Main Generator: Brush TG160-60 Mk2, TG160-60 Mk4 or TM172-50 Mk1.
Traction Motors: Brush TM64-68 Mk1 or Mk1A (axle hung).
Max. Tractive Effort: 267 kN (60000 lbf).
Cont. Tractive Effort: 133 kN (30000 lbf) at 26 m.p.h.

Power At Rail: 1550 kW (2080 hp).	**Length over Buffers:** 19.38 m.
Brake Force: 61 t.	**Wheel Diameter:** 1143 mm.
Design Speed: 95 m.p.h.	**Weight:** 120.5–125 t.
Max. Speed: various.	**RA:** 6 or 7.
Train Brakes: Air & Vacuum.	

Multiple Working: Green Circle (m) or Blue Star (*) Coupling Code. Otherwise not equipped.
ETH Index (47/4, 47/6 and 47/7): 66 (75 Class 47/6).

Non-standard liveries:
47114 is two-tone green with Freightliner lettering and markings.
47145 is dark blue with Railfreight Distribution markings.
47488, 47705, 47710 & 47712 are Waterman Railways black with red and cream lining.
47627 is maroon.
47798 & 47799 are royal train purple with a maroon and gold bodyside stripes.
47803 is grey, red and yellow.
47846 is white.

47701 is owned by Alan & Tracey Lear and managed by Fragonset Railways.
a Vacuum brake isolated.

Formerly numbered 1100–11, 1500–1999 not in order.

Class 47/0. Built with train heating boiler. RA6. Max. Speed 75 m.p.h.

47004	**G**	E	*EW*	IM	Old Oak Common Traction & Rolling Stock Depot

47016	F0	E	EW	IM	ATLAS
47033 am+	RD	E	EW	TI	The Royal Logistics Corps
47049 am+	RD	E	EW	TI	GEFCO
47051 am+	RD	E	EW	TI	
47052	FL	P	FL	CD	
47053 am+	RD	E	EW	TI	Dollands Moor International
47060 a	F	P	FL	CD	
47079	FL	FL	FL	CD	
47085 am+	RD	E	EW	TI	REPTA 1893–1993
47095 am+	RD	E	EW	TI	
47114 am+	0	FL	FL	CD	Freightlinerbulk
47125 am+	RD	E		TI (U)	
47142	FR	P		BL (U)	
47144 am+	F	E		TI (U)	
47145 am+	0	E	EW	TI	
47146 am	RD	E	EW	TI	Loughborough Grammar School
47147	F	P		BL (U)	
47150 am+	RD	FL	FL	CD	
47152 am+	FL	FL	FL	CD	
47156 am+	F	FL		CD (U)	
47157 am	FL	P	FL	CD	Johnson Stevens Agencies
47186 am+	RD	E	EW	TI	Catcliffe Demon
47187	F	P		BL (U)	
47188 am+	RD	E		TI (U)	
47193	F	FL	FL	CD	
47194 am+	F	E	EW	TI	
47197	FL	P	FL	CD	
47200 am+	RD	E	EW	TI	Herbert Austin
47201 am+	RD	E	EW	TI	
47204 am+	FL	FL	FL	CD	
47205 am+	FL	FL	FL	CD	
47206	FL	P	FL	CD	The Morris Dancer
47207	F	FL	FL	CD	
47209 am+	FL	FL	FL	CD	
47210 am+	F	E	EW	TI	
47211 am+	F	E	EW	TI	
47212 +	FL	P	FL	CD	
47213 am+	F	E	EW	TI	Marchwood Military Port
47217 am+	RD	E	EW	TI	
47218 am+	RD	E	EW	TI	United Transport Europe
47219 am+	RD	E	EW	TI	Arnold Kunzler
47221 +	F	E		IM (U)	
47222 am+	F	E		TI (U)	
47223 +	F	E		CD (U)	
47224 +	F	E		IM (U)	
47225	FL	P	FL	CD	
47226 am+	F	E	EW	TI	
47228 am+	RD	E	EW	TI	axial
47229 am+	RD	E	EW	TI	
47231	FL	P	FL	CD	
47234 am+	FL	FL	FL	CD	

47236 am+	**RD**	E	*EW*	TI	ROVER GROUP QUALITY ASSURED
47237 am+	**RD**	E	*EW*	TI	
47238	**F**	E		BS (U)	
47241 am+	**RD**	E	*EW*	TI	Halewood Silver Jubilee 1988
47245 am+	**RD**	E	*EW*	TI	The Institute of Export
47256	**F**	E		DR (U)	
47258 am+	**RD**	FL	*FL*	CD	
47270	**FL**	P	*FL*	CD	Cory Brothers 1842–1992
47276 am+	**F**	E	*EW*	TI	
47277	**F**	E		IM (U)	
47278	**F**	E		SP (U)	
47279 am+	**FL**	P	*FL*	CD	
47280 am+	**F**	E	*EW*	TI	Pedigree
47281 am+	**F**	E	*EW*	TI	
47283	**FL**	P		CD (U)	
47284 am+	**F**	E	*EW*	TI	
47285 am+	**RD**	E	*EW*	TI	
47286 am+	**RD**	E	*EW*	TI	Port of Liverpool
47287 am+	**RD**	FL	*FL*	CD	
47289 am	**FL**	P	*FL*	CD	
47290 am+	**FL**	FL	*FL*	CD	
47291 am+	**F**	E		TI (U)	
47292 am+	**F**	FL	*FL*	CD	
47293 am+	**RD**	E	*EW*	TI	TRANSFESA
47294 +	**F**	E		TO (U)	
47295 +	**F**	FL	*FL*	CD	
47296	**FL**	P	*FL*	CD	
47297 am+	**RD**	E	*EW*	TI	Cobra RAILFREIGHT
47298 am+	**F**	E		TI (U)	Pegasus
47299 am+	**RD**	E		TI (U)	

Class 47/3. Built without Train Heat. (except 47300). RA6. Max. Speed 75 m.p.h.
All equipped with slow speed control.

47300	**C**	E		BS (U)	
47301	**FL**	P	*FL*	CD	Freightliner Birmingham
47302 a	**FR**	FL		TI (U)	
47303 am+	**FL**	FL	*FL*	CD	Freightliner Cleveland
47304 +	**F**	E	*EW*	TI	
47305	**FL**	P	*FL*	CD	
47306 am+	**RD**	E	*EW*	TI	The Sapper
47307 am+	**RD**	E	*EW*	TI	
47308	**C**	E		BS (U)	
47309 am+	**F**	FL	*FL*	CD	The Halewood Transmission
47310 am+	**RD**	E	*EW*	TI	Henry Ford
47312 am+	**RD**	E	*EW*	TI	Parsec of Europe
47313 am+	**F**	E	*EW*	TI	
47314 am+	**F**	E	*EW*	TI	Transmark
47315	**C**	E	*EW*	IM	
47316 am+	**RD**	E	*EW*	TI	
47317	**F**	P		CD (U)	
47319 +	**F**	E		IM (U)	Norsk Hydro

47322		FR	P		CD (U)	
47323	am+	FL	FL	FL	CD	
47326	am+	RD	E	EW	TI	Saltley Depot Quality Approved
47328	am+	F	E		TI (U)	
47329		C	FL	FL	CD	
47330	am+	F	FL	FL	CD	
47331		C	E	EW	IM	
47332		C	FL	FL	CD	
47333		C	E		TO (U)	
47334	a	FL	FL	FL	CD	P&O Nedlloyd
47335	am+	F	E	EW	TI	
47337	am+	FL	P	FL	CD	
47338	am+	RD	E	EW	TI	
47339		FL	P	FL	CD	
47340		C	FL		ZC (U)	
47341		C	E		TO (U)	
47344	am+	RD	E	EW	TI	
47345		FL	P	FL	CD	
47347	a	F	P		CD (U)	
47348	am	RD	E	EW	TI	St. Christopher's Railway Home
47349		FL	P	FL	CD	
47350		FO	FL		CD (U)	
47351	am+	RD	E		TI (U)	
47352		C	E		FH (U)	
47353		FL	FL	FL	CD	
47354	a	FL	P	FL	CD	
47355	am+	F	E	EW	TI	
47356		FO	FL		BL (U)	
47357		C	E		BS (U)	
47358	am	FL	P	FL	CD	
47360	am+	RD	E	EW	TI	
47361	am+	FL	FL	FL	CD	
47362	am+	F	E	EW	TI	
47363	am+	F	E	EW	TI	
47365	am+	RD	E	EW	TI	ICI Diamond Jubilee
47366		C	E		SP (U)	
47367		FL	FL		TO (U)	
47368		F	E		SF (U)	
47369		F	E		IM (U)	
47370	am	FL	FL	FL	CD	Andrew A Hodgkinson
47371		FL	P	FL	CD	
47372		C	FL	FL	CD	
47375	am+	RD	E	EW	TI	Tinsley Traction Depot Quality Approved
47376		FL	P	FL	CD	Freightliner 1995
47377	a	FL	P	FL	CD	
47378	am+	F	E		TI (U)	
47379	am+	F	E	EW	TI	

Class 47/4. Equipped with train heating. RA6. Max. Speed 95 m.p.h.

| 47462 | | R | E | | TO (U) | |

47467		**BR**	E	*EW*	CD	
47471		**I0**	E		CD (U)	
47473		**BR**	FL		ZC (U)	
47474		**R**	E	*EW*	IM	Sir Rowland Hill
47475		**RX**	E	*EW*	IM	Restive
47476		**R**	E	*EW*	IM	Night Mail
47478			E		BS (U)	
47481		**BR**	E		CD (U)	
47484		**G**	E		OC (U)	ISAMBARD KINGDOM BRUNEL
47488		**0**	FG		TM (U)	DAVIES THE OCEAN
47489		**R**	E		BS (U)	
47492		**RX**	E	*EW*	IM	
47501		**R**	E	*EW*	CD	
47513		**BR**	E		CD (U)	Severn
47519	+	**G**	E	*EW*	IM	
47520		**I**	E	*EW*	IM	
47522		**R**	E	*EW*	IM	
47523		**M**	E		TO (U)	
47524		**RX**	E		CD (U)	
47525		**RD**	FL		CD (U)	
47526		**BR**	E	*EW*	IM	
47528		**M**	E	*EW*	IM	The Queen's Own Mercian Yeomanry
47530		**RX**	E		CD (U)	
47532		**RX**	E		CD (U)	
47535		**RX**	E	*EW*	IM	
47536		**RX**	E		CD (U)	
47539		**RX**	E		CD (U)	
47540		**C**	FL		CD (U)	The Institution of Civil Engineers
47543		**R**	E	*EW*	IM	
47547		**N**	E		CD (U)	
47550		**M**	E		IM (U)	
47555		**RD**	E		TI (U)	The Commonwealth Spirit
47565		**RX**	E	*EW*	CD	Responsive
47566		**RX**	E		CD (U)	
47572		**R**	E	*EW*	CD	Ely Cathedral
47574		**R**	E		IM (U)	
47575		**R**	E	*EW*	CD	City of Hereford
47576		**RX**	E		CD (U)	
47584		**RX**	E	*EW*	CD	THE LOCOMOTIVE & CARRIAGE INSTITUTION 1911
47596		**RX**	E	*EW*	CD	
47624		**RX**	E	*EW*	CD	Saint Andrew
47627		**0**	E	*EW*	CD	
47628	j	**RX**	E		CD (U)	
47634		**R**	E	*EW*	CD	Holbeck
47635	j	**R**	E	*EW*	CD	
47640	j	**R**	E	*EW*	CD	University of Strathclyde

Class 47/6. Fitted with high phosphorus brake blocks. RA6. Max. Speed 75 m.p.h.

| 47676 | | **I** | E | | DR (U) | |

Class 47/7. Fitted with an older form of TDM. RA6. Max. Speed 95 m.p.h.
All have twin fuel tanks.

47701	**FG**	FG	*VX*	TM	Waverley
47702	**F**	E	*EW*	IM	County of Suffolk
47703	**FG**	FG	*VX*	TM	
47704	**RX**	E		CD (U)	
47705	**0**	LN	*SS*	CQ	GUY FAWKES
47709	**RX**	FG		TM (U)	
47710	**0**	FG		TM (U)	
47711	**N**	E	*EW*	IM	County of Hertfordshire
47712	**0**	FG	*VX*	TM	
47715	**N**	E		CD (U)	
47716	**RX**	E		CD (U)	
47717	**R**	E		CD (U)	

Class 47/7. Railnet dedicated locos. RA6. Max. Speed 95 m.p.h.
All have twin fuel tanks and are fitted with RCH jumper cables for operating
with propelling control vehicles (PCVs).

47721		**RX**	E	*EW*	CD	Saint Bede
47722	a	**RX**	E	*EW*	CD	The Queen Mother
47725		**RX**	E	*EW*	CD	The Railway Mission
47726		**RX**	E	*EW*	CD	Progress
47727	a	**RX**	E	*EW*	CD	Duke of Edinburgh's Award
47732		**RX**	E	*EW*	CD	Restormel
47733	a	**RX**	E	*EW*	CD	Eastern Star
47734		**RX**	E	*EW*	CD	Crewe Diesel Depot Quality Approved
47736	a	**RX**	E	*EW*	CD	Cambridge Traction & Rolling Stock Depot
47737		**RX**	E	*EW*	CD	Resurgent
47738	a	**RX**	E	*EW*	CD	Bristol Barton Hill
47739	a	**RX**	E	*EW*	CD	Resourceful
47741		**RX**	E	*EW*	CD	Resilient
47742		**RX**	E	*EW*	CD	The Enterprising Scot
47744	a	**E**	E	*EW*	CD	The Cornish Experience
47745		**RX**	E	*EW*	CD	Royal London Society for the Blind
47746	a	**RX**	E	*EW*	CD	The Bobby
47747	a	**RX**	E	*EW*	CD	Res Publica
47749		**RX**	E	*EW*	CD	Atlantic College
47750	a	**RX**	E	*EW*	CD	Royal Mail Cheltenham
47756	a	**RX**	E	*EW*	ML	Royal Mail Tyneside
47757	a	**RX**	E	*EW*	CD	Restitution
47758		**RX**	E	*EW*	CD	
47759		**RX**	E	*EW*	CD	
47760		**RX**	E	*EW*	CD	Restless
47761		**RX**	E	*EW*	CD	
47762		**RX**	E	*EW*	CD	
47763		**RX**	E	*EW*	CD	
47764		**RX**	E	*EW*	CD	Resounding
47765		**RX**	E	*EW*	CD	Ressaldar
47766		**RX**	E	*EW*	CD	Resolute

47767	a	**RX**	E	*EW*	ML	Saint Columba
47768		**RX**	E	*EW*	CD	Resonant
47769		**RX**	E	*EW*	CD	Resolve
47770		**RX**	E	*EW*	CD	Reserved
47771		**RX**	E	*EW*	CD	Heaton Traincare Depot
47772		**RX**	E	*EW*	CD	
47773	a	**RX**	E	*EW*	ML	Reservist
47774		**RX**	E	*EW*	CD	Poste Restante
47775		**RX**	E	*EW*	CD	Respite
47776		**RX**	E	*EW*	CD	Respected
47777		**RX**	E	*EW*	CD	Restored
47778		**RX**	E	*EW*	CD	Irresistible
47779		**RX**	E	*EW*	CD	
47780		**RX**	E	*EW*	CD	
47781		**RX**	E	*EW*	CD	Isle of Iona
47782		**RX**	E	*EW*	CD	
47783		**RX**	E	*EW*	CD	Saint Peter
47784		**RX**	E	*EW*	CD	Condover Hall
47785		**E**	E	*EW*	CD	Fiona Castle
47786	a	**E**	E	*EW*	CD	Roy Castle OBE
47787		**RX**	E	*EW*	CD	Victim Support
47788	a	**RX**	E	*EW*	CD	Captain Peter Manisty RN
47789	a	**RX**	E	*EW*	CD	Lindisfarne
47790	a	**RX**	E	*EW*	ML	Saint David/Dewi Sant
47791	a	**RX**	E	*EW*	CD	VENICE SIMPLON ORIENT EXPRESS
47792		**RX**	E	*EW*	CD	Saint Cuthbert
47793		**RX**	E	*EW*	CD	Saint Augustine

Class 47/4 continued. RA6. Max. Speed 95 m.p.h.

47798	a	**0**	E	*EW*	CD	Prince William
47799	a	**0**	E	*EW*	CD	Prince Henry
47802	+	**I**	E	*EW*	IM	
47803	+	**0**	E		SF (U)	
47805	a+	**I**	P	*VX*	CD	
47806	a+	**V**	P	*VX*	CD	
47807	a+	**PL**	P	*VX*	CD	
47810	a+	**I**	P	*VX*	CD	PORTERBROOK
47811	a+	**I**	P	*GW*	LA	
47812	a+	**I**	P	*VX*	CD	
47813	a+	**I**	P	*GW*	LA	
47814	a+	**V**	P	*VX*	CD	Totnes Castle
47815	a+	**I**	P	*GW*	LA	
47816	a+	**I**	P	*GW*	LA	Bristol Bath Road Quality Approved
47817	a+	**PL**	P	*VX*	CD	
47818	a+	**I**	P	*VX*	CD	
47822	a+	**V**	P	*VX*	CD	
47825	a+	**I**	P	*VX*	CD	Thomas Telford
47826	a+	**I**	P	*VX*	CD	
47827	a+	**I**	P	*VX*	CD	
47828	a+	**I**	P	*VX*	CD	
47829	a+	**I**	P	*VX*	CD	

47830	a+	I	P		ZC (U)	
47831	a+	I	P	VX	CD	Bolton Wanderer
47832	a+	I	P	GW	LA	
47839	a+	I	P	VX	CD	
47840	a+	I	P	VX	CD	NORTH STAR
47841	a+	I	P	VX	CD	The Institution of
						Mechanical Engineers
47843	a+	I	P	VX	CD	
47844	a+	V	P	VX	CD	
47845	a+	V	P	VX	CD	County of Kent
47846	a+	O	P	GW	LA	THOR
47847	a+	I	P	VX	CD	
47848	a+	I	P	VX	CD	
47849	a+	I	P	VX	CD	
47851	a+	I	P	VX	CD	
47853	a+	I	P	VX	CD	
47854	a+	I	P	VX	CD	Women's Royal Voluntary Service
47971	*	BR	E	EW	CD	Robin Hood
47972		CS	E	EW	IM	The Royal Army Ordnance Corps
47976	*	C	E	EW	CD	Aviemore Centre

Class 47/3 continued. RA6. Max. Speed 75 m.p.h.

47981		C	E	EW	IM	

CLASS 50   ENGLISH ELECTRIC TYPE 4   Co–Co

Built: 1967–68 by English Electric at Vulcan Foundry, Newton-le-Willows.
Engine: English Electric 16CVST of 2010 kW (2700 hp) at 850 r.p.m.
Main Generator: English Electric 840/4B.
Traction Motors: English Electric 538/5A.
Max. Tractive Effort: 216 kN (48500 lbf).
Cont. Tractive Effort: 147 kN (33000 lbf) at 23.5 m.p.h.

Power At Rail: 1540 kW (2070 hp).	**Length over Buffers:** 20.88 m.
Brake Force: 59 t.	**Wheel Diameter:** 1092 mm.
Design Speed: 105 m.p.h.	**Weight:** 117 t.
Max. Speed: 90 m.p.h.	**RA:** 6.

Train Brakes: Air & Vacuum.
Multiple Working: Orange Square Coupling Code (within class only).
ETH Index: 66.
All equipped with slow speed control.

50031		BR	FF	SS	KR	Hood

CLASS 55   DELTIC   Co–Co

Built: 1961 by English Electric at Vulcan Foundry, Newton-le-Willows.
Engine: Two Napier-Deltic T18-25 of 1230 kW (1650 h.p.) at 1500 r.p.m.
Main Generators: Two English Electric EE829.
Traction Motors: EE538 axle-hung.
Max. Tractive Effort: 222 kN (50000 lbf).
Cont. Tractive Effort: 136 kN (30500 lbf) at 32.5 m.p.h.
Power At Rail: 1969 kW (2640 hp).   **Length over Buffers:** 17.65 m.
Brake Force: 51 t.   **Wheel Diameter:** 1092 mm.
Design Speed: 100 m.p.h.   **Weight:** 105 t.
Max. Speed: 100 m.p.h.   **RA:** 5.
Train Brakes: Air & Vacuum.   **Multiple Working:** Not equipped.
ETH Index: 66.

55022	**G**	NT	*SS*	BN	ROYAL SCOTS GREY

CLASS 56   BRUSH TYPE 5   Co–Co

Built: 1976–84 by Electroputere at Craiova, Romania (as sub contractors for Brush) or BREL at Doncaster or Crewe Works.
Engine: Ruston Paxman 16RK3CT of 2460 kW (3250 hp) at 900 rpm.
Main Alternator: Brush BA1101A.
Traction Motors: Brush TM73-62.
Max. Tractive Effort: 275 kN (61800 lbf).
Cont. Tractive Effort: 240 kN (53950 lbf) at 16.8 m.p.h.
Power At Rail: 1790 kW (2400 hp).   **Length over Buffers:** 19.36 m.
Brake Force: 60 t.   **Wheel Diameter:** 1143 mm.
Design Speed: 80 m.p.h.   **Weight:** 125 t.
Max. Speed: 80 m.p.h.   **RA:** 7.
Train Brakes: Air.
Multiple Working: Red Diamond Coupling Code.
All equipped with slow speed control.

56003	**LH**	E	*EW*	IM	
56004		E	*EW*	IM	
56006	**LH**	E	*EW*	IM	Ferrybridge 'C' Power Station
56007	**FT**	E	*EW*	IM	
56008		E		IM (U)	
56010	**FT**	E	*EW*	IM	
56011	**F**	E	*EW*	IM	
56012	**F**	E		IM (U)	
56014	**F**	E		IM (U)	
56018	**E**	E	*EW*	IM	
56019	**FR**	E	*EW*	IM	
56021	**LH**	E	*EW*	IM	
56022	**FT**	E	*EW*	IM	
56025	**FT**	E	*EW*	IM	
56027	**LH**	F	*EW*	IM	
56029	**F**	E	*EW*	IM	
56031	**C**	E	*EW*	IM	

56032	E	E	EW	IM	
56033	FT	E	EW	IM	Shotton Paper Mill
56034	LH	E	EW	IM	Castell Ogwr/Ogmore Castle
56035	LH	E	EW	IM	
56036	CT	E	EW	IM	
56037	E	E	EW	IM	
56038	FT	E	EW	IM	Western Mail
56039	LH	E	EW	IM	
56040	FT	E	EW	IM	Oystermouth
56041	E	E	EW	IM	
56043	F	E	EW	IM	
56044	FT	E	EW	IM	Cardiff Canton Quality Assured
56045	LH	E	EW	IM	British Steel Shelton
56046	C	E	EW	IM	
56047	CT	E	EW	IM	
56048	C	E	EW	IM	
56049	CT	E	EW	IM	
56050	LH	E	EW	IM	British Steel Teeside
56051	E	E	EW	IM	
56052	FT	E	EW	IM	The Cardiff Rod Mill
56053	FT	E	EW	IM	Sir Morgannwg Ganol/
					County of Mid Glamorgan
56054	FT	E	EW	IM	British Steel Llanwern
56055	LH	E	EW	IM	
56056	FT	E	EW	IM	
56057	E	E	EW	IM	British Fuels
56058	E	E	EW	IM	
56059	E	E	EW	IM	
56060	E	E	EW	IM	
56061	F	E	EW	IM	
56062	F	E	EW	IM	Mountsorrel
56063	F	E	EW	IM	Bardon Hill
56064	FT	E	EW	IM	
56065	E	E	EW	IM	
56066	FT	E	EW	IM	
56067	E	E	EW	IM	
56068	E	E	EW	IM	
56069	F	E	EW	IM	Thornaby TMD
56070	FT	E	EW	IM	
56071	FT	E	EW	IM	
56072	FT	E	EW	IM	
56073	FT	E	EW	IM	
56074	LH	E	EW	IM	Kellingley Colliery
56075	F	E	EW	IM	West Yorkshire Enterprise
56076	F	E	EW	IM	
56077	LH	E	EW	IM	Thorpe Marsh Power Station
56078	F	E	EW	IM	
56079	FT	E	EW	IM	
56080	F	E	EW	IM	Selby Coalfield
56081	F	E	EW	IM	
56082	F	E	EW	IM	

56083	LH	E	EW	IM	
56084	LH	E	EW	IM	
56085	LH	E	EW	IM	
56086	FT	E	EW	IM	The Magistrates' Association
56087	E	E	EW	IM	ABP Port of Hull
56088	E	E	EW	IM	
56089	E	E	EW	IM	
56090	LH	E	EW	IM	
56091	F	E	EW	IM	Castle Donington Power Station
56092	FT	E	EW	IM	
56093	FT	E	EW	IM	The Institution of Mining Engineers
56094	F	E	EW	IM	Eggborough Power Station
56095	F	E	EW	IM	Harworth Colliery
56096	E	E	EW	IM	
56097	F	E	EW	IM	
56098	F	E	EW	IM	
56099	FT	E	EW	IM	Fiddlers Ferry Power Station
56100	LH	E	EW	IM	
56101	FT	E	EW	IM	Mutual Improvement
56102	LH	E	EW	IM	
56103	E	E	EW	IM	Stora
56104	F	E	EW	IM	
56105	E	E	EW	IM	
56106	LH	E	EW	IM	
56107	LH	E	EW	IM	
56108	F	E	EW	IM	
56109	LH	E	EW	IM	
56110	LH	E	EW	IM	Croft
56111	LH	E	EW	IM	
56112	LH	E	EW	IM	Stainless Pioneer
56113	FT	E	EW	IM	
56114	E	E	EW	IM	Maltby Colliery
56115	FT	E	EW	IM	
56116	LH	E	EW	IM	
56117	E	E	EW	IM	
56118	LH	E	EW	IM	
56119	E	E	EW	IM	
56120	E	E	EW	IM	
56121	F	E	EW	IM	
56123	FT	E	EW	IM	Drax Power Station
56124	F	E	EW	IM	
56125	FT	E	EW	IM	
56126	F	E	EW	IM	
56127	FT	E	EW	IM	
56128	FT	E	EW	IM	
56129	FT	E	EW	IM	
56130	LH	E	EW	IM	Wardley Opencast
56131	F	E	EW	IM	Ellington Colliery
56132	FT	E	EW	IM	
56133	FT	E	EW	IM	Crewe Locomotive Works
56134	F	E	EW	IM	Blyth Power

56135   F   E   *EW*   IM   Port of Tyne Authority

CLASS 58   BREL TYPE 5   Co–Co

Built: 1983–87 by BREL at Doncaster Works.
Engine: Ruston Paxman RK3ACT of 2460 kW (3300 hp) at 1000 rpm.
Main Alternator: Brush BA1101B.
Traction Motors: Brush TM73-62.
Max. Tractive Effort: 275 kN (61800 lbf).
Cont. Tractive Effort: 240 kN (53950 lbf) at 17.4 m.p.h.
Power At Rail: 1780 kW (2387 hp).   **Length over Buffers:** 19.13 m.
Brake Force: 62 t.   **Wheel Diameter:** 1120 mm.
Design Speed: 80 m.p.h.   **Weight:** 130 t.
Max. Speed: 80 m.p.h.   **RA:** 7.
Train Brakes: Air.
Multiple Working: Red Diamond Coupling Code.
All equipped with slow speed control.

58001	**FM**	E	*EW*	TO	
58002	**ML**	E	*EW*	TO	Daw Mill Colliery
58003	**FM**	E	*EW*	TO	Markham Colliery
58004	**FM**	E	*EW*	TO	
58005	**ML**	E	*EW*	TO	Ironbridge Power Station
58006	**F**	E	*EW*	TO	
58007	**FM**	E	*EW*	TO	Drakelow Power Station
58008	**ML**	E	*EW*	TO	
58009	**FM**	E	*EW*	TO	
58010	**FM**	E	*EW*	TO	
58011	**FM**	E	*EW*	TO	Worksop Depot
58012	**FM**	E	*EW*	TO	
58013	**ML**	E	*EW*	TO	
58014	**ML**	E	*EW*	TO	Didcot Power Station
58015	**FM**	E	*EW*	TO	
58016	**E**	E	*EW*	TO	
58017	**FM**	E	*EW*	TO	Eastleigh Depot
58018	**FM**	E	*EW*	TO	High Marnham Power Station
58019	**FM**	E	*EW*	TO	Shirebrook Colliery
58020	**FM**	E	*EW*	TO	Doncaster Works
58021	**ML**	E	*EW*	TO	Hither Green Depot
58022	**FM**	E	*EW*	TO	
58023	**ML**	E	*EW*	TO	Peterborough Depot
58024	**E**	E	*EW*	TO	
58025	**E**	E	*EW*	TO	
58026	**FM**	E	*EW*	TO	
58027	**FM**	E	*EW*	TO	
58028	**FM**	E	*EW*	TO	
58029	**FM**	E	*EW*	TO	
58030	**E**	E	*EW*	TO	
58031	**FM**	E	*EW*	TO	
58032	**ML**	E	*EW*	TO	Thoresby Colliery
58033	**E**	E	*EW*	TO	

58034	**FM**	E	*EW*	TO	Bassetlaw
58035	**FM**	E	*EW*	TO	
58036	**ML**	E	*EW*	TO	
58037	**E**	E	*EW*	TO	
58038	**ML**	E	*EW*	TO	
58039	**E**	E	*EW*	TO	
58040	**FM**	E	*EW*	TO	Cottam Power Station
58041	**FM**	E	*EW*	TO	Ratcliffe Power Station
58042	**ML**	E	*EW*	TO	Petrolea
58043	**FM**	E	*EW*	TO	
58044	**FM**	E	*EW*	TO	Oxcroft Opencast
58045	**FM**	E	*EW*	TO	
58046	**ML**	E	*EW*	TO	Asfordby Mine
58047	**E**	E	*EW*	TO	
58048	**E**	E	*EW*	TO	
58049	**E**	E	*EW*	TO	Littleton Colliery
58050	**E**	E	*EW*	TO	

CLASS 59   GENERAL MOTORS TYPE 5   Co–Co

Built: 1985 (59001–4), 1989 (59005) by General Motors, La Grange, Illinois, U.S.A. or 1990 (59101–4), 1994 (59201) and 1995 (59202–6) by General Motors, London, Ontario, Canada.
Engine: General Motors 645E3C two stroke of 2460 kW (3300 hp) at 900 rpm.
Main Alternator: General Motors AR11 MLD-D14A.
Traction Motors: General Motors D77B.
Max. Tractive Effort: 506 kN (113 550 lbf).
Cont. Tractive Effort: 291 kN (65 300 lbf) at 14.3 m.p.h.
Power At Rail: 1889 kW (2533 hp).   **Length over Buffers:** 21.35 m.
Brake Force: 69 t.   **Wheel Diameter:** 1067 mm.
Weight: 121 t.   **RA:** 7.
Design Speed: 60 m.p.h. (75 m.p.h. Class 59/2).
Max. Speed: 60 m.p.h. (75 m.p.h. Class 59/2).

Class 59/0. Owned by Foster-Yeoman Ltd.

59001	**FY**	FY	*MD*	MD	YEOMAN ENDEAVOUR
59002	**FY**	FY	*MD*	MD	ALAN J DAY
59004	**FY**	FY	*MD*	MD	PAUL A HAMMOND
59005	**FY**	FY	*MD*	MD	KENNETH J. PAINTER

Class 59/1. Owned by ARC Limited.

59101	**AC**	AC	*MD*	WH	Village of Whatley
59102	**AC**	AC	*MD*	WH	Village of Chantry
59103	**AC**	AC	*MD*	WH	Village of Mells
59104	**AC**	AC	*MD*	WH	Village of Great Elm

Class 59/2. Owned by National Power.

59201	**NP**	NP	*NP*	FB	Vale of York
59202	**NP**	NP	*NP*	FB	Vale of White Horse
59203	**NP**	NP	*NP*	FB	Vale of Pickering
59204	**NP**	NP	*NP*	FB	Vale of Glamorgan

| 59205 | **NP** | NP | *NP* | FB | Vale of Evesham |
| 59206 | **NP** | NP | *NP* | FB | Pride of Ferrybridge |

CLASS 60   BRUSH TYPE 5   Co–Co

Built: 1989–1993 by Brush Traction at Loughborough.
Engine: Mirrlees MB275T of 2310 kW (3100 hp) at 1000 rpm.
Main Alternator: Brush
Traction Motors: Brush separately excited.
Max. Tractive Effort: 500 kN (106500 lbf).
Cont. Tractive Effort: 336 kN (71570 lbf) at 17.4 m.p.h.
Power At Rail: 1800 kW (2415 hp).   **Length over Buffers:** 21.34 m.
Brake Force: 74 t.   **Wheel Diameter:** 1118 mm.
Design Speed: 62 m.p.h.   **Weight:** 129 t.
Max. Speed: 60 m.p.h.   **RA:** 7.
Multiple Working: Within class only.
All equipped with slow speed control.

Non-standard liveries:
60006 & 60033 are British Steel Trafalgar blue.

60001		**E**	E	*EW*	TO	
60002	+	**E**	E	*EW*	TO	
60003	+	**E**	E	*EW*	TO	FREIGHT TRANSPORT ASSOCIATION
60004	+	**E**	E	*EW*	TO	
60005		**E**	E	*EW*	TO	
60006		**0**	E	*EW*	TO	Scunthorpe Ironmaster
60007	+	**LH**	E	*EW*	TO	GYPSUM QUEEN II
60008		**LH**	E	*EW*	TO	GYPSUM QUEEN II
60009	+	**E**	E	*EW*	TO	
60010		**E**	E	*EW*	TO	
60011		**ML**	E	*EW*	TO	
60012	+	**E**	E	*EW*	TO	
60013		**F**	E	*EW*	TO	Robert Boyle
60014		**E**	E	*EW*	TO	
60015	+	**FT**	E	*EW*	TO	Bow Fell
60016		**E**	E	*EW*	TO	
60017	+	**E**	E	*EW*	TO	Shotton Works Centenary Year 1996
60018		**E**	E	*EW*	TO	
60019		**E**	E	*EW*	TO	
60020	+	**E**	E	*EW*	TO	
60021	+	**F**	E	*EW*	TO	Pen-y-Ghent
60022	+	**E**	E	*EW*	TO	
60023	+	**E**	E	*EW*	TO	
60024		**E**	E	*EW*	TO	
60025	+	**LH**	E	*EW*	TO	
60026	+	**E**	E	*EW*	TO	
60027	+	**E**	E	*EW*	TO	
60028	+	**E**	E	*EW*	TO	
60029		**E**	E	*EW*	TO	
60030		**E**	E	*EW*	TO	
60031		**F**	E	*EW*	TO	

60032	**FT**	E	*EW*	TO	William Booth
60033	**0**	E	*EW*	TO	Tees Steel Express
60034	**FT**	E	*EW*	TO	Carnedd Llewelyn
60035	**FT**	E	*EW*	TO	Florence Nightingale
60036	**E**	E	*EW*	TO	
60037 +	**E**	E	*EW*	TO	Aberddawan/Aberthaw
60038 +	**LH**	E	*EW*	TO	
60039	**E**	E	*EW*	TO	
60040	**E**	E	*EW*	TO	
60041 +	**E**	E	*EW*	TO	
60042	**E**	E	*EW*	TO	
60043	**E**	E	*EW*	TO	
60044	**ML**	E	*EW*	TO	
60045	**E**	E	*EW*	TO	The Permanent Way Institution
60046	**E**	E	*EW*	TO	
60047 +	**E**	E	*EW*	TO	
60048	**E**	E	*EW*	TO	Eastern
60049	**E**	E	*EW*	TO	
60050	**E**	E	*EW*	TO	
60051 +	**E**	E	*EW*	TO	
60052 +	**E**	E	*EW*	TO	
60053	**E**	E	*EW*	TO	Nordic Terminal
60054 +	**F**	E	*EW*	TO	Charles Babbage
60055	**FT**	E	*EW*	TO	Thomas Barnardo
60056 +	**FT**	E	*EW*	TO	William Beveridge
60057	**F**	E	*EW*	TO	Adam Smith
60058	**FT**	E	*EW*	TO	John Howard
60059 +	**LH**	E	*EW*	TO	Swinden Dalesman
60060	**F**	E	*EW*	TO	James Watt
60061	**FT·**	E	*EW*	TO	Alexander Graham Bell
60062	**FT**	E	*EW*	TO	Samuel Johnson
60063	**FT**	E	*EW*	TO	James Murray
60064 +	**FH**	E	*EW*	TO	Back Tor
60065	**FT**	E	*EW*	TO	Kinder Low
60066	**FT**	E	*EW*	TO	John Logie Baird
60067	**F**	E	*EW*	TO	James Clerk-Maxwell
60068	**F**	E	*EW*	TO	Charles Darwin
60069	**F**	E	*EW*	TO	Humphry Davy
60070 +	**FH**	E	*EW*	TO	John Loudon McAdam
60071 +	**FM**	E	*EW*	TO	Dorothy Garrod
60072	**FM**	E	*EW*	TO	
60073	**FM**	E	*EW*	TO	
60074	**FM**	E	*EW*	TO	Braeriach
60075	**FM**	E	*EW*	TO	
60076	**FM**	E	*EW*	TO	
60077 +	**FM**	E	*EW*	TO	
60078	**ML**	E	*EW*	TO	
60079	**FM**	E	*EW*	TO	Foinaven
60080 +	**FT**	E	*EW*	TO	Kinder Scout
60081 +	**FT**	E	*EW*	TO	
60082	**F**	E	*EW*	TO	Mam Tor

60083		E	E	EW	TO	
60084		FT	E	EW	TO	Cross Fell
60085		FT	E	EW	TO	
60086		FM	E	EW	TO	Schiehallion
60087		FM	E	EW	TO	Slioch
60088		FM	E	EW	TO	Buachaille Etive Mor
60089		FT	E	EW	TO	Arcuil
60090	+	F	E	EW	TO	Quinag
60091		F	E	EW	TO	An Teallach
60092		FT	E	EW	TO	Reginald Munns
60093		FT	E	EW	TO	Jack Stirk
60094		FM	E	EW	TO	Tryfan
60095		F	E	EW	TO	
60096	+	E	E	EW	TO	Ben Macdui
60097		FT	E	EW	TO	Pillar
60098	+	E	E	EW	TO	Charles Francis Brush
60099		FM	E	EW	TO	Ben More Assynt
60100		FM	E	EW	TO	Boar of Badenoch

CLASS 66   GENERAL MOTORS TYPE 5   Co–Co

Built: 1998 onwards by General Motors, London, Ontario, Canada.
Engine: General Motors 12N-710-3GB-EC two stroke of 2460 kW (3300 hp) at 900 rpm.
Main Alternator: General Motors AR8.
Traction Motors: General Motors D43TR.
Max. Tractive Effort: 399 kN (89800 lbf).
Cont. Tractive Effort: 253 kN (57000 lbf) at 15.8 m.p.h.

Power At Rail:	**Length over Buffers:** 21.39 m.
Brake Force:	**Wheel Diameter:** 1120 mm.
Weight:	**RA:** 7.
Design Speed: 75 m.p.h.	
Max. Speed: 75 m.p.h.	

66001
66002
66003
66004
66005
66006
66007
66008
66009
66010
66011
66012
66013
66014
66015
66016
66017

66018
66019
66020
66021
66022
66023
66024
66025
66026
66027
66028
66029
66030
66031
66032
66033
66034
66035
66036
66037
66038
66039
66040
66041
66042
66043
66044
66045
66046
66047
66048
66049
66050
66051
66052
66053
66054
66055
66056
66057
66058
66059
66060
66061
66062
66063
66064
66065
66006
66067
66068

66069
66070
66071
66072
66073
66074
66075
66076
66077
66078
66079
66080
66081
66082
66083
66084
66085
66086
66087
66088
66089
66090
66091
66092
66093
66094
66095
66096
66097
66098
66099
66100

On order to 66250.

1.2. ELECTRIC LOCOMOTIVES

CLASS 73/0   ELECTRO-DIESEL   Bo–Bo

Built: 1962 by BR at Eastleigh Works.
Supply System: 660–850 V d.c. from third rail.
Engine: English Electric 4SRKT of 447 kW (600 hp) at 850 rpm.
Main Generator: English Electric 824/3D.
Traction Motors: English Electric 542A.
Max. Tractive Effort: Electric 187 kN (42000 lbf). Diesel 152 kN (34100 lbf).
Continuous Rating: Electric 1060 kW (1420 hp) giving a tractive effort of 43 kN (9600 lbf) at 55.5 m.p.h.
Cont. Tractive Effort: Diesel 72 kN (16100 lbf) at 10 m.p.h.
Maximum Rail Power: Electric 1830 kW (2450 hp) at 37 m.p.h.
Brake Force: 31 t.   **Length over Buffers:** 16.36 m.
Design Speed: 80 m.p.h.   **Weight:** 76.5 t.
Max. Speed: 60 m.p.h.   **RA:** 6.
Wheel Diameter: 1016 mm.   **ETH Index (Elec. power):** 66.
Train Brakes: Air, Vacuum and electro-pneumatic.
Multiple Working: Within sub-class, with Class 33/1 and various 750 V d.c. EMUs.
Couplings: Drop-head buckeye.

Formerly numbered E 6002/5.

73002	**BR**	ME	Kirkdale (U)
73005		ME	Kirkdale (U)

CLASS 73/1 & 73/2   ELECTRO-DIESEL   Bo–Bo

Built: 1965–67 by English Electric Co. at Vulcan Foundry, Newton le Willows.
Supply System: 660–850 V d.c. from third rail.
Engine: English Electric 4SRKT of 447 kW (600 hp) at 850 rpm.
Main Generator: English Electric 824/5D.
Traction Motors: English Electric 546/1B.
Max. Tractive Effort: Electric 179 kN (40000 lbf). Diesel 160 kN (36000 lbf).
Continuous Rating: Electric 1060 kW (1420 hp) giving a tractive effort of 35 kN (7800 lbf) at 68 m.p.h.
Cont. Tractive Effort: Diesel 60 kN (13600 lbf) at 11.5 m.p.h.
Maximum Rail Power: Electric 2350 kW (3150 hp) at 42 m.p.h.
Brake Force: 31 t.   **Length over Buffers:** 16.36 m.
Design Speed: 90 m.p.h.   **Weight:** 77 t.
Max. Speed: 60 (90*) m.p.h.   **RA:** 6.
Wheel Diameter: 1016 mm.   **ETH Index (Elec. power):** 66.
Train Brakes: Air, Vacuum and electro-pneumatic.
Multiple Working: Within sub-class, with Class 33/1 and various 750 V d.c. EMUs.
Couplings: Drop-head buckeye.

a Vacuum brake isolated.

Formerly numbered E 6001–20/22–26/28–49 (not in order).

73101		**PC**	E	*EW*	HG	The Royal Alex'
73103		**IO**	E	*EW*	HG	
73104		**IO**	E	*EW*	HG	
73105		**C**	E	*EW*	HG	
73106		**D**	E	*EW*	HG	
73107		**C**	E	*EW*	HG	Redhill 1844–1994
73108		**C**	E	*EW*	HG	
73109	*	**ST**	SW	*SW*	BM	Battle of Britain 50th Anniversary
73110		**C**	E	*EW*	HG	Stewarts Lane
73114		**ML**	E	*EW*	HG	Traction Maintenance Depot
73117		**IO**	E	*EW*	HG	University of Surrey
73118	c	**EP**	LC	*ES*	OC	
73119		**C**	E	*EW*	HG	Kentish Mercury
73126		**N**	E		OC (U)	
73128		**E**	E	*EW*	HG	
73129		**N**	E	*EW*	HG	City of Winchester
73130	c	**EP**	LC	*ES*	OC	
73131		**E**	E	*EW*	HG	
73132		**IO**	E	*EW*	HG	
73133		**ML**	E	*EW*	HG	The Bluebell Railway
73134		**IO**	E	*EW*	HG	Woking Homes 1885–1985
73136		**ML**	E	*EW*	HG	Kent Youth Music
73138		**C**	E	*EW*	HG	
73139		**IO**	E	*EW*	HG	
73140		**IO**	E	*EW*	HG	
73141		**IO**	E	*EW*	HG	
73201	a*	**GX**	P	*GX*	SL	Broadlands
73202	a*	**GX**	P	*GX*	SL	Royal Observer Corps
73203	a*	**GX**	P	*GX*	SL	
73204	a*	**GX**	P	*GX*	SL	Stewarts Lane 1860–1985
73205	a*	**GX**	P	*GX*	SL	
73206	a*	**GX**	P	*GX*	SL	Gatwick Express
73207	a*	**GX**	P	*GX*	SL	County of East Sussex
73208	a*	**GX**	P	*GX*	SL	Croydon 1883–1983
73209	a*	**GX**	P	*GX*	SL	
73210	a*	**GX**	P	*GX*	SL	Selhurst
73211	a*	**GX**	P	*GX*	SL	
73212	a*	**GX**	P	*GX*	SL	Airtour Suisse
73213	a*	**GX**	P	*GX*	SL	University of Kent at Canterbury
73235	a*	**GX**	P	*GX*	SL	

CLASS 73/9   ELECTRO-DIESEL   Bo–Bo

For details see Class 73/0. Sandite fitted locos.

Formerly numbered E 6001/6.

73901	**MD**	ME	*ME*	BD
73906	**MD**	ME	*ME*	BD

NOTES FOR CLASSES 86–91.

The following common features apply to all locos of Classes 86–91.

Supply System: 25 kV a.c. from overhead equipment.
Communication Equipment: Driver–guard telephone.
Multiple Working: Time division multiplex system.

a vacuum brake isolated.

Class 86 were formerly numbered E 3101–3200 (not in order).

CLASS 86/1   BR DESIGN   Bo–Bo

Built: 1965–66 by English Electric Co. at Vulcan Foundry, Newton le Willows or BR at Doncaster Works. Rebuilt with Class 87 type bogies and motors. Tap changer control.
Traction Motors: GEC G412AZ frame mounted.
Max. Tractive Effort: 258 kN (58000 lbf).
Continuous Rating: 3730 kW (5000 hp) giving a tractive effort of 95 kN (21300 lbf) at 87 m.p.h.
Maximum Rail Power: 5860 kW (7860 hp) at 50.8 m.p.h.

Brake Force: 40 t.	**Length over Buffers:** 17.83 m.
Design Speed: 110 m.p.h.	**Weight:** 87 t.
Max. Speed: 110 m.p.h.	**RA:** 6.
ETH Index: 74.	**Wheel Diameter:** 1150 mm.
Train Brakes: Air & Vacuum.	**Electric Brake:** Rheostatic.

86101	I	F	ZC (U)	Sir William A Stanier FRS
86102	I	F	ZC (U)	Robert A Riddles
86103	I	F	ZC (U)	André Chapelon

CLASS 86/2   BR DESIGN   Bo–Bo

Built: 1965–66 by English Electric Co. at Vulcan Foundry, Newton le Willows or BR at Doncaster Works. Later rebuilt with resilient wheels and flexicoil suspension. Tap changer control.
Traction Motors: AEI 282BZ.
Max. Tractive Effort: 207 kN (46500 lbf).
Continuous Rating: 3010 kW (4040 hp) giving a tractive effort of 85 kN (19200 lbf) at 77.5 m.p.h.
Maximum Rail Power: 4550 kW (6100 hp) at 49.5 m.p.h.

Brake Force: 40 t.	**Length over Buffers:** 17.83 m.
Design Speed: 125 m.p.h.	**Weight:** 85 t–86 t.
Max. Speed: 100 (110*) m.p.h.	**RA:** 6.
ETH Index: 66 (75§).	**Wheel Diameter:** 1156 mm.
Train Brakes: Air & Vacuum.	**Electric Brake:** Rheostatic.

86204		I	F	VX	LG	City of Carlisle
86205	a	I	F	VX	LG	City of Lancaster
86206	a	I	F	VX	LG	City of Stoke on Trent
86207	a	I	F	VW	WN	City of Lichfield

86208		I	E	*EW*	CE	City of Chester
86209		I	F	*VW*	WN	City of Coventry
86210		**RX**	E	*EW*	CE	C.I.T. 75th Anniversary
86212	a	I	F	*VX*	LG	Preston Guild 1328–1992
86213		I	F	*VX*	LG	Lancashire Witch
86214	a	I	F	*VX*	LG	Sans Pareil
86215		I	F	*AR*	NC	
86216	a	I	F	*VW*	WN	Meteor
86217	a	I	F	*AR*	NC	City University
86218		I	F	*AR*	NC	YEAR OF OPERA & MUSICAL THEATRE 1997
86219	a	I	F		ZH (U)	Phoenix
86220	a§	I	F	*AR*	NC	The Round Tabler
86221		I	F	*AR*	NC	B.B.C. Look East
86222	a	I	F	*VX*	LG	Clothes Show Live
86223	a§	I	F	*AR*	NC	Norwich Union
86224	a	I	F	*VW*	WN	Caledonian
86225	a	I	F	*VW*	WN	Hardwicke
86226	a	I	F	*VX*	LG	CHARLES RENNIE MACKINTOSH
86227	a§	I	F	*VX*	LG	Sir Henry Johnson
86228		I	F	*AR*	NC	Vulcan Heritage
86229	a	I	F	*VX*	LG	Sir John Betjeman
86230		I	F	*AR*	NC	
86231	*	I	F	*VW*	WN	Starlight Express
86232	a	I	F	*AR*	NC	Norfolk and Norwich Festival
86233	a	I	F	*VX*	LG	Laurence Olivier
86234	a	I	F	*VX*	LG	J B Priestley OM
86235		I	F	*AR*	NC	Crown Point
86236	a	I	F	*VW*	WN	Josiah Wedgwood MASTER POTTER 1736–1795
86237		I	F	*AR*	NC	University of East Anglia
86238	a	I	F	*AR*	NC	European Community
86240		I	F	*VW*	WN	Bishop Eric Treacy
86241		**RX**	E	*EW*	CE	Glenfiddich
86242		I	F	*VW*	WN	James Kennedy GC
86243		**RX**	E	*EW*	CE	
86244	a	I	F	*VX*	LG	The Royal British Legion
86245	a	I	F	*VW*	WN	Dudley Castle
86246	a	I	F	*AR*	NC	Royal Anglian Regiment
86247	a	I	F	*VX*	LG	Abraham Darby
86248		I	F	*VW*	WN	Sir Clwyd/County of Clwyd
86249	a	I	F		ZC (U)	County of Merseyside
86250	a	I	F	*AR*	NC	The Glasgow Herald
86251		I	F	*VW*	WN	The Birmingham Post
86252	a	I	F	*VX*	LG	The Liverpool Daily Post
86253		I	F	*VW*	WN	The Manchester Guardian
86254		**RX**	E	*EW*	CE	
86255	a	I	F	*VX*	LG	Penrith Beacon
86256		I	F	*VW*	WN	Pebble Mill
86257	a	I	F		NC (U)	Snowdon
86258		I	F	*VW*	WN	Talyllyn–The First Preserved Railway

86259	a	**I**	F	*VX*	LG	Greater MANCHESTER
						THE LIFE & SOUL OF BRITAIN
86260	a	**I**	F	*VX*	LG	Driver Wallace Oakes G.C.
86261		**E**	E	*EW*	CE	THE RAIL CHARTER PARTNERSHIP

CLASS 86/4 & 86/6 — BR DESIGN — Bo-Bo

Built: 1965–66 by English Electric Co. at Vulcan Foundry, Newton le Willows or BR at Doncaster Works. Later rebuilt with resilient wheels and flexicoil suspension. Tap changer control.
Traction Motors: AEI 282AZ.
Max. Tractive Effort: 258 kN (58000 lbf).
Continuous Rating: 2680 kW (3600 hp) giving a tractive effort of 89 kN (20000 lbf) at 67 m.p.h.
Maximum Rail Power: 4400 kW (5900 hp) at 38 m.p.h.
Brake Force: 40 t.   **Length over Buffers:** 17.83 m.
Design Speed: 100 m.p.h.   **Weight:** 83 t–84 t.
Max. Speed: 100 (75*) m.p.h.   **RA:** 6.
ETH Index: 74 (66§).   **Wheel Diameter:** 1156 mm.
Train Brakes: Air & Vacuum.   **Electric Brake:** Rheostatic.

Class 86/6 have the ETH equipment isolated.

86401		**E**	E	*EW*	CE	
86602	a*	**F**	FL	*FL*	CE	
86603	a*	**FL**	FL	*FL*	CE	
86604	a*	**FL**	FL	*FL*	CE	
86605	a*	**FL**	FL	*FL*	CE	
86606	a*	**FL**	FL	*FL*	CE	
86607	a*	**F**	FL	*FL*	CE	
86608	a*	**RD**	FL	*FL*	CE	
86609	a*	**F**	FL	*FL*	CE	
86610	a*	**F**	FL	*FL*	CE	
86611	a*	**FL**	FL	*FL*	CE	Airey Neave
86612	a*	**FL**	P	*FL*	CE	Elizabeth Garrett Anderson
86613	a*	**F**	P	*FL*	CE	County of Lancashire
86614	a*	**FL**	P	*FL*	CE	Frank Hornby
86615	a*	**F**	P	*FL*	CE	Rotary International
86416		**RX**	E	*EW*	CE	
86417		**RX**	E	*EW*	CE	
86618	a*	**FL**	P	*FL*	CE	
86419		**RX**	E	*EW*	CE	
86620	a*	**F**	P	*FL*	CE	
86621	a*	**F**	P	*FL*	CE	London School of Economics
86622	a*	**FL**	P	*FL*	CE	
86623	a*	**FL**	P	*FL*	CE	
86424		**RX**	E	*EW*	CE	
86425		**RX**	E	*EW*	CE	Saint Mungo
86426		**E**	E	*EW*	CE	
86627	a*	**F**	P	*FL*	CE	The Industrial Society
86628	a*	**FL**	P	*FL*	CE	Aldaniti
86430		**RX**	E	*EW*	CE	Saint Edmund

86631	a*	F	P	*FL*	CE	
86632	a*	F	P	*FL*	CE	Brookside
86633	a*	FL	P	*FL*	CE	Wulfruna
86634	a*	F	P	*FL*	CE	
86635	a*	F	P	*FL*	CE	
86636	a*	F	P	*FL*	CE	
86637	a*	FL	P	*FL*	CE	
86638	a*	FL	P	*FL*	CE	
86639	a*	F	P	*FL*	CE	

CLASS 87   BR DESIGN   Bo–Bo

Built: 1973–75 by BREL at Crewe Works.
Traction Motors: GEC G412AZ frame mounted (87/0), G412BZ (87/1).
Max. Tractive Effort: 258 kN (58000 lbf).
Continuous Rating: 3730 kW (5000 hp) giving a tractive effort of 95 kN (21300 lbf) at 87 m.p.h. (Class 87/0), 3620 kW (4850 hp) giving a tractive effort of 96 kN (21600 lbf) at 84 m.p.h. (Class 87/1).
Maximum Rail Power: 5860 kW (7860 hp) at 50.8 m.p.h.

Brake Force: 40 t.	**Length over Buffers:** 17.83 m.
Design Speed: 110 m.p.h.	**Weight:** 83.5 t.
Max. Speed: 110 (100*) m.p.h.	**RA:** 6.
ETH Index: 95 (75§).	**Wheel Diameter:** 1150 mm.
Train Brakes: Air.	**Electric Brake:** Rheostatic.

Class 87/0. Standard Design. Tap Changer Control.

87001	I		P	*VW*	WN	Royal Scot
87002	I		P	*VW*	WN	Royal Sovereign
87003	I		P	*VW*	WN	Patriot
87004	I		P	*VW*	WN	Britannia
87005	I		P	*VW*	WN	City of London
87006	V		P	*VW*	WN	George Reynolds
87007	I		P	*VW*	WN	City of Manchester
87008	I		P	*VW*	WN	City of Liverpool
87009	§	V	P	*VW*	WN	
87010	I		P	*VW*	WN	King Arthur
87011	I		P	*VW*	WN	The Black Prince
87012	I		P	*VW*	WN	The Royal Bank of Scotland
87013	I		P	*VW*	WN	John O' Gaunt
87014	I		P	*VW*	WN	Knight of the Thistle
87015	I		P	*VW*	WN	Howard of Effingham
87016	V		P	*VW*	WN	Willesden Intercity Depot
87017	I		P	*VW*	WN	Iron Duke
87018	I		P	*VW*	WN	Lord Nelson
87019	I		P	*VW*	WN	Sir Winston Churchill
87020	I		P	*VW*	WN	North Briton
87021	I		P	*VW*	WN	Robert The Bruce
87022	I		P	*VW*	WN	Cock o' the North
87023	I		P	*VW*	WN	Velocity
87024	I		P	*VW*	WN	Lord of the Isles
87025	I		P	*VW*	WN	County of Cheshire

87026	I	P	VW	WN	Sir Richard Arkwright
87027	I	P	VW	WN	Wolf of Badenoch
87028	I	P	VW	WN	Lord President
87029	§ I	P	VW	WN	Earl Marischal
87030	I	P	VW	WN	Black Douglas
87031	I	P	VW	WN	Hal o' the Wynd
87032	I	P	VW	WN	Kenilworth
87033	I	P	VW	WN	Thane of Fife
87034	I	P	VW	WN	William Shakespeare
87035	I	P	VW	WN	Robert Burns

Class 87/1. Thyristor Control.

| 87101 | * | E | EW | CE | STEPHENSON |

CLASS 89   BRUSH DESIGN   Co–Co

Built: 1987 by BREL at Crewe Works.
Traction Motors: Brush design frame mounted.
Max. Tractive Effort: 205 kN (46000 lbf).
Continuous Rating: 2390 kW (3200 hp) giving a tractive effort of 105 kN (23600 lbf) at 92 m.p.h.
Maximum Rail Power:
Brake Force: 40 t.
Design Speed: 125 m.p.h.
Max. Speed: 125 m.p.h.
ETH Index: 95.
Train Brakes: Air.
Couplings: Drop-head buckeye.

Length over Buffers: 18.80 m.
Weight: 104 t.
RA: 6.
Wheel Diameter: 1150 mm.
Electric Brake: Rheostatic.

| 89001 | GN | SL | GN | BN |

CLASS 90   GEC DESIGN   Bo–Bo

Built: 1987–90 by BREL at Crewe Works. Thyristor control.
Traction Motors: GEC G412CY separately excited frame mounted.
Max. Tractive Effort: 258 kN (58000 lbf).
Continuous Rating: 3730 kW (5000 hp) giving a tractive effort of 95 kN (21300 lbf) at 87 m.p.h.
Maximum Rail Power: 5860 kW (7860 hp) at 68.3 m.p.h.
Brake Force: 40 t.
Design Speed: 110 m.p.h.
Max. Speed: 110 (75*) m.p.h.
ETH Index: 95.
Train Brakes: Air.
Couplings: Drop-head buckeye (removed on Class 90/1).

Length over Buffers: 18.80 m.
Weight: 84.5 t.
RA: 7.
Wheel Diameter: 1156 mm.
Electric Brake: Rheostatic.

Non-standard Liveries:
90128 is in SNCB/NMBS (Belgian Railways) electric loco livery.
90129 is in DB (German Federal Railways) 'neurot' (new red) livery.
90130 is in SNCF (French Railways) 'Sybic' livery.
90136 is **RD**, but with full yellow ends and roof and red 'Railfreight Distribution' lettering.

Class 90/0. As Built.

90001	**I**	P	*VW*	WN	BBC Midlands Today
90002	**V**	P	*VW*	WN	Mission:Impossible
90003	**I**	P	*VW*	WN	THE HERALD
90004	**V**	P	*VW*	WN	
90005	**I**	P	*VW*	WN	Financial Times
90006	**I**	P	*VW*	WN	High Sheriff
90007	**I**	P	*VW*	WN	Lord Stamp
90008	**I**	P	*VW*	WN	The Birmingham Royal Ballet
90009	**I**	P	*VW*	WN	The Economist
90010	**I**	P	*VW*	WN	275 Railway Squadron (Volunteers)
90011	**I**	P	*VW*	WN	The Chartered Institute of Transport
90012	**V**	P	*VW*	WN	British Transport Police
90013	**I**	P	*VW*	WN	The Law Society
90014	**V**	P	*VW*	WN	
90015	**V**	P	*VW*	WN	
90016	**RX**	E	*EW*	CE	
90017	**RX**	E	*EW*	CE	Rail express systems Quality Assured
90018	**RX**	E	*EW*	CE	
90019	**RX**	E	*EW*	CE	Penny Black
90020	**E**	E	*EW*	CE	Sir Michael Heron
90021	**RD**	E	*EW*	CE	
90022	**RD**	E	*EW*	CE	Freightconnection
90023	**RD**	E	*EW*	CE	
90024	**RD**	E	*EW*	CE	

Class 90/1. ETH equipment isolated.

90125	*	**RD**	E	*EW*	CE	
90126	*	**RD**	E	*EW*	CE	Crewe International Electric Maintenance Depot
90127	*	**F**	E	*EW*	CE	Allerton T&RS Depot Quality Approved
90128	*	**0**	E	*EW*	CE	Vrachtverbinding
90129	*	**0**	E	*EW*	CE	Frachtverbindungen
90130	*	**0**	E	*EW*	CE	Fretconnection
90131	*	**RD**	E	*EW*	CE	Intercontainer
90132	*	**RD**	E	*EW*	CE	Cerestar
90133	*	**RD**	E	*EW*	CE	
90134	*	**RD**	E	*EW*	CE	
90135	*	**RD**	E	*EW*	CE	Crewe Basford Hall
90136	*	**0**	E	*EW*	CE	
90137	*	**F**	E	*EW*	CE	
90138	*	**RD**	E	*EW*	CE	
90139	*	**F**	E	*EW*	CE	
90140	*	**F**	E	*EW*	CE	
90141	*	**FL**	P	*FL*	CE	
90142	*	**FL**	P	*FL*	CE	
90143	*	**FL**	P	*FL*	CE	Freightliner Coatbridge
90144	*	**FL**	P	*FL*	CE	
90145	*	**FL**	P	*FL*	CE	

90146	*	**FL**	P	*FL*	CE
90147	*	**FL**	P	*FL*	CE
90148	*	**FL**	P	*FL*	CE
90149	*	**FL**	P	*FL*	CE
90150	*	**FL**	P	*FL*	CE

CLASS 91   GEC DESIGN   Bo–Bo

Built: 1988–91 by BREL at Crewe Works. Thyristor control.
Traction Motors: GEC G426AZ.
Continuous Rating: 4540 kW (6090 hp).
Maximum Rail Power: 4700 kW (6300 hp).
Brake Force: 45 t.   **Length over Buffers:** 19.40 m.
Design Speed: 140 m.p.h.   **Weight:** 84 t.
Max. Speed: 140 m.p.h.   **RA:** 7.
ETH Index: 95.   **Wheel Diameter:** 1000 mm.
Train Brakes: Air.   **Electric Brake:** Rheostatic.
Couplings: Drop-head buckeye.

91001	**GN**	F	*GN*	BN	
91002	**GN**	F	*GN*	BN	
91003	**GN**	F	*GN*	BN	
91004	**GN**	F	*GN*	BN	
91005	**GN**	F	*GN*	BN	
91006	**GN**	F	*GN*	BN	
91007	**GN**	F	*GN*	BN	
91008	**GN**	F	*GN*	BN	
91009	**GN**	F	*GN*	BN	The Samaritans
91010	**GN**	F	*GN*	BN	
91011	**GN**	F	*GN*	BN	
91012	**GN**	F	*GN*	BN	
91013	**GN**	F	*GN*	BN	
91014	**GN**	F	*GN*	BN	
91015	**GN**	F	*GN*	BN	
91016	**GN**	F	*GN*	BN	
91017	**GN**	F	*GN*	BN	
91018	**GN**	F	*GN*	BN	
91019	**GN**	F	*GN*	BN	
91020	**GN**	F	*GN*	BN	
91021	**GN**	F	*GN*	BN	
91022	**GN**	F	*GN*	BN	
91023	**GN**	F	*GN*	BN	
91024	**GN**	F	*GN*	BN	
91025	**GN**	F	*GN*	BN	
91026	**GN**	F	*GN*	BN	
91027	**GN**	F	*GN*	BN	
91028	**GN**	F	*GN*	BN	
91029	**GN**	Γ	*GN*	BN	
91030	**GN**	F	*GN*	BN	
91031	**GN**	F	*GN*	BN	

CLASS 92   BRUSH DESIGN   Co–Co

Built: 1993–5 by Brush Traction at Loughborough. Thyristor control.
Supply System: 25 kV a.c. from overhead equipment and 750 V d.c. third rail.
Electrical equipment: ABB Transportation, Zürich, Switzerland.
Traction Motors: Brush design.
Max. Tractive Effort: 400 kN (90 000 lbf).
Continuous Rating at Motor Shaft: 5040 kW (6760 hp).
Maximum Rail Power (25 kV a.c.): 5000 kW (6700 hp).
Maximum Rail Power (750 V d.c.): 4000 kW (5360 hp).

Brake Force: t.	**Length over Buffers:** 21.34 m.
Design Speed: 140 km/h (87½ m.p.h.).	**Weight:** 126 t.
Max. Speed: 140 km/h (87½ m.p.h.).	**RA:** 8.
ETH Index: 108.	**Wheel Diameter:** 1160 mm.

Train Brakes: Air.
Electric Brake: Rheostatic & regenerative.
Multiple Working: Time division multiplex system.
Communication Equipment: Driver–guard telephone.
Cab Signalling: Fitted with TVM430 cab signalling for Channel Tunnel.

92001	**EP**	E	*EW*	CE	Victor Hugo
92002	**EP**	E	*EW*	CE	H G Wells
92003	**EP**	E	*EW*	CE	Beethoven
92004	**EP**	E	*EW*	CE	Jane Austen
92005	**EP**	E	*EW*	CE	Mozart
92006	**EP**	CF	*EW*	CE	Louis Armand
92007	**EP**	E	*EW*	CE	Schubert
92008	**EP**	E	*EW*	CE	Jules Verne
92009	**EP**	E	*EW*	CE	Elgar
92010	**EP**	CF	*EW*	CE	Molière
92011	**EP**	E	*EW*	CE	Handel
92012	**EP**	E	*EW*	CE	Thomas Hardy
92013	**EP**	E	*EW*	CE	Puccini
92014	**EP**	CF	*EW*	CE	Emile Zola
92015	**EP**	E	*EW*	CE	D H Lawrence
92016	**EP**	E	*EW*	CE	Brahms
92017	**EP**	E	*EW*	CE	Shakespeare
92018	**EP**	CF	*EW*	CE	Stendhal
92019	**EP**	E	*EW*	CE	Wagner
92020	**EP**	LC	*EW*	CE	Milton
92021	**EP**	LC	*EW*	CE	Purcell
92022	**EP**	E	*EW*	CE	Charles Dickens
92023	**EP**	CF	*EW*	CE	Ravel
92024	**EP**	E	*EW*	CE	J S Bach
92025	**EP**	E	*EW*	CE	Oscar Wilde
92026	**EP**	E	*EW*	CE	Britten
92027	**EP**	E	*EW*	CE	George Eliot
92028	**EP**	CF	*EW*	CE	Saint Saëns
92029	**EP**	E	*EW*	CE	Dante
92030	**EP**	E	*EW*	CE	Ashford

92031	**EP**	E	*EW*	CE	
92032	**EP**	LC	*EW*	CE	César Franck
92033	**EP**	CF	*EW*	CE	Berlioz
92034	**EP**	E	*EW*	CE	Kipling
92035	**EP**	E	*EW*	CE	Mendelssohn
92036	**EP**	E	*EW*	CE	Bertolt Brecht
92037	**EP**	E	*EW*	CE	Sullivan
92038	**EP**	CF	*EW*	CE	Voltaire
92039	**EP**	E	*EW*	CE	Johann Strauss
92040	**EP**	LC	*EW*	CE	Goethe
92041	**EP**	E	*EW*	CE	Vaughan Williams
92042	**EP**	E	*EW*	CE	Honegger
92043	**EP**	CF	*EW*	CE	Debussy
92044	**EP**	LC	*EW*	CE	Couperin
92045	**EP**	LC	*EW*	CE	Chaucer
92046	**EP**	LC	*EW*	CE	Sweelinck

KEEP RIGHT UP TO DATE WITH....

TODAY'S
RAILWAYS
A DIFFERENT RAILWAY MAGAZINE

Today's Railways is the PLATFORM 5 monthly magazine containing news and features from Britain and the rest of Europe. **Today's Railways** is the most comprehensive modern railway magazine available in Britain today.

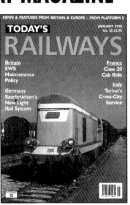

Every month there are sections on:

- News from Great Britain

- News from the Continent of Europe

- British Stock News, including the official Platform 5 stock changes

- European Stock News, including stock changes from Austria, Belgium, France, Germany, Italy, the Netherlands, Switzerland and other countries

- Light Rail and Metros

- Heritage News

Today's Railways is the only magazine which provides comprehensive stock change information in the Platform 5 format used in our pocket books, instead of the gobbledygook coding used in other publishers' magazines, e.g. Entry in Todays Railways:

92937 *SR* IS – *oou* PC

i.e. vehicle 92937 operated by Scotrail at Inverness has been taken out of use and is now stored at Polmadie. The same entry in a fortnightly magazine from another publisher reads:

92937 IS–HQ HAIS–SAXH

SAVE MONEY ON PLATFORM 5 BOOKS WITH TODAY'S RAILWAYS!

Every new subscriber to **Today's Railways** receives a £3.00 discount voucher which may be used against any PLATFORM 5 book which has a value of at least £9.95 when ordered from our Mail Order Department. This includes BRITISH RAILWAYS LOCOMOTIVES & COACHING STOCK 1998.

Today's Railways is published monthly and costs £2.95. It is available from good newsagents or on subscription at £35.40 for 12 issues.

For further details please contact our magazine subscription department at the address below:

Today's Railways (LCS), Platform 5 Publishing., 3 Wyvern House, Sark Road, SHEFFIELD, S2 4HG, ENGLAND.

TEL: 0114-255-2625   FAX: 0114-255-2471

DON'T MISS OUT - CALL OUR SUBSCRIPTION DEPARTMENT TODAY!

1.3. EUROTUNNEL LOCOMOTIVES

CLASS 0   MaK   Bo–Bo

These general purpose diesel locomotives are the same basic design as the
Netherlands Railways 6400 Class.
Built: 1992–3 by Krupp-MaK/ABB at Kiel, Germany. (Type DE1004)
Engine: MaK 940 kW (1280 hp) at 1800 r.p.m.
Traction Motor: Four ABB three-phase traction motors.
Max. Tractive Effort: 305 kN.
Continuous Tractive Effort: 140 kN at 20 m.p.h.
Power at Rail: 750 kW.
Brake Force: 120 kN.   **Length over Couplers:** 16.50 m.
Weight: 84 tonnes.   **Wheel Diameter:** 1000 mm.
Max. Speed: 120 (75 km/h).   **Train Brakes:** Air.
Communication Equipment: Cab to shore radio.
Couplings: High and Low level Sharfenberg plus UIC screw.
Cab Signalling: TVM 430.
Livery: Standard NS grey and yellow (Netherlands Railways).

0001	ET	*ET*	CU
0002	ET	*ET*	CU
0003	ET	*ET*	CU
0004	ET	*ET*	CU
0005	ET	*ET*	CU

CLASS 0   B

Schöma rebuilds of Hunslet 900 mm gauge locos.
Built: 1989–90. Rebuilt 1993–4.
Engine: Deutz 270 kW (200 hp).
Transmission: Mechanical.
Max. Speed: 30 m.p.h. (120 km/h when gears disengaged).
Train Brakes: Air.

0031	ET	*ET*	CU	FRANCES
0032	ET	*ET*	CU	ELIZABETH
0033	ET	*ET*	CU	SILKE
0034	ET	*ET*	CU	AMANDA
0035	ET	*ET*	CU	MARY
0036	ET	*ET*	CU	LAWRENCE
0037	ET	*ET*	CU	LYDIE
0038	ET	*ET*	CU	JENNIE
0039	ET	*ET*	CU	DIGITA
0040	ET	*ET*	CU	JILL
0041	ET	*ET*	CU	KIM
0042	ET	*ET*	CU	NICOLLE

CLASS 9   BRUSH EUROSHUTTLE   Bo–Bo–Bo

A.C. electric locomtives which are used on the Eurotunnel shuttle trains between Cheriton and Coquelles.
Built: 1992–4 (9001–38) and 1998 (9040, 9101–4) by Brush/ABB at Loughborough.
Supply System: 25 kV a.c. from overhead equipment.
Traction Motors: 6 x ABB 6PH.
Control System: GTO thyristor.
Max. Tractive Effort: 400 kN (90 000 lbf).
Continuous Rating: 5760 kW (7725 hp) giving a tractive effort of 310 kN at 65 km/h.

Brake Force: 50 t.	**Length over Buffers:** 22.00 m.
Design Speed: 175 km/h (110 m.p.h.).	**Weight:** 132 t.
Max. Speed: 160 km/h (100 m.p.h.).	**RA:** Channel Tunnel only.
ETH Index:	**Wheel Diameter:** 1090 mm.
Train Brakes: Air	**Electric Brake:** Regenerative.

Multiple Working: Time division multiplex system. RC232 data Bus.
Couplings: High level Sharfenberg plus UIC screw links.
Communication Equipment: Cab to shore radio and in-train system.
Cab Signalling: TVM 430.
Livery: Two-tone grey and white with green, blue bands.

Class 9/0. Standard Design.

9001	ET	*ET*	CU	LESLEY GARRETT
9002	ET	*ET*	CU	STUART BURROWS
9003	ET	*ET*	CU	BENJAMIN LUXON
9004	ET	*ET*	CU	VICTORIA DE LOS ANGELES
9005	ET	*ET*	CU	JESSYE NORMAN
9006	ET	*ET*	CU	REGINE CRESPIN
9007	ET	*ET*	CU	DAME JOAN SUTHERLAND
9008	ET	*ET*	CU	ELISABETH SODERSTROM
9009	ET	*ET*	CU	FRANCOIS POLLET
9010	ET	*ET*	CU	JEAN-PHILIPPE COURTIS
9011	ET	*ET*	CU	JOSE VAN DAM
9012	ET	*ET*	CU	LUCIANO PAVAROTTI
9013	ET	*ET*	CU	MARIA CALLAS
9014	ET	*ET*	CU	LUCIA POPP
9015	ET	*ET*	CU	LÖTCHBERG
9016	ET	*ET*	CU	WILLARD WHITE
9017	ET	*ET*	CU	JOSE CARRERAS
9018	ET	*ET*	CU	WILHELMINIA FERNANDEZ
9019	ET	*ET*	CU	MARIA EWING
9020	ET	*ET*	CU	NICOLAI GHIAUROV
9021	ET	*ET*	CU	TERESA BERGANZA
9022	ET	*ET*	CU	DAME JANET BAKER
9023	ET	*ET*	CU	DAME ELISABETH LEGGE-SCHWARZKOPF
9024	ET	*ET*	CU	GOTTHARD
9025	ET	*ET*	CU	JUNGFRAUJOCH
9026	ET	*ET*	CU	FURKATUNNEL

9027	ET	*ET*	CU	BARBARA HENDRICKS
9028	ET	*ET*	CU	DAME KIRI TE KANAWA
9029	ET	*ET*	CU	THOMAS ALLEN
9031	ET	*ET*	CU	PLACIDO DOMINGO
9032	ET	*ET*	CU	RENATA TIBALDI
9033	ET	*ET*	CU	MONSERRAT CABALLE
9034	ET	*ET*	CU	MIRELLA FRENI
9035	ET	*ET*	CU	NICOLAI GEDDA
9036	ET	*ET*	CU	ALAIN FONDARY
9037	ET	*ET*	CU	GABRIEL BAQUIER
9038	ET	*ET*	CU	HILDEGARD BEHRENS
9040				

Class 9/1. For use on freight shuttles only.

9101
9102
9103
9104

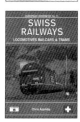

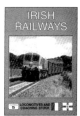

1.4. LOCOMOTIVES AWAITING DISPOSAL

03079		IL	Sandown
03179	N	IL	Ryde T&RSMD
08419		E	ADtranz Crewe Works
08473		E	Leicester LIP
08515		E	Gateshead WRD
08609		E	Willesden TMD
08618		E	Gateshead WRD
08634		E	Stratford TMD
08666		E	Allerton TMD
08673	IO	E	Allerton TMD
08677		E	Willesden TMD
08733		E	Motherwell TMD
08755		E	Millerhill Wagon Wks
08793	O	E	RFS(E) Ltd., Doncaster
08829		E	Toton TMD
08855		E	RFS(E) Ltd., Doncaster
08880		E	Allerton TMD
08895		E	Margam LIP
08898		E	RFS(E) Ltd., Doncaster
20073		E	Bescot Yard
20113	O	D	Brush, Loughborough
20119		E	Toton TMD
20154		E	Toton TMD
20175	O	D	Brush, Loughborough
20177		E	Toton TMD
25083		E	Crewe Brook Sidings
31105	FT	E	Bescot Yard
31112	CT	E	Bescot Yard
31168		E	Bescot Yard
31180	FR	E	Toton Yard
31184	FO	E	Toton Yard
31196	C	E	Stratford TMD
31209	F	E	Toton Yard
31217	F	E	Toton TMD
31282	FR	E	Bescot Yard
31283	O	E	Stratford TMD

31286		E	Bescot Yard
31289		E	Bescot Yard
31290	C	E	Toton Yard
31296	F	E	Crewe Brook Sidings
31299	FO	E	Stratford TMD
31320		E	Stratford TMD
31402		E	Bescot Yard
31403		E	Toton Yard
31428		E	Bescot Yard
31442		E	Crewe Brook Sidings
31460		E	Bescot Yard
31547	C	E	Toton Yard
31553	C	E	Toton Yard
31569	C	E	Toton Yard
33038		E	Stratford TMD
33205	F	E	Hither Green TMD
37252	F	E	Doncaster TMD
45015		E	Toton TMD
47096		E	Tinsley TMD
47102		E	Tinsley TMD
47190	F	E	Tinsley TMD
47214	F	E	Tinsley TMD
47249	FR	E	Tinsley TMD
47318	FO	E	Bescot Yard
47321	F	E	Tinsley TMD
47325	FO	E	Tinsley TMD
47515	M	E	Crewe Coal Siding
47707	RX	E	Crewe Basford Hall Yd
47714	RX	E	Crewe Basford Hall Yd
56009	F	E	Brush, Loughborough
56013	F	E	Toton TMD
56023	F	E	Toton TMD
56028	F	E	Margam WRD
56030	F	E	Margam WRD
56122	F	E	Toton TMD
97653	O	E	Reading T&RSMD

Non-Standard Liveries:
08793 is in London & North Eastern Railway apple green.
20113 & 20175 are RFS grey with blue and yellow bodyside stripes and carry numbers 2003 & 2007 respectively.
31283 is BR blue with large numbers.
97653 is Departmental yellow.

1.5. LOCOMOTIVES TOPS POOL CODES

A list of codes as used by TOPS is shown here for reference purposes:

SERCO

CDJD   Derby Etches Park Class 08

EWS (FORMERLY RAILFREIGHT DISTRIBUTION)

DAAN   Allerton Class 08
DADC   Crewe Electric Class 92 (Dollands Moor–Wembley)
DAEC   Crewe Electric Class 92 (Non-Operational)
DAET   Tinsley Class 47
DAIC   Crewe Electric Class 92 (EPS Testing)
DAMC   Crewe Electric Class 87/1 & 90
DASY   Tinsley Class 08/09 (Saltley)
DATI   Tinsley Class 08
DAWE   Allerton Class 08 (Wembley/Dagenham)
DAXT   Locomotives Awaiting Repair
DAYX   Stored Locomotives

FREIGHTLINER

DFLC   Crewe Electric Class 90/1
DFLM   Crewe Diesel Class 47 (Multiple Working Fitted)
DFLS   Crewe Diesel Class 08
DFLT   Crewe Diesel Class 47
DFNC   Crewe Electric Class 86/6
DFYX   Stored Locomotives
DHLT   Crewe Diesel Class 47 (Holding Pool)

EWS (FORMERLY MAINLINE)

ENAN   Toton Class 60
ENBN   Toton Class 58
ENSN   Toton Class 08/09 (Toton/Peterborough)
ENTN   Toton Class 37
ENXX   Stored Locomotives
ENZX   Locomotives For Withdrawal
EWDB   Eastleigh Class 37
EWDS   Eastleigh Class 33
EWEB   Hither Green Class 73
EWEH   Eastleigh Class 08
EWHG   Hither Green Class 09
EWOC   Old Oak Common Class 08/09
EWRB   Hither Green Class 73 (Restricted Use)
EWSF   Stratford Class 08/09
EWSU   Selhurst Class 08/09
EWSX   Stored/Reserve Shunters

EWS (FORMERLY LOADHAUL)

FDBI   Immingham Class 56

FDCI	Immingham Class 37
FDSD	Doncaster Class 08/09
FDSI	Immingham Class 08
FDSK	Knottingley Class 08/09
FDSX	Stored Shunters
FDYX	Stored Locomotives
FMSY	Thornaby Class 08/09

EUROSTAR (UK)

GPSN	Old Oak Common Class 73 (North Pole)
GPSS	Old Oak Common Class 08 (North Pole)
GPSV	Old Oak Common Class 37/6

TRAIN OPERATING COMPANIES

HASS	ScotRail Railways – Inverness Class 08
HBSH	Great North Eastern Railway – Bounds Green/Craigentinny Class 08
HEBD	Merseyrail Electrics – Birkenhead North Class 73
HFSL	Virgin West Coast – Longsight Class 08
HFSN	Virgin West Coast – Willesden Class 08
HGSS	Central Trains – Tyseley Class 08
HISE	Midland Mainline – Derby Etches Park Class 08
HISL	Midland Mainline – Neville Hill Class 08
HJSE	Great Western Trains – Landore Class 08
HJSL	Great Western Trains – Laira Class 08
HJXX	Great Western Trains – Old Oak Common/St Phillips Marsh Class 08
HLSV	Cardiff Railway Co. – Cardiff Canton Class 08
HSSN	Anglia Railways – Norwich Crown Point Class 08
HWSU	Connex South Central – Selhurst Class 09
HYSB	South Western Trains – Bournemouth Class 73/1
IANA	Anglia Railways – Norwich Crown Point Class 86/2
ICCA	Virgin Cross Country – Longsight Class 86/2
ICCP	Virgin Cross Country – Laira Class 43
ICCS	Virgin Cross Country – Edinburgh Craigentinny Class 43
IECA	Great North Eastern Railway – Bounds Green Class 91
IECB	Great North Eastern Railway – Bounds Green Class 89
IECP	Great North Eastern Railway – Craigentinny/Neville Hill Class 43
ILRA	Virgin Cross Country – Crewe Diesel Class 47/8
ILRB	Virgin Cross Country – Crewe Diesel Class 47/8 (Spot Hire)
IMLP	Midland Mainline – Neville Hill Class 43
IVGA	Gatwick Express – Stewarts Lane Class 73
IWCA	Virgin West Coast – Willesden Class 87/90
IWCP	Virgin West Coast – Manchester Longsight Class 43
IWLA	Great Western Trains – Laira Class 47
IWLX	Great Western Trains – Laira Class 47 (Reserve)
IWPA	Virgin West Coast – Willesden Class 86
IWRP	Great Western Trains – Laira/St Phillips Marsh Class 43

EWS (FORMERLY TRANSRAIL)

LBBS	Bescot Class 08/09
LCWX	Strategic Reserve Locomotives
LCXX	Stored Locomotives

LGBM   Motherwell Class 37
LGHM   Motherwell Class 37/4 (West Highland)
LGML   Motherwell Class 08/09
LNCF   Cardiff Canton Class 08/09
LNCK   Cardiff Canton Class 37 (Wales)
LNLK   Cardiff Canton Class 37 (St Blazey)
LNSK   Cardiff Canton Class 37 (Sandite Fitted)
LNWK   Cardiff Canton Class 08 (Allied Steel & Wire)
LWCW   Immingham Class 47
LWMC   Crewe Diesel Class 37/4 (North West Passenger)
LWNW   Bescot Class 31
LWSP   Crewe Diesel Class 08

HERITAGE LOCOMOTIVES

MBDL   Diesel Locomotives

EWS (FORMERLY RES)

PXLB   Crewe Diesel Class 47 (Extended Range)
PXLC   Crewe Diesel Class 47
PXLD   Motherwell Class 47
PXLE   Crewe Electric Class 86/90
PXLK   Crewe Diesel Class 47/9
PXLP   Crewe Diesel Class 47 (VIP Fleet)
PXLS   Crewe Diesel/Heaton/Old Oak Common Class 08/09
PXXA   Stored Locomotives

FORWARD TRUST RAIL

SAXL   Locomotives Off Lease

PORTERBROOK LEASING COMPANY

SBXL   Locomotives Off Lease

ANGEL TRAIN CONTRACTS

SCXL   Locomotives Off Lease

FRAGONSET RAILWAYS

SDFR   Locomotives For Hire

OTHER OPERATORS

XHSD   Direct Rail Services
XHSS   Direct Rail Services Stored Locomotives
XYPA   ARC Class 59/1
XYPD   Hunslet-Barclay Class 20/9
XYPN   National Power Class 59/2
XYPO   Foster-Yeoman Class 59/0
XYPS   Hunslet-Barclay Stored Locomotives

It will be possible, in most cases, for you to identify which pool a loco belongs to by its class and the owner, operation and depot codes shown. There are, however, a few exceptions to the rule and these are listed below.

Class 08 & 09

E owned locos whose operation code is *EW* and depot code is AN are DAAN except for 08389/93/482/694/825/44/72/913 which are DAWE.

E owned locos whose operation code is *EW* and depot code is CD are LWSP except for 08701/802/73/921, 09012 which are PXLS.

E owned locos whose operation code is *EW* and depot code is CF are LNCF except for 08651/819/900/32/42/54/5/93 which are LNWK.

E owned locos whose operation code is *EW* and depot code is TI are DASY except for 08879 which is DATI.

Class 37

E owned locos whose operation code is *EW* and depot code is CF are LNCK except for 37521/668–74/96 which are LNLK and 37178/97/229/30/54/63/75 which are LNSK.

E owned locos whose operation code is *EW* and depot code is ML are LGBM except for 37401/4/5/6/9/10/3/24/5/8/30 which are LGHM.

Class 47

E owned locos whose operation code is *EW* and depot code is CD are PXLB except for 47467/501/65/72/5/84/96/624/7/34/5/40 which are PXLC, 47971/6 which are PXLK and 47798/9 which are PXLP.

FL and P owned locos whose operation code is *FL* and depot code is CD are DFLT except 47114/50/2/7/204/5/9/34/58/79/87/9/90/2/303/9/23/30/7/58/61/70 which are DFLM.

P owned locos whose operation code is *GW* and depot code is LA are IWLA except for 47813/32 which are IWLX.

P owned locos whose operation code is *VX* and depot code is CD are ILRA except for 47825 which is ILRB.

Class 73

E owned locos whose operation code is *EW* and depot code is HG are EWEB except for 73128/31/2/9/40/1 which are EWRB.

Class 92

All locos are DADC except for 92005/8/14/7/8/21/3/5/7/36/40/3–6 which are DAEC and 92032 which is DAIC.

COACHING STOCK - INTRODUCTION

LAYOUT OF BOOK

Coaches are listed in batches, according to their class, with lot number information for the various batches being shown above the listings. Where a coach has been renumbered, the former number is shown in parentheses. If the coach has been renumbered more than once, both the original number and last number are shown in parentheses. Where the old number of a coach due to be converted or renumbered is known and the conversion or renumbering has not yet taken place, the coach is listed both under its old number with its depot allocation, and under its new number without an allocation. This book now includes 'preserved' coaches in the main section, as their is now no official distinction between those and other coaches registered to operate on Railtrack metals, all coaches in use now being privately-owned.

NUMBERING SYSTEMS

Six different numbering systems were in use on BR. These were the BR series, the four pre-nationalisation companies' series' and the Pullman Car Company's series. BR number series coaches and former Pullman Car Company series ones are listed separately. There is also a separate listing of 'Saloon' type vehicles which are registered to run on Railtrack metals. Please note that the Mark 2 Manchester Pullman vehicles were ordered after the Pullman Car Company had been nationalised and are therefore numbered in the BR series.

DETAILED INFORMATION AND CODES

After the heading, the following details are shown:

*   Diagram code. This consists of the first three characters of the TOPS code followed by two numbers which relate to the particular design of vehicle.
*   'Mark' of coach.
*   Number of first class seats , standard class seats and lavatory compartments shown as F/S nT respectively.
*   Bogie type (see below).
*   Additional features.
*   ETH Index.

BOGIE TYPES

BR Mk 1 (BR1). Standard double bolster leaf spring bogie. Generally 90 m.p.h. but certain vehicles were allowed to run at 100 m.p.h. with special maintenance. Weight: 6.1 t.
BR Mk 2 (BR2). Single bolster leaf-spring bogie used on certain types of non-passenger stock and suburban stock (all now withdrawn). Weight: 5.3 t.

COMMONWEALTH (C). Heavy, cast steel coil spring bogie. 100 m.p.h. Weight: 6.75 t.

B4. Coil spring fabricated bogie for 100 m.p.h. Weight: 5.2 t. Note: B4 bogies are allowed to run at 110 m.p.h. provided that they have a special maintenance regime. This applies to certain BGs (NHA).

B5. Heavy duty version of B4. 100 m.p.h. Weight: 5.3 t.

B5 (SR). A bogie originally used on ex Southern Region EMUs, similar in design to the B5 above. Now also used on locomotive hauled coaches. 100 m.p.h.

BT10. A fabricated bogie designed for 125 m.p.h. Air suspension.

T4. A 125 m.p.h. bogie from BREL (now ADtranz).

BT41. Fitted to Mark 4 vehicles. Manufactured by the Swiss firm of SIG. At present limited to 125 m.p.h. but designed for 140 m.p.h.

BRAKE TYPES

The standard form of braking on British main line trains is now air braking. Exceptions are shown as follows:

v	Vacuum braked.
x	Dual braked (air and vacuum).

HEATING

All heating on British main-line trains is now electric. Certain coaches for use on charter trains may, however have steam heating facilities also.

PUBLIC ADDRESS

It is assumed that all coaches are now fitted with public address, although certain stored coaches may not have the feature. In addition, it is assumed that all vehicles with a guard's compartment have public address transmission facilities, as have catering vehicles.

ADDITIONAL FEATURE CODES

d	Secondary door locking provided.
f	Facelifted or fluorescent lighting provided.
k	Composition brake blocks (instead of cast iron).
n	Day/night lighting.
p	Fitted with public telephone.
pg	Public address transmission and driver-guard communication.
pt	Public address transmission facility.
q	Fitted with catering staff to shore telephone.
r	Refurbished.
w	Fitted with wheelchair space.
z	Fitted with disabled persons' toilet and wheelchair space.

Note: Standard class coaches with wheelchair space also have one tip-up seat per space.

NOTES ON ETH INDICES

The sum of ETH indices in a train must not be more than that of the locomotive. The normal voltage on BR is 1000. Suffix 'X' denotes 600 amp wiring instead of 400 amp. Trains whose ETH index comes to more than 66 must be formed completely with 600 amp wired stock. Class 55 locomotives can not provide an ETH supply for Mark 2E or 2D FO 3192/3202, FK 13585–13607 & BFK 17163–17172. Class 33 locomotives can not provide an ETH supply for Mark 2D, Mark 2E, Mark 2F, Mark 3, Mark 3A, Mark 3B or Mark 4 stock.

LAYOUT

The layout in this section consists of number, original and last numbers in parentheses if applicable, notes (if any), livery, owner code, operation code and depot. For off-loan vehicles, the last storage location is given where known. Thus the layout is as follows:

No.	Notes	Livery	Owner	Operation	Depot
3131	x	**CC**	RS	*SS*	BN

THE DEVELOPMENT OF BR STANDARD COACHES

The standard BR coach built from 1951 to 1963 is the Mark 1. This has a separate underframe and body. The underframe is normally 64'6" long, but certain vehicles were built on short (57') frames. Tungsten lighting is standard and until 1961, BR Mark 1 bogies were generally provided. In 1959 TSOs to lot No. 30525 appeared with fluorescent lighting and melamine interior panels and from 1961 onwards Commonwealth bogies were fitted in an attempt to improve the quality of ride which became very poor when the tyre profiles on the wheels of the Mark 1 bogies became worn. The further batches of TSOs and BSOs retained the features of lot 30525, but the BSKs, SKs, BCKs and CKs, whilst utilising melamine panelling in standard class, still retained tungsten lighting. Wooden interior finish was retained in first class compartments. The FOs had fluorescent lighting with wooden panelling except for lot No. 30648 which had tungsten lighting. In later years many Mark 1s had their mark 1 bogies replaced by B4s.

In 1964, a new train was introduced. Known as "XP64", it featured new seat designs, pressure ventilation, aluminium compartment doors and corridor partitions, foot pedal operated toilets, and B4 bogies. The vehicles were on standard mark 1 underframes. Folding doors were fitted but these proved troublesome and were later replaced with hinged doors. All XP64 coaches have now been withdrawn, but some have been preserved.

The prototype Mark 2 vehicle (W 13252) was produced in 1963. This was an FK of semi-integral construction and was pressure ventilated. Tungsten lighting was provided and B4 bogies. This vehicle has been preserved by the National Railway Museum. The production build was similar, but wider windows were used. The standard class open vehicles used the new seat design similar to that in the XP64 and fluorescent lighting was provided. Interior finish reverted to wood. Mark 2s were built from 1964–66.

The Mark 2As, built 1967–68, incorporated the rest of the novel features first used in the XP64 set, i.e. foot pedal operated toilets (except BSOs), new first class seat design, aluminium compartment doors and partitions together with fluorescent lighting in first class compartments. Folding gangway doors (lime green coloured) were used instead of the traditional variety. The following list summarises the changes made in the later Mark 2 variants:

Mk 2B: Wide wrap round doors, no centre doors, slightly longer body. In standard class, one toilet at each end instead of two at one end as previously. Red gangway doors.

Mk 2C: Lowered ceiling with twin strips of fluorescent lighting, ducting for air conditioning, but no air conditioning.

Mk 2D: Air conditioning. No opening lights in windows.

Mk 2E: Smaller toilets with luggage racks opposite. Fawn gangway doors.

Mk 2F: Plastic interior panels. Inter-City 70 seats. Modified air conditioning system.

The Mark 3 coach has BT10 bogies, is 75' (23 m) long and is of fully integral construction with Inter-City 70 seats. Gangway doors were yellow (red in RFB) when new, although these are being altered on refurbishment. Loco-hauled coaches are classified Mark 3A, Mark 3 being reserved for HST trailers. A new batch of FOs and BFOs classified Mark 3B was built in 1985 with APT style seating and revised lighting. The last vehicles in the Mark 3 series were the driving brake vehicles (officially called driving van trailers) which have been built for West Coast Main Line services.

The Mark 4 coach built by Metro-Cammell for the East Coast Main Line electrification scheme features a body profile suitable for tilting trains, although tilt is not fitted, and is not intended to be. They are suitable for 140 m.p.h. running, although are restricted to 125 m.p.h. pending the installation of a more advanced signalling system on the East Coast Main Line.

2.1. BR NUMBER SERIES STOCK

TOPS CODES

TOPS codes for loco-hauled passenger stock are made up as follows:

(1) Two letters denoting the layout of the vehicle as follows:

AA	Gangwayed Corridor
AB	Gangwayed Corridor Brake
AC	Gangwayed Open (2+2 seating)
AD	Gangwayed Open (2+1 seating)
AE	Gangwayed Open Brake
AF	Gangwayed Driving Open Brake
AG	Micro-Buffet
AH	Brake Micro-Buffet
AI	As 'AC' but fitted with drop-head buckeye and no gangway at one end
AJ	Restaurant Buffet with Kitchen
AK	Kitchen Car
AL	As 'AC' but with disabled person's toilet (Mark 4 only)
AN	Miniature Buffet
AP	Pullman First with Kitchen
AQ	Pullman Parlour First
AR	Pullman Brake First
AS	Sleeping Car
AT	Royal Train Coach
AU	Sleeping Car with Pantry
AX	Generator Van
AY	Eurostar Barrier Vehicle
AZ	Special saloon
GF	DMU/EMU/Mark 4 Barrier Vehicle
GS	HST Barrier Vehicle
GX	Generator Van (three-phase supply)
NM	Sandite Coach

(2) A digit for the class of passenger accommodation:

1	First		4	Unclassified
2	Standard (formerly second)		5	None
3	Composite			

(3) A suffix relating to the build of coach:

1	Mark 1	A	Mark 2A	C	Mark 2C	E	Mark 2E	G	Mark 3	H	Mark 3B
Z	Mark 2	B	Mark 2B	D	Mark 2D	F	Mark 2F		or Mark 3A	J	Mark 4

OPERATOR CODES

The normal operator codes are given in brackets after the TOPS codes. These are as follows. Various other letters are in use and the meaning of these can be ascertained by referring to the titles at the head of each class:

B	Brake		O	Open
C	Composite		S	Standard (formerly second)
F	First		K	Side corridor with lavatory

AJ11 (RF)   RESTAURANT FIRST

Dia. AJ106. Mark 1. Gas cooking. 24/–. B5 bogies. ETH 2. This coach spent most of its life as a Royal train vehicle and was numbered 2907 for a time.

Lot No. 30633 Ashford/Swindon 1961. 41 t.

325   **WV** RS *SS*   BN

AP1Z (PK)   PULLMAN FIRST WITH KITCHEN

Dia. AP101. Mark 2. Pressure Ventilated. Electric cooking. 18/– 2T. B5 bogies. ETH 6.

Lot No. 30755 Derby 1966. 40 t.

Non-Standard Livery: Maroon & beige.

504	**0**	W	*SS*	CS	THE WHITE ROSE
506	**0**	W	*SS*	CS	THE RED ROSE

AQ1Z (PC)   PULLMAN PARLOUR FIRST

Dia. AQ101. Mark 2. Pressure Ventilated. 36/– 2T. B4 bogies. ETH 5.

Lot No. 30754 Derby 1966. 35 t.

Non-Standard Livery: Maroon & beige.

546	**0**	W	*SS*	CS	CITY OF MANCHESTER
548	**0**	W	*SS*	CS	ELIZABETHAN
549	**0**	W	*SS*	CS	PRINCE RUPERT
550	**0**	W	*SS*	CS	GOLDEN ARROW
551	**0**	W	*SS*	CS	CALEDONIAN
552	**0**	W	*SS*	CS	SOUTHERN BELLE
553	**0**	W	*SS*	CS	KING ARTHUR

AR1Z (PB)   PULLMAN BRAKE FIRST

Dia. AR101. Mark 2. Pressure Ventilated. 30/– 2T. B4 bogies. ETH 4.

 Lot No. 30753 Derby 1966. 35 t.

Non-Standard Livery: Maroon & beige.

586	**0**	W	*SS*	CS	TALISMAN

AJ1F (RFB)   BUFFET OPEN FIRST

Dia. AJ104. Mark 2F. Air conditioned. Electric cooking. Converted 1988–9/91 at BREL, Derby from Mark 2F FOs. 1200/1/3/6/11/14–17/20/21/50/2/5/6/9 have Stones equipment, others have Temperature Ltd. 25/– 1T plus wheelchair space (except 1217 and 1253 which are 26/– 1T). B4 bogies. p. q. Secondary door locks. ETH 6X.

1200/3/6/11/14/16/20/52/5/6. Lot No. 30845 Derby 1973. 33 t.
1201/4/5/7/8/10/12/13/15/17–9/21/50/1/4/7/9. Lot No. 30859 Derby 1973–4. 33 t.
1202/9/53/8. Lot No. 30873 Derby 1974–5. 33 t.

r Refurbished with new seat trim also new ma sets (except 1203, 1256).

1200	(3287, 6459)			F	VX	DY
1201	(3361, 6445)			F	VX	DY
1202	(3436, 6456)	r	V	F	VX	DY
1203	(3291)	r		F	VX	DY
1204	(3401)			F	VX	MA
1205	(3329, 6438)			F	VX	DY
1206	(3319)			F	VX	MA
1207	(3328, 6422)			F	VX	MA
1208	(3393)			F	VX	MA
1209	(3437, 6457)	r	V	F	VX	DY
1210	(3405, 6462)			F	VX	DY
1211	(3305)			F	VX	MA
1212	(3427, 6453)			F	VX	DY
1213	(3419)			F	VX	DY
1214	(3317, 6433)	r	V	F	VX	MA
1215	(3377)	r	V	F	VX	MA
1216	(3302)			F	VX	DY
1217	(3357, 6444)			F		ZH
1218	(3332)			F	VX	DY
1219	(3418)			F	VX	MA
1220	(3315, 6432)			F	VX	MA
1221	(3371)			F	VX	MA
1250	(3372)	r	V	F	VX	MA
1251	(3383)	r	V	F	VX	DY
1252	(3280)	r	V	F	VX	DY
1253	(3432)	r	V	F	VX	MA
1254	(3391)	r	V	F	VX	DY
1255	(3284)			F	VX	MA
1256	(3296)	r		F	VX	MA
1258	(3322)			F	VX	DY
1259	(3439)	r	V	F	VX	DY
1260	(3378)	r	V	F	VX	MA

AK51 (RKB)   KITCHEN BUFFET

Dia. AK502. Mark 1. Gas Cooking. No seats. B5 bogies. ETH 1.

Lot No. 30624 Cravens 1960–1. 41 t.

1566	G	VS	SS	SL	

AJ41 (RBR)   RESTAURANT BUFFET

Dia. AJ403. Mark 1. Gas cooking. Built with 23 loose chairs (Dia. AJ402). All remaining vehicles refurbished with 23 (21 w) fixed polypropylene chairs and fluorescent lighting. ETH 2 (2X*).

r Further refurbished with 21 chairs, payphone, wheelchair space and carpets (Dia. AJ417).

1644–1699. Lot No. 30628 Pressed Steel 1960–61. Commonwealth bogies. 39 t.
1730. Lot No. 30512 BRCW 1960–61. B5 bogies. 37 t.

1691 is leased to the Venice Simplon Orient Express.

1644		E		OM	1674			RS *SS* EC
1645		RS *SS*		EC	1679		**G**	RS *SS* BN
1649 w		F		KI	1680 x*w	**WV**	RS *SS* BN
1650 w		E		OM	1683 r	**FD**	F *SS* CP
1652 w		E		OM	1684 x*	**BG**	E	KM
1653 w		CC		Ferme Park	1686 r		F	LM
1655		E		KM	1688 w	**BG**	E	OM
1658		RS *SS*		BN	1689 r		F	LM
1659 x		MH *SS*		RL	1691 r	**G**	F *SS* SL
1663 x*		E		OM	1692 xr	**CH**	RV *SS* CP
1666 x*		E		OM	1693 x*		CC	Ferme Park
1667 x		RS *SS*		BN	1696	**G**	RS *SS* BN
1670 x*w **BG**		E		OM	1697 r		F	CP
1671 x*		RS *SS*		BN	1698	**WV**	RS *SS* BN
1672 x*		E		BN	1699 r		F	CP
1673 w		F		KI	1730 x	**M**	SP *SS* BO

AN21 (RMB)   MINIATURE BUFFET CAR

Dia. AN203. Mark 1. Gas cooking. –/44 2T. These vehicles are basically an open standard with two full window spaces removed to accommodate a buffet counter, and four seats removed to allow for a stock cupboard. All remaining vehicles now have fluorescent lighting. All vehicles have Commonwealth bogies except 1850 (B5). ETH 3.

1813–1832. Lot No. 30520 Wolverton 1960. 38 t.
1840–1850. Lot No. 30507 Wolverton 1960. 37 t (1850 is 36 t).
1853–1863. Lot No. 30670 Wolverton 1961–2. 38 t.
1871–1882. Lot No. 30702 Wolverton 1962. 38 t.

1842/50/71 have been been refurbished and are fitted with a microwave oven and payphone. Dia. AN208.

1813 x	**CC**	RS *SS*	BN	1859 x	**M**	SP *SS*	BO
1832 x		RS *SS*	EC	1860 x	**M**	W *SS*	CS
1840 v	**G**	MH *SS*	RL	1861 x	**M**	MH *SS*	RL
1842 x		F	*AR*	NC	1863 x	**CH**	RV *SS*	CP
1850		F	*AR*	NC	1871 x		F	*AR*	NC
1853 x		RS *SS*	BN	1882 x	**M**	W *SS*	CS

AJ41 (RBR)   RESTAURANT BUFFET

Dia. AJ414. Mark 1. Gas cooking. This vehicle was built as an unclassified restaurant (RU). It was rebuilt with buffet counter and 23 fixed polypropylene chairs (RBS), then further refurbished by fitting fluorescent lighting and reclassified RBR. B4/B5 bogies. ETH 2X.

Lot No. 30575 Ashford/Swindon 1960. 36.5 t.

1953 x **RC** RA *SS* CP

AS41   FIRST CLASS SLEEPING CAR

Dia. AS101. Mark 1. Pressure Ventilated. 11 single-berth compartments plus an attendant's compartment with gas cooking. ETH 3 (3X*).

2013. Lot No. 30159 Wolverton 1958. B5 bogies. 39 t.
2127. Lot No. 30687 Wolverton 1961. Commonwealth bogies. 41 t.

2013 was numbered 2908 for a time when in use with the Royal Train.

2013   **M** FS   SZ | 2127 *   **M** GS   CS

AU51   CHARTER TRAIN STAFF COACHES

Dia. AU501. Mark 1. Converted from BCKs in 1988. Commonwealth bogies. ETH 2.

Lot No. 30732 Derby 1964. 37 t.

2833 (21270)   RS *SS* BN
2834 (21267)   **WV** RS *SS* BN

AT5G   HM THE QUEENS SALOON

Dia. AT525. Mark 3. Converted from a FO built 1972. Consists of a lounge, bedroom and bathroom for HM The Queen, and a combined bedroom and bathroom for the Queen's dresser. One entrance vestibule has double doors. Air conditioned. BT10 bogies. ETH 9X.

Lot No. 30886 Wolverton 1977. 36 t.

2903 (11001)   **RP** RT *RT* ZN

AT5G   HRH THE DUKE OF EDINBURGH'S SALOON

Dia. AT526. Mark 3. Converted from a TSO built 1972. Consists of a combined lounge/dining room, a bedroom and a shower room for the Duke, a kitchen and a valet's bedroom and bathroom. Air conditioned. BT10 bogies. ETH 15X.

Lot No. 30887 Wolverton 1977. 36 t.

2904 (12001)   **RP** RT *RT* ZN

AT5B ROYAL STAFF COUCHETTES

Dia. AT527. Mark 2B. Converted from a BFK built 1969. Consists of luggage accommodation, guard's compartment, 350 kW diesel generator and Staff sleeping accommodation. Pressure ventilated. B5 bogies. ETH 5X.

Lot No. 30888 Wolverton 1977. 46 t.

2905 (14105) **RP** RT *RT* ZN

Dia. AT528. Mark 2B. Converted from a BFK built 1969. Consists of luggage accomodation, guards compartment and staff accomodation. Pressure ventilated. B5 bogies. ETH 4X.

Lot No. 30889 Wolverton 1977. 35.5 t.

2906 (14112) **RP** RT *RT* ZN

AT5G ROYAL STAFF SLEEPING CARS

Dia. AT531. Mark 3A Details as for 10646–732 except that controlled emission toilets are not fitted. ETH11X.

Lot No. 31002 Derby/Wolverton 1985. 42.5 t (44 t*).

2914   **RP** RT *RT* ZN
2915 * **RP** RT *RT* ZN

AT5G ROYAL KITCHEN/DINING CARS

Dia AT537 (AT539*). Mark 3. Converted from HST TRUKs built 1976/7. BT10 bogies. ETH 13X.

Lot No. 31059 (31084*) Wolverton 1988 (1990*). 43 t.

2916 (40512)   **RP** RT *RT* ZN
2917 (40514) * **RP** RT *RT* ZN

AT5G ROYAL HOUSEHOLD CARS

Dia. AT538 (AT540*). Mark 3. Converted from HST TRUKs built 1976/7. BT10 bogies. ETH 10X.

Lot Nos. 31083 (31085*) Wolverton 1989. 41.05 t.

2918 (40515)   **RP** RT *RT* ZN
2919 (40518) * **RP** RT *RT* ZN

AT5B ROYAL STAFF COUCHETTES

Dia. AT536 (AT541*). Mark 2B. Converted from BFKs built 1969. Pressure ventilated. B5 bogies. ETH2X (ETH 7X*).

Lot No. 31044 Wolverton 1986. With generator (similar to 2905). 48 t.

2920 (14109, 17109) **RP** RT *RT* ZN

Lot No. 31086 Wolverton 1990. 41.5 t.

2921 (14107, 17107)   * **RP**   RT   *RT*   ZN

AT5G HRH THE PRINCE OF WALES'S SLEEPING CAR

Dia. AT534. Mark 3B. BT10 bogies.

Lot No. 31035 Derby/Wolverton 1987.

2922   **RP**   RT   *RT*   ZN

AT5G   HRH THE PRINCE OF WALES'S SALOON

Dia. AT535. Mark 3B. BT10 bogies.

Lot No. 31036 Derby/Wolverton 1987.

2923   **RP**   RT   *RT*   ZN

AD11 (FO)   OPEN FIRST

Dia. AD103. Mark 1. 42/– 2T. ETH 3. Many now fitted with table lights.

3063–3069. Lot No. 30169 Doncaster 1955. B4 bogies. 33 t.
3096–3100. Lot No. 30576 BRCW 1959. B4 bogies. 33 t.

3064 and 3068 were numbered DB 975607 and DB 975606 for at time when in
departmental service for British Rail.

3063	**BG**	VS		SL	3096	x **M**	SP *SS*	BO
3064	**BG**	VS		SL	3097	**WV**	RS *SS*	BN
3066	**G**	VS *SS*		SL	3098	x **CH**	RV *SS*	CP
3068	**G**	VS *SS*		SL	3100	x **CC**	RS *SS*	BN
3069	**G**	VS *SS*		SL				

Later design with fluorescent lighting, aluminium window frames and Commonwealth bogies.

3105–3128. Lot No. 30697 Swindon 1962–3. 36 t.
3130–3150. Lot No. 30717 Swindon 1963. 36 t.

3128/36/41/3/4/6/7/8 were renumbered 1058/60/3/5/6/8/9/70 when reclassified
RUO, then 3600/5/8/9/2/6/4/10 when declassified, but have now regained their
original numbers.

3105	x **M**	W *SS*	CS	3119	x	RS *SS*	BN
3107	x	RS *SS*	BN	3120	**WV**	RS *SS*	BN
3110	x **CC**	RS *SS*	BN	3121	**WV**	RS *SS*	BN
3111	x	CC	Ferme Park	3122	x **CH**	RV *SS*	CP
3112	x **CH**	RV *SS*	CP	3123	**G**	RS *SS*	BN
3113	x **M**	W *SS*	CS	3124		CC	Ferme Park
3114	**G**	RS *SS*	BN	3125	**RC**	RA *SS*	CP
3115	x	RS *SS*	BN	3127	**G**	RS *SS*	BN
3117	x **M**	W *SS*	CS	3128	x **M**	W *SS*	CS
3118	x	RV	CQ	3130	v **M**	W	CS

3131	x **CC**	RS	SS	BN	3143		FS	SZ
3132	x **CC**	RS	SS	BN	3144	x **CC**	RS	SS	BN
3133	x **CC**	RS	SS	BN	3146	**WV**	RS	SS	BN
3134	x	CC		Ferme Park	3147	**WV**	RS	SS	BN
3136		RS	SS	EC	3148		RS	SS	BN
3140	x	CC		Ferme Park	3149		RS	SS	EC
3141	**WV**	RS	SS	BN	3150	**G**	RS	SS	BN

AD1D (FO)   OPEN FIRST

Dia. AD105. Mark 2D. Air conditioned. 3172–88 have Stones equipment. 3192/
3202 have Temperature Ltd and require at least 800 V train heating supply.
42/– 2T. B4 bogies. ETH 5.

Lot No. 30821 Derby 1971–2. 32.5 t.

3172		CC	DY	3186		CC	DY
3174		CC	Ferme Park	3187		E	KM
3178		CC	DY	3188	**RC**	RA	SS	CP
3181	**RC**	RA	SS	CP	3192		E	DY
3182		CC	DY	3202		E	KM

AD1E (FO)   OPEN FIRST

Dia. AD106. Mark 2E. Air conditioned. Stones equipment. Require at least 800
V train heating supply. 42/– 2T (41/– 2T w). B4 bogies. ETH 5.

* Seats removed to accommodate catering module. 40F 1T.
§ Fitted with power supply for Mk. 1 RBR.

Lot No. 30843 Derby 1972–3. 32.5 t.

3221	w	F		ZH	3244	dw	F		ZH
3223		CC		OM	3246	w	CC		Ferme Park
3225		E		KI	3247		CC		Hornsey Up CS
3226		E		KI	3248		CC		DY
3227		CC		Ferme Park	3251	*	CC		Ferme Park
3228	d§	F		ZH	3252	w	F		LM
3229	d	F		ZH	3256	w	F		LM
3230		CC		DY	3257	w	CC		Ferme Park
3231		CC		Ferme Park	3258	n	E		KI
3232	dw	F		LM	3259	d*	F		ZP
3233	**BG**	E		BK	3261	dw	F		CP
3234	w	CC		Ferme Park	3267	**CH**	RV	SS	CP
3235	§	F		LM	3268		CC		KI
3237		CC		BN	3269	d	F		CP
3239		CC		Ferme Park	3270		CC		Ferme Park
3240	**CH**	RV	SS	CP	3272		CC		Hornsey Up CS
3241	d	F		LM	3273	**CH**	RV	SS	CP
3242	w§	F		LM	3275		CC		Hornsey Up CS

AD1F (F0)   OPEN FIRST

Dia. AD107. Mark 2F. Air conditioned. 3277–3318/58–81 have Stones equipment, others have Temperature Ltd. 42/– 2T (41/– 2T w). All now refurbished with power-operated vestibule doors, new panels and new seat trim. B4 bogies. Secondary door locks. ETH 5X.

3277–3318. Lot No. 30845 Derby 1973. 33 t.
3325–3428. Lot No. 30859 Derby 1973–4. 33 t.
3429–3438. Lot No. 30873 Derby 1974–5. 33 t.

§ Fitted with power supply for Mk. 1 RBR.
r Further refurbished with table lamps and new burgundy seat trim.

3403 was numbered 6450 for a time when declassified.

3277		F	AR	NC	3356	r	F	VW	OY
3278	r V F	VW	OY	3358		F	AR	NC
3279	§ F	AR	NC	3359		F	VW	OY
3285	w F	VW	OY	3360		F	VW	OY
3290	F	AR	NC	3362		F	VW	OY
3292	F	AR	NC	3363		F	VW	OY
3293	F	VW	OY	3364	r	F	VW	OY
3295	F	AR	NC	3366		F	VW	OY
3299	F	VW	OY	3368		F	AR	NC
3300	r	F	VW	OY	3369		F	VW	OY
3303	F	AR	NC	3373		F	AR	NC
3304	F	VW	OY	3374		F	VW	OY
3309	F	AR	NC	3375		F	AR	NC
3312	F	VW	OY	3379	§	F	AR	NC
3313	r V F	VW	OY	3381		F	AR	NC
3314	F	VW	OY	3384		F	VW	OY
3318	F	AR	NC	3385		F	VW	OY
3325	r V F	VW	OY	3386		F	VW	OY
3326	r V F	VW	OY	3387		F	VW	OY
3330	F	VW	OY	3388		F	AR	NC
3331	F	AR	NC	3389		F	VW	OY
3333	F	VW	OY	3390		F	VW	OY
3334	F	AR	NC	3392	r V F	VW	OY
3336	§	F	AR	NC	3395		F	VW	OY
3337	F	VW	OY	3397	r V F	VW	OY
3338	§	F	AR	NC	3399	§	F	AR	NC
3340	F	VW	OY	3400		F	AR	NC
3344	F	VW	OY	3402		F	VW	OY
3345	r V F	VW	OY	3403		F	VW	OY
3348	r	F	VW	OY	3408		F	VW	OY
3350	r V F	VW	OY	3411		F	VW	OY
3351	F	AR	NC	3414		F	AR	NC
3352	F	VW	OY	3416		F	AR	NC
3353	F	VW	OY	3417		F	AR	NC
3354	F	VW	OY	3424		F	AR	NC

3425		F	*VW*	OY		3431	r	**V** F	*VW*	OY
3426	r	F	*VW*	OY		3433		F	*VW*	OY
3428		F	*VW*	OY		3434		F	*VW*	OY
3429		F	*VW*	OY		3438		F	*VW*	OY

AG1E (FOt)   OPEN FIRST (PANTRY)

Dia. AG101. Mark 2E. Air conditioned. Converted from FO. Fitted with pantry, microwave oven and payphone for use on sleeping car services. 36/– 1T. B4 bogies. Secondary door locks. ETH 5X.

Lot No. 30843 Derby 1972–3. 32.5 t.

3520	(3253)	F	*GW*	LA		3523	(3238)	F	*SR*	IS
3521	(3271)	F	*GW*	LA		3524	(3254)	F	*SR*	IS
3522	(3236)	F	*GW*	LA		3525	(3255)	F		CP

AC21 (TSO)   OPEN STANDARD

Dia. AC204. Mark 1. This vehicle has 2+2 seating and is classified TSO ('Tourist second open'–a former LNER designation). It has narrower seats than later vehicles. –/64 2T. Commonwealth bogies (originally built with BR Mark 1 bogies). ETH 4.

Lot No. 30079 York 1953. 36 t.

3766	x **M**	W	*SS*	CS

AC21 (TSO)   OPEN STANDARD

Dia. AC201. Mark 1. These vehicles are a development of Dia. AC204 with fluorescent lighting and modified design of seat headrest. Built with BR Mark 1 bogies. –/64 2T. ETH 4.

4831–4836. Lot No. 30506 Wolverton 1959. Commonwealth bogies. 33 t.
4842–4880. Lot No. 30525 Wolverton 1959–60. B4 bogies. 33 t.

4831	x	**M**	SP	*SS*	BO		4860	x	**M**	W		CS
4832	x	**M**	SP	*SS*	BO		4866	x	**RR**	F	*NW*	LL
4836	x	**M**	SP	*SS*	BO		4869	x		CC		Ferme Park
4842	x		CC		Ferme Park		4873	x	**RR**	F	*NW*	LL
4849	x	**RR**	F	*NW*	LL		4875	x	**RR**	F	*NW*	LL
4854	x	**RR**	F	*NW*	LL		4876	x	**RR**	F	*NW*	LL
4856	x	**M**	SP	*SS*	BO		4880	x	**RR**	F	*NW*	LL
4858	x		E		KM							

Lot No. 30646 Wolverton 1961. Built with Commonwealth bogies, but BR Mark 1 bogies substituted by the SR on 4902/5/9/10/12/15/16. All now re-rebogied. 34 t B4, 36 t C.

4902	x B4	**CH**	RV	*SS*	CP		4912	x C	**M**	W	*SS*	CS
4905	v C	**M**	MH	*SS*	RL		4915	x B4	**CC**	RS	*SS*	BN
4909	x B4		CC		Ferme Park		4916	x B4		CC		Ferme Park
4910	v C	**M**	MH	*SS*	RL		4917	x C	**RR**	F	*NW*	LL

This lot has Commonwealth bogies and aluminium window frames.

f Facelifted.

Lot No. 30690 Wolverton 1961–2. 37 t.

4925		G	RS	SS	BN	4997 v	BG	W	CS
4927 f			RV		CP	4998		RS SS	BN
4931 v	M	W		SS	CS	4999		RS SS	BN
4932 v	N	W			CS	5002	W	RS SS	EC
4936 v	M	W			CS	5005	W	RS SS	BN
4938	W	RS	SS	BN	5007 f	G	RS SS	BN
4939		RS	SS	BN	5008 x	CC	RS SS	BN
4940 v	M	MH	SS	RL	5009		RV	CP
4946 x	CC	RS	SS	BN	5010		RV	CP
4949		RS	SS	BN	5023	G	RS SS	BN
4951 v	M	MH	SS	RL	5025 x	CH	RV SS	CP
4954 v	M	W	SS	CS	5027	G	RS SS	BN
4956		RS	SS	BN	5028 x	M	SP SS	BO
4958 v	M	W	SS	CS	5029 x	CH	RV SS	CP
4959		RS	SS	BN	5030 x	CH	RV SS	CP
4960 v	M	MH	SS	RL	5032 x	M	W SS	CS
4963 x	CH	RV	SS	CP	5033 x	M	W SS	CS
4973 v	M	MH	SS	RL	5035 x	M	W SS	CS
4977		RS	SS	BN	5037	G	RS SS	BN
4984 v	M	MH	SS	RL	5038 x		E	OM
4986	G	RS	SS	BN	5040 x	CH	RV SS	CP
4991	W	RS	SS	EC	5041		CC	Wolsingham
4993		RS	SS	EC	5042 x		CC	Ferme Park
4994 v	M	MH	SS	RL	5044 x	M	W SS	CS
4996 xf	CC	RS	SS	BN				

AC2Z (TSO)   OPEN STANDARD

Dia. AC205. Mark 2. Pressure ventilated. –/64 2T. B4 bogies. ETH 4.

Lot No. 30751 Derby 1965–7. 32 t.

5125 v	G	MH	SS	RL	5180 v	RR	F		LT
5132 v	H	F		LT	5183 v	RR	F		LT
5135 v	RR	F		LT	5186 v	RR	F		LT
5148 v	RR	F		LT	5191 v	H	F		LT
5154 v	H	F		LT	5193 v	H	F		LT
5156 v	RR	F		LT	5194 v	RR	F		LT
5157 v	RR	F		LT	5198 v	RR	F		LT
5158 v	RR	F		LT	5200 v	G	MH SS	RL
5161 v	RR	F		LT	5207 v	RR	F		LT
5163 v	RR	F		LT	5209 v	RR	F		LT
5166 v	H	F		LT	5212 v	H	F		LT
5167 v	RR	F		LT	5213 v	RR	F		LT
5174 v	RR	F		LT	5221 v	RR	F		LT
5177 v	RR	F		LT	5222 v	G	MH SS	RL
5179 v	RR	F		LT	5225 v	RR	F		LT

5226 v	**RR**	F		LT

AD2Z (SO)   OPEN STANDARD

Dia. AD203. Mark 2. Pressure ventilated. –/48 2T. B4 bogies. ETH 4.

Lot No. 30752 Derby 1966. 32 t.

5237 v	**G**	MH	*SS*	RL		5254	**BG**	F		DY
5249 v	**G**	MH	*SS*	RL						

AC2A (TSO)   OPEN STANDARD

Dia. AC206. Mark 2A. Pressure ventilated. –/64 2T (–/62 2T w). B4 bogies. ETH 4.

5265–5345. Lot No. 30776 Derby 1967–8. 32 t.
5350–5433. Lot No. 30787 Derby 1968. 32 t.

5265	**RR**	F		LM		5337	**BG**	F		LM
5266	**RR**	F		LM		5341	**RR**	F		CP
5267	**RR**	F		LM		5345	**RR**	F	*NW*	LL
5271	**RR**	F		LM		5350	**NR**	F		CP
5272	**RR**	F		CP		5353	**RR**	F		LM
5275	**FD**	F	*SS*	CP		5354	**RR**	F		LM
5276	**RR**	F	*NW*	LL		5364	**RR**	F		CP
5277	**BG**	F		LM		5365	**FD**	F	*SS*	CP
5278	**RR**	F	*NW*	LL		5366	**RR**	F		LM
5279	**BG**	F		LM		5373	**FD**	F	*SS*	CP
5282	**RR**	F		LM		5376	**NR**	F		CP
5290	**NR**	F		LM		5378	**NR**	F		CP
5291	**RR**	F		ZB		5379	**RR**	F		LM
5292	**RR**	F		CP		5381 w	**RR**	F	*NW*	LL
5293	**NR**	F		LM		5384	**N**	F		LM
5299	**M**	W	*SS*	CS		5386 w	**RR**	F	*NW*	LL
5300	**BG**	F		LM		5389 w	**RR**	F	*NW*	LL
5304	**RR**	F		LM		5392	**BG**	F		LM
5307	**RR**	F	*SS*	CP		5393	**RR**	F		LM
5309	**RR**	F	*NW*	LL		5396	**RR**	F		LM
5314	**BG**	F		LM		5401	**RR**	F		LM
5316	**RR**	F		LM		5410	**N**	F		LM
5322	**RR**	F	*NW*	LL		5412 w	**RR**	F	*NW*	LL
5323	**RR**	F		LM		5419 w	**RR**	F	*NW*	LL
5331	**RR**	F	*NW*	LL		5420 w	**RR**	F	*NW*	LL
5335	**RR**	F	*NW*	LL		5433 w	**RR**	F	*NW*	LL

AC2B (TSO)   OPEN STANDARD

Dia. AC207. Mark 2B. Pressure ventilated. –/62 2T. B4 bogies. ETH 4.

Lot No. 30791 Derby 1969. 32 t.

Non-Standard Livery: 5453, 5478 and 5491 are royal blue with white lining.

5439	**N**	F		LM		5443	**N**	F		LM

5446	N	F		LM	5471	N	F		LM
5447	N	F		LM	5472	N	F		LM
5449	N	CC		CB	5475	N	F		LM
5450	N	F		LM	5476	BG	RR		NL
5453 d	0	W	WW	CF	5478 d	0	W	WW	CF
5454	N	F		LM	5480	N	F		LM
5462	N	CC		CB	5487 d	M	W	WW	CF
5463 d	M	W	WW	CF	5491 d	0	W	WW	CF
5464	N	CC		CB	5494	N	CC		CB

AC2C (TSO)   OPEN STANDARD

Dia. AC208. Mark 2C. Pressure ventilated. –/62 2T. B4 bogies. ETH 4.

Lot No. 30795 Derby 1969–70. 32 t.

5505	RR	W		CS	5574	BG	RR		NL
5520	RR	RR		Melmerby	5585	BG	RR		NL
5533	BG	RR		NL	5595	BG	RR		NL
5554	RR	F		LM	5600	M	W	SS	CS
5569 d	M	W	WW	CF	5614	RR	F		LM

AC2D (TSO)   OPEN STANDARD

Dia. AC209. Mark 2D. Air conditioned. Stones equipment. –/62 2T. B4 bogies.
ETH 5.

Non-Standard Livery: 5630, 5647, 5732 & 5739 are Waterman VIP without lining.

Lot No. 30822 Derby 1971. 33 t.

5616		CC		Ferme Park	5663		F		KI
5617		E		OM	5665		F		LM
5618		F		LM	5669 d		F		LT
5620		F		LT	5674		F		KI
5623		F		LT	5675		E		OM
5628		F		LM	5676 d		F		LM
5629		F		LT	5679 d		F		LM
5630	0	CC	CA	CF	5685		F		LM
5631 d		F		LM	5686		F		LM
5632 d		F		LM	5687		F		KI
5633		E		OM	5690		F		LT
5634		F		LM	5692		F		LM
5636 d		F		LM	5693		F		LM
5640		F		LT	5694		F		KI
5642		W		CS	5699		F		LM
5645		W		CS	5700 d		F		LM
5647	0	CC	CA	CF	5701		F		KI
5648		E		LM	5704	M	W	SS	CS
5650		F		LT	5709	BG	W		CS
5657		F		LM	5710 d		F		LM
5661		F		KI	5711		F		LT
5662		F		LM	5712		W		CS

Pullman Stock. Umber and cream liveried Pullman Parlour First No. 301 'PERSEUS' is pictured at Southall, West London on 13th June 1996. The vehicle is fitted with Gresley bogies.

Kevin Conkey

▲ Carrying a maroon and beige livery, Mark 2 Pullman First with Kitchen No. 504 'THE WHITE ROSE' is pictured in the formation of a Carnforth bound empty coaching stock train at Preston on 10th August 1997. This vehicle is one of a number which once formed the 'Manchester Pullman' set. **Martyn Hilbert**

▼ **Mark 1 Stock.** Chocolate and cream liveried Open First No. 3098 is pictured in the formation of a charter service at Elgin on 4th March 1997. **Colin J. Marsden**

VSOE-owned Kitchen Buffet No. 1566 is pictured at Okehampton on 24th May 1997. The vehicle carries Southern Railway green livery.

Colin J. Marsden

Originally used as a generator for use with HST trailers, Generator Van No. 6310 is pictured at Wessex Traincare's Eastleigh works on 4th November 1997 after receiving a repaint into Porterbrook livery. **Brian Morrison**

▲ Waterman Railways VIP liveried Corridor Brake Composite No. 21269 is pictured at Worcester Shrub Hill on 11th March 1997. **Stephen Widdowson**

▼ **Mark 2A Stock.** Regional Railways liveried Open Standard No. 5322 in the formation of the 12.18 to Holyhead at Crewe on 7th September 1996. **Peter Fox**

▲ **Mark 2C Stock.** Some of the vehicles hired to South Wales & West Railway (since renamed Wales & West Railway) from West Coast Railway Company carry a royal blue livery with 'South Wales & West' branding. One such vehicle, Open Standard No. 5453 is pictured at Weymouth on 18th January 1997.

Denise Johnson

▼ **Mark 2D Stock.** Maroon liveried Open Standard No. 5704 is seen on the rear of a Blackburn to Carnforth empty coaching stock working at Preston on 1st June 1997.

Martyn Hilbert

▲ **Mark 2E Stock.** Riviera Trains-owned Open First No. 3240 'PENDENNIS' at Aberdeen on 2nd March 1997.   **Colin J. Marsden**

▼ **Mark 2F Stock.** Virgin Trains liveried Open Brake Standard No. 9516 is pictured at Sheffield on 1st August 1997. The service is the 09.00 Poole–York.

Peter Fox

Mark 3 Stock. Royal Household Car No. 2918 is pictured passing through Malvern Wells in the formation of the royal train on 3rd May 1996. **Stephen Widdowson**

Midland Mainline liveried Trailer First No. 41061 at London St Pancras on 29th July 1997. Note absence of yellow stripes at roof level on this company's first class vehicles. **Denise Johnson**

▲ Great Western Trains liveried Trailer Buffet First No. 40733 is pictured at Par on 23rd August 1997. **Denise Johnson**

▼ **Mark 3A Stock.** Intercity liveried Sleeping Car with Pantry No. 10507 is pictured in Railcare's Glasgow works after a C3 overhaul. The date is 29th August 1996. **Colin J. Marsden**

Mark 3B Stock. A Virgin Trains liveried Open First, No. 11084 is seen in the formation of the 11.20 Preston–London Euston at Watford Junction on 18th October 1997.

Kevin Conkey

Mark 4 Stock. Great North Eastern Railway liveried Restaurant Buffet First No. 10304 arrives at London King's Cross with the 14.05 from Leeds on 20th September 1997. Many coaches in this livery have been fitted with a cast GNER crest located centrally on the lower bodyside.

Kevin Conkey

Nightstar Stock. The Nightstar stock, built for overnight services between Britain and continental Europe, has still not entered service. A rake of Nightstar stock with some unidentified seating cars at the front is pictured passing over Shap behind Class 92 No. 92032 whilst on trial. The date is 10th April 1997.

Dave McAlone

▲ **NPCCS Stock.** Royal Mail Stowage Van No. 80400 is seen between duties at Bristol Barton Hill on 18th March 1997. **Colin J. Marsden**

▼ The 16.35 London Euston–Carlisle 'Lancashire Pullman' is pictured passing the site of Standon Bridge station, Staffordshire on 27th August 1997. Driving Van Trailer No. 82152 is at the rear of the formation which is in Virgin Trains livery. **Hugh Ballantyne**

GNER Driving Van Trailer No. 82200 arrives at King's Cross, leading the 10.00 arrival from Glasgow Central, on 23rd September 1997.
Kevin Conkey

▲ Propelling Control Vehicle No. 94345 is pictured at rear of Carlisle station on 31st May 1997. The vehicle is painted in Rail Express Systems livery.
Kevin Conkey

▼ Rail Express Systems liveried High Security Brake Van No. 94478 at Carlisle on 31st May 1997.
Kevin Conkey

5714	**M**	W	*SS*	CS
5715		F		LT
5716		F		KI
5718		F		KI
5722		E		KM
5724		F		LT
5726		F		LM
5727	**M**	W	*SS*	CS
5728		F		LM

5729			CC		OM
5731			F		KI
5732	**0**		CC	*CA*	CF
5735			F		LM
5737	d		F		LM
5738			F		KI
5739	**0**		CC	*CA*	CF
5740	d		F		LM

AC2E (TSO)　　　　　　　　　　OPEN STANDARD

Dia. AC210. Mark 2E. Air conditioned. Stones equipment. –/64 2T (–/62 2T w).
B4 bogies. Require at least 800 V train heat supply to oper TE ir conditioning (as
built). Secondary door locks (except 5756, 5879, 5885, 5904 & 5907). ETH 5.

5744–5803. Lot No. 30837 Derby 1972. 33.5 t.
5810–5907. Lot No. 30844 Derby 1972–3. 33.5 t.

f　Facelifted with centre luggage stack seating –/60 2T (–/58 + 2T w), modified
design of seat headrest and new claret seat trim.
§　Facelifted with modified design of seat headrest and new red seat trim.
*　Refurbished with new green seat trim.
†　Refurbished with new green seat trim, centre luggage stack and pt. –/58 2T (w).

5744			F	*VX*	MA
5745	f		F	*VX*	DY
5746			F	*VX*	MA
5748	w		F	*VX*	MA
5750	f		F	*VX*	DY
5751	§w		F	*VW*	OY
5752	w		F	*VX*	DY
5754	w		F	*VX*	MA
5756		**M**	W	*SS*	CS
5760	§		F	*VX*	DY
5764			F		ZP
5766			F		ZP
5769			F	*VX*	MA
5772	w		F	*VX*	DY
5773	f		F	*VX*	DY
5775	f		F	*VX*	DY
5776			F	*VX*	MA
5778	w		F	*VX*	MA
5779			F	*VX*	MA
5780	w		F	*VX*	MA
5781	w		F	*VX*	MA
5784			F	*VX*	MA
5787	f		F	*VX*	DY
5788	w		F	*VX*	MA
5789			F	*VX*	MA
5791	w		F	*VX*	MA
5792			F	*VX*	MA
5793	fw		F	*VX*	DY

5794	w		F	*VX*	MA
5796	w		F	*VX*	MA
5797			F	*VX*	MA
5799			F	*VW*	OY
5800		**W**	F		ZH
5801			F	*VX*	MA
5803			F		ZH
5810	f		F	*VX*	DY
5812	w		F	*VX*	MA
5814			F	*VX*	MA
5815	fw		F	*VX*	DY
5816			F	*VX*	MA
5821			F	*VX*	MA
5822	†	**V**	F	*VX*	MA
5824	w		F	*VX*	MA
5826	§		F	*VX*	DY
5827	w		F	*VX*	MA
5828	w		F	*VX*	MA
5831			F		ZH
5833			F	*VX*	MA
5836			F		ZH
5840			F	*VW*	OY
5843	w		F	*VX*	DY
5845	w		F	*VX*	MA
5847	w		F	*VX*	MA
5851	§		F	*VW*	OY
5852		**W**	F		ZH
5853			F	*AR*	NC

5854	F	VX	MA	5888 w	F	VX	MA
5859	F	VX	MA	5889	F	VX	MA
5863	F		ZH	5890 §	F	VW	OY
5866	F	VX	MA	5892	F	VX	DY
5868	F	VX	MA	5893	F	VX	MA
5869	F	AR	NC	5897	F	VX	MA
5871 §	F	VX	DY	5899	F	VX	MA
5874 w	F	AR	NC	5900 f	F	VX	DY
5875	F		ZP	5901 f	F	VX	DY
5876	F	VX	MA	5902	F	VX	MA
5879	E		OM	5903	F	VX	MA
5881 fw	F	VX	DY	5904	F		PC
5885	F		ZP	5905	F	VX	MA
5886 f	F	VX	DY	5906 *	F	VX	MA
5887 *w	F	VX	MA	5907	F		ZP

AC2F (TSO)   OPEN STANDARD

Dia. AC211. Mark 2F. Air conditioned. Temperature Ltd. equipment. –/64 2T. (–/62 2T w) Inter-City 70 seats. All were refurbished in the 1980s with power-operated vestibule doors and new panels and new seat trim. B4 bogies. Secondary door locks. ETH 5X.

5908–5958. Lot No. 30846 Derby 1973. 33 t.
5959–6170. Lot No. 30860 Derby 1973–4. 33 t.
6171–6184. Lot No. 30874 Derby 1974–5. 33 t.

* Early Mark 2 style seats.
These vehicles are now undergoing a second refurbishment with carpets, new motor-alternator sets and new seat trim .

r Standard refurbished vehicles.

Cross-Country vehicles:

§ Fitted with centre luggage stack seating –/60 2T.
‡ Fitted with centre luggage stack seating –/60 2T plus pt.
† Fitted with centre luggage stack and wheelchair space seating –/58 2T.

West Coast vehicles:

• Refurbished vehicles with two wheelchair spaces seating –/60 2T.

5908		F	VW	OY	5921		F	AR	NC
5910 w		F	VW	OY	5922		F	AR	NC
5911 §	V	F	VX	MA	5924		F	AR	NC
5912		F	VX	DY	5925 r		F	VX	MA
5913 r		F	VX	DY	5926		F	AR	NC
5914 •	V	F	VW	OY	5927		F	AR	NC
5915 r	V	F	VW	OY	5928		F	AR	NC
5916 rw		F	VX	DY	5929		F	AR	NC
5917 §	V	F	VX	DY	5930 †	V	F	VX	DY
5918 w		F	VX	DY	5931 w		F	VW	OY
5919 ‡	V	F	VX	DY	5932		F	VW	OY
5920		F	VW	OY	5933		F	VW	OY

5934			F	VW	OY		5994	*		F	VX	MA
5935			F	AR	NC		5995			F	VX	DY
5936			F	AR	NC		5996	‡	V	F	VX	DY
5937			F	VW	OY		5997			F	VW	OY
5939	r	V	F	VW	OY		5998			F	AR	NC
5940	w		F	VW	OY		5999	§	V	F	VX	DY
5941			F	VW	OY		6000	†	V	F	VX	DY
5943	w		F	VW	OY		6001	w		F	VW	OY
5944	w		F	AR	NC		6002			F	VW	OY
5945	w		F	VW	OY		6005	*		F	VX	MA
5946			F	VW	OY		6006			F	AR	NC
5947			F	VX	MA		6008	§	V	F	VX	MA
5948	w		F	VW	OY		6009	r	V	F	VW	OY
5949	•	V	F	VW	OY		6010	§	V	F	VX	DY
5950			F	AR	NC		6011	§	V	F	VX	DY
5951			F	VX	MA		6012	*		F	VW	OY
5952			F	VW	OY		6013	r		F	VX	DY
5953	r		F	VW	OY		6014	r		F	VX	DY
5954			F	AR	NC		6015	†	V	F	VX	DY
5955	r	V	F	VW	OY		6016			F	VW	OY
5956			F	AR	NC		6018	†	V	F	VX	DY
5957			F	VW	OY		6021			F	VW	OY
5958	r		F	VX	MA		6022	§	V	F	VX	DY
5959	n		F	AR	NC		6024	§	V	F	VX	MA
5960	§	V	F	VX	DY		6025	w*		F	VX	MA
5961			F	VX	DY		6026	§	V	F	VX	DY
5962	‡	V	F	VX	DY		6027	w		F	VW	OY
5963	*		F	VW	OY		6028			F	AR	NC
5964			F	AR	NC		6029			F	VW	OY
5965	rw		F	VX	DY		6030	†	V	F	VX	MA
5966			F	AR	NC		6031			F	VW	OY
5967	†	V	F	VX	DY		6034			F	AR	NC
5968			F	AR	NC		6035	rw		F	VX	DY
5969	w		F	VW	OY		6036	*		F	AR	NC
5971			F	VX	MA		6037			F	AR	NC
5973			F	AR	NC		6038			F	VX	DY
5975	§	V	F	VX	DY		6041	§	V	F	VX	DY
5976	†	V	F	VX	MA		6042			F	AR	NC
5977	r	V	F	VW	OY		6043			F	VW	OY
5978	r*	V	F	VW	OY		6045	w		F	VW	OY
5980			F	VW	OY		6046	§	V	F	VX	DY
5981	r		F	VX	MA		6047	n*		F	VW	OY
5983			F	VX	DY		6049			F	VW	OY
5984	*		F	VW	OY		6050	r		F	VX	DY
5985			F	AR	NC		6051	*		F	VW	OY
5986			F	VW	OY		6052	rw		F	VX	MA
5987	*		F	VW	OY		6053	*		F	AR	NC
5988	r	V	F	VW	OY		6054			F	VW	OY
5989	w		F	VX	MA		6055			F	VW	OY
5991			F	VX	MA		6056			F	VW	OY
5993	w*		F	AR	NC		6057	r		F	VW	OY

6059	§	**V**	F	*VX*	DY		6144	*		F	*VW*	OY
6060	•		F	*VW*	OY		6145	*		F	*VX*	DY
6061	‡	**V**	F	*VX*	DY		6146	*		F	*AR*	NC
6062			F	*VW*	OY		6147	r		F	*VW*	OY
6063	w		F	*VW*	OY		6148	r		F	*VX*	DY
6064	§	**V**	F	*VX*	MA		6149	•*	**V**	F	*VW*	OY
6065			F	*VW*	OY		6150	r*		F	*VX*	DY
6066	r		F	*VX*	DY		6151	*		F	*VW*	OY
6067	‡	**V**	F	*VX*	MA		6152	*		F	*AR*	NC
6073	§	**V**	F	*VX*	DY		6153	r		F	*VW*	OY
6100	*		F	*VW*	OY		6154	r*		F	*VX*	DY
6101			F	*VW*	OY		6155	*		F	*AR*	NC
6102	r		F	*VW*	OY		6157	§	**V**	F	*VX*	MA
6103			F	*AR*	NC		6158	r*	**V**	F	*VW*	OY
6104	r	**V**	F	*VW*	OY		6159	‡	**V**	F	*VX*	DY
6105	†pt	**V**	F	*VX*	DY		6160	*		F	*AR*	NC
6106	r		F	*VW*	OY		6161	*		F	*VW*	OY
6107			F	*VW*	OY		6162	‡	**V**	F	*VX*	MA
6110	w		F	*AR*	NC		6163			F	*VW*	OY
6111			F	*VW*	OY		6164			F	*VW*	OY
6112	‡	**V**	F	*VX*	DY		6165	r	**V**	F	*VW*	OY
6113			F	*VW*	OY		6166			F	*AR*	NC
6115	r		F	*VX*	DY		6167			F	*AR*	NC
6116			F	*VW*	OY		6168	r		F	*VX*	DY
6117	†	**V**	F	*VX*	DY		6170	§	**V**	F	*VX*	MA
6119	w		F	*VX*	DY		6171			F	*VW*	OY
6120	§	**V**	F	*VX*	DY		6172	§	**V**	F	*VX*	DY
6121			F	*VW*	OY		6173	§	**V**	F	*VX*	DY
6122	§	**V**	F	*VX*	DY		6174			F	*AR*	NC
6123			F	*AR*	NC		6175	r	**V**	F	*VW*	OY
6124	r		F	*VX*	DY		6176	w		F	*VX*	DY
6134			F	*VW*	OY		6177			F	*VX*	DY
6135	r		F	*VX*	DY		6178	r		F	*VX*	MA
6136			F	*VW*	OY		6179	r	**V**	F	*VW*	OY
6137			F	*VX*	MA		6180	w		F	*VW*	OY
6138			F	*VW*	OY		6181	wn		F	*VW*	OY
6139	n*		F	*AR*	NC		6182	§	**V**	F	*VX*	DY
6141	w		F	*VW*	OY		6183			F	*VX*	MA
6142	*		F	*VW*	OY		6184	§	**V**	F	*VX*	DY

AC2D (TSO)   OPEN STANDARD

Dia. AC217. Mark 2D. Air conditioned. Stones. –/58 2T. (–/58 1T*). B4 bogies. ETH 5X. Rebuilt from FO with new style 2+2 seats.

Lot No. 30821 Derby 1971–2. 33.5 t.

* One toilet converted to store room for use of attendant on sleeping car services.

6203/7/12/19 are leased to Rail Charter Services Ltd.

6200	(3198)	d		F *GW*	LA

6202	(3191)	d*	F		CP
6203	(3180)	d	F	SS	EC
6206	(3183)	d	F	GW	LA
6207	(3204)	d	F	SS	EC
6212	(3176)	d	F	SS	EC
6213	(3208)	d	F	GW	LA
6216	(3179)		F		LM
6219	(3213)	d	F	SS	EC
6221	(3173)	d	F		CP
6226	(3203)	d	F	GW	LA
6233	(3206)		F		LM

GX51   GENERATOR VAN

Dia. GX501. Renumbered 1989 from BR departmental series. Converted from NDA in 1973 to three-phase supply generator van for use with HST trailers. Currently used to test overhauled HST trailers. B4 bogies.

Lot No. 30400 Pressed Steel 1958.

6310	(81448, ADB 975325)	**PL**	P	PL	ZG

AX51   GENERATOR VAN

Dia. AX501. Converted from NDA in 1992 to generator vans for use with pairs of Class 37s on Anglo-Scottish sleeping car services. Now normally used on tours hauled by steam locomotives. B4 bogies. ETH75.

6311. Lot No. 30162 Pressed Steel 1958. 37.25 t.
6312. Lot No. 30224 Cravens 1956. 37.25 t.
6313. Lot No. 30484 Pressed Steel 1958. 37.25 t.

Non-Standard Livery: 6311 is purple.

6313 is leased to the Venice Simplon Orient Express.

6311	(80903, 92911)	**0**	RS	SS	BN
6312	(81023, 92925)		FS		SK
6313	(81553, 92167)	**PC**	P	SS	SL

GS5 (HSBV)   HST BARRIER VEHICLE

Various diagrams. Renumbered from departmental stock, or converted from various types. B4 bogies (Commonwealth bogies *).

6330. Lot No. 30786 Derby 1968. 32 t.
6334. Lot No. 30400 Pressed Steel 1957–8. 31.5 t.
6335. Lot No. 30775 Derby 1967–8. 32 t.
6336/8/44. Lot No. 30715 Gloucester 1962. 31 t.
6340. Lot No. 30669 Swindon 1962. 36 t.
6343. Lot No. 30795 Derby 1969/70. 32 t.
6346. Lot No. 30777 Derby 1967. 31.5 t.
6347. Lot No. 30787 Derby 1968. 31.5 t.
6348. Lot No. 30163 Pressed Steel 1957. 31.5 t.

6330	(14084, 975629)		A	*AT*	LA
6334	(81478, 92128)	PL	P	*PL*	NL
6335	(14065, 975655)	BG	P		LA
6336	(81591, 92185)		A	*AT*	LA
6338	(81581, 92180)		A	*AT*	LA
6340	(21251, 975678)	*	A	*AT*	LA
6343	(5522)		RR		HT
6344	(81263, 92080)		A	*AT*	EC
6346	(9422)		A	*AT*	EC
6347	(5395)		A	*AT*	LA
6348	(81233, 92963)		A	*AT*	LA

GF5 (MFBV)   MARK 4 BARRIER VEHICLE

Various diagrams. Renumbered from departmental stock, or converted from
FK, BSO or BG. B4 bogies.

6351. Lot No. 30091 Doncaster 1954. 33 t.
6352/3. Lot No. 30774 Derby 1968. 33 t.
6354–6. Lot No. 30820 Derby 1970. 32 t.
6357. Lot No. 30798 Derby 1970. 32 t.
6358–9. Lot No. 30788 Derby 1968. 31.5 t.
6390. Lot No. 30136 Metro-Cammell 1955. 31.5 t.

6351	(3050, 975435)		F	*GN*	EC
6352	(13465, 19465)	BG	F	*GN*	BN
6353	(13478, 19478)	BG	F	*GN*	EC
6354	(9459)		F	*GN*	BN
6355	(9477)	BG	F	*GN*	BN
6356	(9455)	BG	F	*GN*	BN
6357	(9443)	BG	F	*GN*	BN
6358	(9432)	BG	F	*GN*	BN
6359	(9429)	BG	F	*GN*	BN
6390	(80723, 92900)		F	*GN*	BN

GF5 (BV)   DMU/EMU* BARRIER VEHICLE

Various diagrams. Converted from BFK, BSO or BG. B4 (BR Mark 1*) bogies.

6360. Lot No. 30777 Derby 1967. 31.5 t.
6361–2. Lot No. 30820 Derby 1970. 32 t.
6363. Lot No. 30796 Derby 1970. 32 t.
6364. Lot No. 30039 Derby 1954. 32 t.
6365. Lot No. 30323 Pressed Steel 1957. 32 t.

6360	(9420)		RR	P	*PL*	NL
6361	(9460)		RR	P	*PL*	NL
6362	(9467)		RR	RR		LL
6363	(14117, 17117)		RR	RR		LL
6364	(80565)	*	RR	P	*PL*	TS
365	(81296, 84296)	*	RR	P	*PL*	TS

AY5 (BV)   EUROSTAR BARRIER VEHICLE

Dia. AY501. Converted from GUVs. Bodies removed. B4 bogies.

6380–6382/9. Lot No. 30417 Pressed Steel 1958–9. 40 t.
6383. Lot No. 30565 Pressed Steel 1959. 40 t.
6384/6/7. Lot No. 30616 Pressed Steel 1959–60. 40 t.
6385. Lot No. 30343 York 1957. 40 t.
6388. Lot No. 30403 York/Glasgow 1958–60. 40 t.

6380	(86386, 93386)	**B**	LC	*ES*	PI
6381	(86187, 93187)	**B**	LC	*ES*	PI
6382	(86295, 93295)	**B**	LC	*ES*	PI
6383	(86664, 93664)	**B**	LC	*ES*	PI
6384	(86955, 93955)	**B**	LC	*ES*	PI
6385	(86515, 93515)	**B**	LC	*ES*	PI
6386	(86859, 93859)	**B**	LC	*ES*	PI
6387	(86973, 93973)	**B**	LC	*ES*	PI
6388	(86562, 93562)	**B**	LC	*ES*	PI
6389	(86135, 93135)	**B**	LC	*ES*	PI

GS5 (HSBV)   HST BARRIER VEHICLE

Dia. Converted from BG. B4 bogies.

6392. Lot No. 30715 Gloucester 1962. 29.5 t.
6393/6/7. Lot No. 30716 Gloucester 1962. 29.5 t.
6394. Lot No. 30162 Pressed Steel 1956–7. 30.5 t.
6395. Lot No. 30484 Pressed Steel 1958. 30.5 t.
6398/9. Lot No. 30400 Pressed Steel 1957–8. 30.5 t.

6392	(81588, 92183)	**PL**	P	*PL*	NL
6393	(81609, 92196)	**PL**	P	*PL*	NL
6394	(80878, 92906)	**PL**	P	*PL*	NL
6395	(81506, 92148)	**PL**	P	*PL*	NL
6396	(81606, 92195)	**PL**	P	*PL*	NL
6397	(81600, 92190)	**PL**	P	*PL*	NL
6398	(81471, 92126)	**PL**	P	*PL*	NL
6399	(81367, 92994)	**PL**	P	*PL*	NL

AD2C (FO)   OPEN SECOND

Dia. AD205. Mark 2C. Pressure ventilated. Declassified open firsts. –/42 2T. B4
bogies. ETH 4.

Lot No. 30810 Derby 1970. 33 t.

6411 and 6416 were numbered DB 977547 and DB 977546 for a time when
in departmental service with British Rail.

6400	(3167)	**BG** RR	Heysham	6415	(3155)	**BG** RR	Melmerby
6411	(3152)	**M** RR	CP	6416	(3165)	**BG** RR	Mossend Yd
6414	(3161)	**BG** RR	Melmerby				

AG2C (TSOT)   OPEN STANDARD (TROLLEY)

Dia. AG201. Mark 2C. Converted from TSO by removal of one seating bay and
replacing this by a counter with a space for a trolley. Adjacent toilet removed
and converted to steward's washing area/store. Pressure ventilated. –/54 1T.
B4 bogies. ETH 4.

Lot No. 30795 Derby 1969–70. 32.5 t.

6510	(5518)	**BG**	RR	Melmerby	6523	(5569)	**BG** W		CS
6513	(5538)	**N**	F	OM	6528	(5592)	**M**	W *SS*	CS
6517	(5499)	**N**	F	OM					

AG2D (TSOT)   OPEN STANDARD (TROLLEY)

Dia. AG202. Mark 2D. Converted from TSO by removal of one seating bay and
replacing this by a counter with a space for a trolley. Adjacent toilet removed
and converted to steward's washing area/store. Air conditioned. Stones
equipment. –/54 1T. B4 bogies. ETH 5.

Lot No. 30822 Derby 1971. 33 t.

6609	(5698)	F	KI		6619	(5655)	F	KI

AN2D (RMBT)   MINIATURE BUFFET CAR

Dia. AN207. Mark 2D. Converted from TSOT by the removal of another seating
bay and fitting a proper buffet counter with boiler and microwave oven. Air
conditioned. Stones equipment. –/46 1T. B4 bogies. p. q. Secondary door
locks. ETH 5.

Lot No. 30822 Derby 1971. 33 t.

6652	(5622, 6602)	F	LM
6660	(5627, 6610)	F	CP
6661	(5736, 6611)	F	CP
6662	(5641, 6612)	F	LM
6665	(5721, 6615)	F	LM

AN1F (RLO)   SLEEPER RECEPTION CAR

Dia. AN101 (AN102*). Mark 2F. Converted from FO, these vehicles consist of a
pantry, microwave cooking facilities, seating area for passengers, telephone
booth and staff toilet. 6703–8 also have a bar. Converted at RTC, Derby (6700),
Ilford (6701–5) and Derby (6706–8). Air conditioned. 26/– 1T. B4 bogies. p. q.
Secondary door locks. ETH 5X.

6700–2/4/8. Lot No. 30859 Derby 1973–4. 33.5 t.
6703/5–7. Lot No. 30845 Derby 1973. 33.5 t.

6700	(3347)		F	*SR*	IS
6701	(3346)	*	F	*SR*	IS
6702	(3421)	*	F	*SR*	IS
6703	(3308)		F	*SR*	IS

6704	(3341)	F	*SR*	IS
6705	(3310, 6430)	F	*SR*	IS
6706	(3283, 6421)	F	*SR*	IS
6707	(3276, 6418)	F	*SR*	IS
6708	(3370)	F	*SR*	IS

AC2F (TSO)   OPEN STANDARD

Dia. AC224. Mark 2F. Renumbered 1985–6 from FO. Converted 1990 to TSO with mainly unidirectional seating and power-operated sliding doors. Air conditioned. B4 bogies. –/74 2T + one tip-up seat. 6800–14 were converted by BREL Derby and have Temperature Ltd. air conditioning. 6815–29 were converted by RFS Industries Doncaster and have Stones air conditioning. Secondary door locks. ETH 5X.

6800–07. 6810–12. 6813–14. 6819/22/28. Lot No. 30859 Derby 1973–4. 33 t.
6808–6809. Lot No. 30873 Derby 1974–5. 33.5 t.
6815–18. 6820–21. 6823–27. 6829. Lot No. 30845 Derby 1973. 33 t.

6800	(3323, 6435)	F	*AR*	NC
6801	(3349, 6442)	F	*AR*	NC
6802	(3339, 6439)	F	*AR*	NC
6803	(3355, 6443)	F	*AR*	NC
6804	(3396, 6449)	F	*AR*	NC
6805	(3324, 6436)	F	*AR*	NC
6806	(3342, 6440)	F	*AR*	NC
6807	(3423, 6452)	F	*AR*	NC
6808	(3430, 6454)	F	*AR*	NC
6809	(3435, 6455)	F	*AR*	NC
6810	(3404, 6451)	F	*AR*	NC
6811	(3327, 6437)	F	*AR*	NC
6812	(3394, 6448)	F	*AR*	NC
6813	(3410, 6463)	F	*AR*	NC
6814	(3422, 6465)	F	*AR*	NC
6815	(3282, 6420)	F	*AR*	NC
6816	(3316, 6461)	F	*AR*	NC
6817	(3311, 6431)	F	*AR*	NC
6818	(3298, 6427)	F	*AR*	NC
6819	(3365, 6446)	F	*AR*	NC
6820	(3320, 6434)	F	*AR*	NC
6821	(3281, 6458)	F	*AR*	NC
6822	(3376, 6447)	F	*AR*	NC
6823	(3289, 6424)	F	*AR*	NC
6824	(3307, 6429)	F	*AR*	NC
6825	(3301, 6460)	F	*AR*	NC
6826	(3294, 6425)	F	*AR*	NC
6827	(3306, 6428)	F	*AR*	NC
6828	(3380, 6464)	F	*AR*	NC
6829	(3288, 6423)	F	*AR*	NC

NM51 — MERSEYRAIL SANDITE COACH

Dia. NM504. Mark 1. Former Class 501 750 V d.c. third rail EMU driving trailers converted for use as Sandite/de-icing coaches. Mark 1 Bogies.

Lot No. 30328 Ashford/Eastleigh 1958.   . t.

| 6910 | (75178, 977346) | **MD** RT | *SA* | BD |
| 6911 | (75180, 977348) | **MD** RT | *SA* | BD |

AH2Z (BSOT) — OPEN BRAKE STANDARD (MICRO-BUFFET)

Dia. AH203. Mark 2. Converted from BSO by removal of one seating bay and replacing this by a counter with a space for a trolley. Adjacent toilet removed and converted to a steward's washing area/store. –/23 0T. ETH 4.

Lot No. 30757 Derby 1966. 31 t.

9100	(9405)	v	**RR** F		LT
9101	(9398)	v	**RR** F		LT
9104	(9401)	v	**G** MH	*SS*	RL
9105	(9404)	v	**RR** F		LT

AE21 (BSO) — OPEN BRAKE STANDARD

Dia. AE201. Mark 1. –/39 1T. BR Mark 1 bogies. ETH 3.

Lot No. 30170 Doncaster 1955. 34 t.

9227 xk **M** SP *SS*   BO

AE2Z (BSO) — OPEN BRAKE STANDARD

Dia. AE203. Mark 2. These vehicles use the same body shell as the mark 2 BFK and have first class seat spacing and wider tables. Pressure ventilated. –/31 1T. B4 bogies. ETH 4.

Lot No. 30757 Derby 1966. 31.5 t.

| 9385 | v | **H** F | LT | | 9414 | v | **H** F | LT |
| 9388 | v | **H** F | LT | | | | | |

AE2A (BSO) — OPEN BRAKE STANDARD

Dia. AE204. Mark 2A. These vehicles use the same body shell as the mark 2A BFK and have first class seat spacing and wider tables. Pressure ventilated. –/31 1T. B4 bogies. ETH 4.

9417–9424. Lot No. 30777 Derby 1967. 31.5 t.
9428–9438. Lot No. 30788 Derby 1968. 31.5 t.

9417	**RR** F	CP		9421	**RR** F	LM
9418	**RR** F	LM		9424	**RR** F	CP
9419	**RR** F	LM		9428	**RR** F	CP

9431	**RR** F	LM	9435	**RR** F	LM
9434	**RR** F	ZB	9438	**RR** F	CQ

AE2C (BSO)   OPEN BRAKE STANDARD

Dia. AE205. Mark 2C. Pressure ventilated. –/31 1T. B4 bogies. ETH 4.

9440–48. Lot No. 30798 Derby 1970. 32 t.
9458–70. Lot No. 30820 Derby 1970. 32 t.

Non-Standard Livery: 9440 is in Royal blue with white lining.

9470 was numbered DB 977577 for a time when in departmental service with British Rail.

9440	d **0**	W	*WW*	CF	9458	**RR** F		ZB
9444	**BG** RR			Norwich Goods	9470	**BG** RR		CB
9448	d **M**	W	*WW*	CF				

AE2D (BSO)   OPEN BRAKE STANDARD

Dia. AE206. Mark 2D. Air conditioned (Stones). –/31 1T. B4 bogies. pg. ETH 5.

Lot No. 30824 Derby 1971. 33 t.

9479	d **V**	F	*VX*	MA	9486		F		LM
9480	d	F	*VX*	MA	9488	d	F		CP
9481	d	F	*GW*	LA	9489	d	F	*VX*	MA
9482		F		NL	9490	d	F		LT
9483		F		LM	9492	d	F	*GW*	LA
9484	d	F		LT	9493	d	F	*VX*	DY
9485		F		LT	9494	d	F		CP

AE2E (BSO)   OPEN BRAKE STANDARD

Dia. AE207. Mark 2E. Air conditioned (Stones). –/32 1T. B4 bogies. Secondary door locks (except 9499 which is not in service). pg. ETH 5.

Lot No. 30838 Derby 1972. 33 t.

f facelifted with modified design of seat headrest and new claret seat trim.
r Refurbished with new green seat trim.

9496		F	*VX*	MA	9503 f		F	*VX*	DY
9497		F	*VX*	MA	9504		F	*VX*	MA
9498		F	*VX*	DY	9505 r		F	*VX*	MA
9499		F		ZH	9506		F	*VX*	MA
9500		F	*VX*	MA	9507 r	**V**	F	*VX*	MA
9501		F	*GW*	LA	9508		F	*VX*	DY
9502 f		F	*VX*	DY	9509		F	*VX*	MA

AE2F (BSO)   OPEN BRAKE STANDARD

Dia. AE208. Mark 2F. Air conditioned (Temperature Ltd.). All now refurbished with power-operated vestibule doors, new panels and seat trim. –/32 1T. B4 bogies. Secondary door locks. pg. ETH 5X.

Lot No. 30861 Derby 1974. 34 t.

r Further refurbished with new green seat trim.

9513	r	**V**	F	*VX*	DY	9526	r	F	*VX*	DY
9516	r	**V**	F	*VX*	DY	9527		F	*VX*	DY
9520	r	**V**	F	*VX*	MA	9529		F	*VX*	MA
9521	r	**V**	F	*VX*	DY	9531	r	**V**	F	*VX*	MA
9522	r	**V**	F	*VX*	DY	9537	n	F	*VX*	DY
9523		F	*VX*	MA	9538	r	**V**	F	*VX*	DY
9524	n	F	*VX*	DY	9539		F	*VX*	DY
9525	r	**V**	F	*VX*	DY					

AF2F (DBSO)   DRIVING OPEN BRAKE STANDARD

Dia. AF201. Mark 2F. Air conditioned (Temperature Ltd.). Push & pull (t.d.m. system). Converted from BSO, these vehicles originally had half cabs at the brake end. They have since been refurbished and have had their cabs widened and the outer gangways removed. Fitted with cowcatchers. Cab to shore communication. BR Cellnet phone and data transmitter. Secondary door locks. –/30 + wheelchair space 1T. B4 bogies. pg.ETH 5X.

9701–9710. Lot No. 30861 Derby 1974. Converted 1979. Disc brakes. 34 t.
9711–9713. Lot No. 30861 Derby 1974. Converted Glasgow 1985. 34 t.
9714. Lot No. 30861 Derby 1974. Converted Glasgow 1986. Disc brakes. 34 t.

9701	(9528)	F	*AR*	NC	9709	(9515)	F	*AR*	NC
9702	(9510)	F	*AR*	NC	9710	(9518)	F	*AR*	NC
9703	(9517)	F	*AR*	NC	9711	(9532)	F	*AR*	NC
9704	(9512)	F	*AR*	NC	9712	(9534)	F	*AR*	NC
9705	(9519)	F	*AR*	NC	9713	(9535)	F	*AR*	NC
9707	(9511)	F	*AR*	NC	9714	(9536)	F	*AR*	NC
9708	(9530)	F	*AR*	NC					

AJ1G (RFM)   RESTAURANT BUFFET FIRST (MODULAR)

Dia. AJ103 (10200/1 are Dia. AJ101). Mark 3A. Air conditioned. Converted from HST TRFKs, RFBs and FOs. 22/– (24/– *). BT10 bogies. p. q. Secondary door locks. ETH 14X.

10200–10211. Lot No. 30884 Derby 1977.
10212–10229. Lot No. 30878 Derby 1975–6. 39.8 t.
10230–10260. Lot No. 30890 Derby 1979. 39.8 t.

r Refurbished with table lamps and new burgundy set trim.

10200	(40519)	*	P		ZD	10202	(40504)	r	**V**	P	*VW*	MA
10201	(40520)		P	*VW*	OY	10203	(40506)	*	P		*AR*	NC

10204	(40502)			P	VW	OY
10205	(40503)			P	VW	OY
10206	(40507)	r	V P	VW	MA
10207	(40516)			P	VW	MA
10208	(40517)	r	V P	VW	MA
10209	(40508)			P	VW	PC
10210	(40509)			P	VW	PC
10211	(40510)			P	VW	PC
10212	(11049)			P	VW	PC
10213	(11050)			P	VW	PC
10214	(11034)	*		P	AR	NC
10215	(11032)			P	VW	MA
10216	(11041)	*		P	AR	NC
10217	(11051)	r	V P	VW	MA
10218	(11053)			P	VW	PC
10219	(11047)	r	V P	VW	MA
10220	(11056)	r	V P	VW	OY
10221	(11012)			P	VW	PC
10222	(11063)	r	V P	VW	MA
10223	(11043)	*		P	AR	NC
10224	(11062)			P	VW	PC
10225	(11014)	r		P	VW	OY
10226	(11015)	r	V P	VW	MA
10227	(11057)			P	VW	PC
10228	(11035)	*		P	AR	NC
10229	(11059)	r	V P	VW	MA
10230	(10021)			P	VW	PC

10231	(10016)			P	VW	MA
10232	(10027)	r		P	VW	OY
10233	(10013)			P	VW	PC
10234	(10004)			P	VW	PC
10235	(10015)	r		P	VW	PC
10236	(10018)			P	VW	PC
10237	(10022)	r		P	VW	MA
10238	(10017)	r	V P	VW	OY
10240	(10003)			P	VW	OY
10241	(10009)	*		P	AR	NC
10242	(10002)	r		P	VW	OY
10245	(10019)			P	VW	PC
10246	(10014)	r	V P	VW	PC
10247	(10011)	*		P	AR	NC
10248	(10005)			P	VW	OY
10249	(10012)			P	VW	PC
10250	(10020)			P	VW	OY
10251	(10024)	r	V P	VW	OY
10252	(10008)			P	VW	OY
10253	(10026)			P	VW	PC
10254	(10006)			P	VW	PC
10255	(10010)	r	V P	VW	OY
10256	(10028)			P	VW	PC
10257	(10007)			P	VW	PC
10258	(10023)			P	VW	PC
10259	(10025)			P	VW	OY
10260	(10001)			P	VW	MA

AJ1J (RFM)   RESTAURANT BUFFET FIRST (MODULAR)

Dia. AJ105. Mark 4. Air conditioned. 20/– 1T. BT41 bogies. ETH 6X.

Lot No. 31045 Metro-Cammell 1989 onwards. 45.5 t.

10300	**GN**	F	GN	BN
10301	**GN**	F	GN	BN
10302	**GN**	F	GN	BN
10303	**GN**	F	GN	BN
10304	**GN**	F	GN	BN
10305	**GN**	F	GN	BN
10306	**GN**	F	GN	BN
10307	**GN**	F	GN	BN
10308	**GN**	F	GN	BN
10309	**GN**	F	GN	BN
10310	**GN**	F	GN	BN
10311	**GN**	F	GN	BN
10312	**GN**	F	GN	BN
10313	**GN**	F	GN	BN
10314	**GN**	F	GN	BN
10315	**GN**	F	GN	BN
10316	**GN**	F	GN	BN
10317	**GN**	F	GN	BN
10318	**GN**	F	GN	BN
10319	**GN**	F	GN	BN
10320	**GN**	F	GN	BN
10321	**GN**	F	GN	BN
10322	**GN**	F	GN	BN
10323	**GN**	F	GN	BN
10324	**GN**	F	GN	BN
10325	**GN**	F	GN	BN
10326		F	GN	BN
10327	**GN**	F	GN	BN
10328	**GN**	F	GN	BN
10329	**GN**	F	GN	BN
10330	**GN**	F	GN	BN
10331	**GN**	F	GN	BN
10332	**GN**	F	GN	BN
10333	**GN**	F	GN	BN

AU4G (SLEP)   SLEEPING CAR WITH PANTRY

Dia. AU401. Mark 3A. Air conditioned. 12 compartments with a fixed lower berth and a hinged upper berth, plus an attendants compartment with 2T (controlled emission). BT10 bogies. ETH 7X.

Lot No. 30960 Derby 1981–3.

† Attendants and adjacent two sleeping compartments converted to generator room. 10569 is leased to the Venice Simplon Orient Express.

10500		SL		EC	10551	d	P	*SR*	IS
10501	d	P	*SR*	IS	10553	d	P	*SR*	IS
10502	d	P	*SR*	IS	10554	d	P		ZD
10503		SL		EC	10555	d	P		KI
10504	d	P	*SR*	IS	10556	d †	GS *SS*		EN
10506	d	P	*SR*	IS	10557	d	P		ZD
10507	d	P	*SR*	IS	10558	d	P		PC
10508	d	P	*SR*	IS	10559	d	P		KI
10510	d	P	*SR*	IS	10560	d	P		ZD
10512	d	P		ZG	10561	d	P	*SR*	IS
10513	d	P	*SR*	IS	10562	d	P	*SR*	IS
10514		SL		EC	10563	d	P	*GW*	LA
10515	d	P	*SR*	IS	10565	d	P	*SR*	IS
10516	d	P	*SR*	IS	10566	d	P		ZD
10519	d	P	*SR*	IS	10567	d	P		ZG
10520	d	P	*SR*	IS	10569	d **PC**	P	*SS*	SL
10522	d	P	*SR*	IS	10570		P		KI
10523	d	P	*SR*	IS	10571		SL		Ferme Park
10526	d	P	*SR*	IS	10572	d	P		ZD
10527	d	P	*SR*	IS	10573	d	P		ZD
10529	d	P	*SR*	IS	10574		RS		BN
10530	d	P		ZD	10575		SL		EC
10531	d	P	*SR*	IS	10577	**BG**	P		ZD
10532	d	P	*GW*	LA	10578		P		KI
10533	d	P		ZD	10579	**BG**	P		KI
10534	d	P	*GW*	LA	10580	d	P	*SR*	IS
10535	d	P		ZD	10582	d	P		ZD
10536	d	P		KI	10583	d	P	*GW*	LA
10537	d	P		ZD	10584	d	P	*GW*	LA
10538	d	P		KI	10586	d	P		KI
10539	d	P		KI	10588	d	P	*GW*	LA
10540	d	P		ZD	10589	d	P	*GW*	LA
10541	d †	GS *SS*		EN	10590	d	P	*GW*	LA
10542	d	P	*SR*	IS	10591		P		KI
10543	d	P	*SR*	IS	10592		P		KI
10544	d	P	*SR*	IS	10593	d	P		KI
10546		P		ZD	10594	d	P	*GW*	LA
10547	d	P	*SR*	IS	10595	**BG**	P		KI
10548	d	P	*SR*	IS	10506	d	P		KI
10549	d	P		ZD	10597	d	P	*SR*	IS
10550	d	P		ZD	10598	d	P	*SR*	IS

10599		P		KI
10600 d		P	SR	IS
10601		P		ZD
10602		P		ZD
10603		P		KI
10604		P		ZD
10605 d		P	SR	IS
10606		P		KI
10607 d		P	SR	IS

10608	BG	P		ZN
10609	BG	P		ZG
10610 d		P	SR	IS
10612 d		P	GW	LA
10613 d		P	SR	IS
10614 d		P	SR	IS
10616 d		P	GW	LA
10617 d		P	SR	IS

AS4G (SLE)   SLEEPING CAR

Dia. AS403. Mark 3A. Air conditioned. 13 compartments with a fixed lower berth and a hinged upper berth. 2T (controlled emission). BT10 bogies. ETH 6X.

Lot No. 30961 Derby 1980–4.

10646 d		FS		BN
10647 d		P		KI
10648 d		P	SR	IS
10649 d		P		KI
10650 d		P	SR	IS
10651 d		P		ZD
10653 d		P		ZD
10654 d		P		ZD
10655		SL		EC
10656		P		KI
10657		SL		EC
10658 d		P		KI
10660 d		P		ZD
10662		P		ZD
10663 d		P	SR	IS
10665	BG	P		ZG
10666 d		P	SR	IS
10668 d		P		ZD
10670		P		KI
10672 d		P		ZG
10674 d		P		ZG
10675 d		P	SR	IS
10678	BG	P		KI
10679	BG	P		KI
10680 d		P	SR	IS
10682 d		P		ZD
10683 d		P	SR	IS
10684	BG	P		KI
10685 d		P		ZH
10686 d		P		ZD
10687 d		P		ZD
10688 d		P	SR	IS
10689 d		P	SR	IS
10690 d		P	SR	IS

10691 d		P		ZD
10692 d		P		ZD
10693 d		P	SR	IS
10696 d		P		KI
10697 d		P		KI
10699 d		P	SR	IS
10700	BG	P		KI
10701 d		P		KI
10702		SL		EC
10703 d		P	SR	IS
10704 d		SO	TE	ZA
10706 d		P	SR	IS
10707		P		ZG
10708 d		P		ZD
10709 d		P		ZD
10710 d		P		KI
10711 d		P		ZD
10712 d		P		ZD
10713 d		P		ZD
10714 d		P	SR	IS
10715 d		P		ZD
10716 d		P		ZD
10717 d		P		ZD
10718 d		P	SR	IS
10719 d		P	SR	IS
10720		P		KI
10722 d		P	SR	IS
10723 d		P	SR	IS
10724		SL		Ferme Park
10725		SL		EC
10726		SL		EC
10727		SL		EC
10728		P		ZN
10729		SL		Ferme Park

10730 d	P	ZD	10732 d	P	KI
10731 d	P	KI			

AD1G (FO)   OPEN FIRST

Dia. AD108. Mark 3A. Air conditioned. All now refurbished with new upholstery, carpets etc. 11005–7 have regained their original numbers, having being converted back from open composites 11905–7. 48/– 2T (47/– 2T w). BT10 bogies. Secondary door locks. ETH 6X.

Lot No. 30878 Derby 1975–6. 34.3 t.

r Further refurbished with table lamps and new burgundy seat trim.
* Disabled toilet, but no wheelchair space.

11005			P	VW	PC	11031	rw	V	P	VW	MA
11006			P	VW	PC	11033			P	VW	PC
11007	w		P	VW	PC	11036	rw	V	P	VW	MA
11011	r*	V	P	VW	MA	11037			P	VW	PC
11013			P	VW	PC	11038	w		P	VW	PC
11016			P	VW	PC	11039	w		P	VW	PC
11017			P	VW	PC	11040			P	VW	MA
11018	rw	V	P	VW	MA	11042	rw	V	P	VW	MA
11019			P	VW	PC	11044			P	VW	MA
11020	r		P	VW	MA	11045			P	VW	PC
11021	w		P	VW	PC	11046	w		P	VW	MA
11023	w		P	VW	PC	11048			P	VW	PC
11024	r	V	P	VW	MA	11052	r	V	P	VW	MA
11026	r		P	VW	MA	11054			P	VW	MA
11027	r		P	VW	PC	11055			P	VW	PC
11028	r		P	VW	MA	11058	r	V	P	VW	MA
11029	r	V	P	VW	MA	11060			P	VW	PC
11030	rw	V	P	VW	MA						

AD1H (FO)   OPEN FIRST

Dia. AD109. Mark 3B. Air conditioned. Inter-City 80 seats. 48/– 2T (47/– 2T w). BT10 bogies. Secondary door locks. ETH 6X.

Lot No. 30982 Derby 1985. 36.46 t.

r Refurbished with table lamps and new burgundy seat trim.

11064			P	VW	PC	11075			P	VW	PC
11065			P	VW	MA	11076			P	VW	PC
11066			P	VW	PC	11077			P	VW	MA
11067	rw	V	P	VW	MA	11078			P	VW	MA
11068			P	VW	PC	11079	r	V	P	VW	PC
11069			P	VW	MA	11080	rw	V	P	VW	MA
11070	w		P	VW	PC	11081			P	VW	PC
11071	r	V	P	VW	PC	11082	w		P	VW	PC
11072			P	VW	PC	11083	pr	V	P	VW	MA
11073			P	VW	PC	11084	prw	V	P	VW	MA
11074	rw	V	P	VW	MA	11085	p		P	VW	PC

11086	p	P	*VW*	PC	11094	pw	P	*VW*	PC
11087	p	P	*VW*	PC	11095	p	P	*VW*	PC
11088	p	P	*VW*	PC	11096	p	P	*VW*	PC
11089	p	P	*VW*	PC	11097	pr **V**	P	*VW*	MA
11090	p	P	*VW*	PC	11098	p	P	*VW*	PC
11091	p	P	*VW*	PC	11099	pr **V**	P	*VW*	MA
11092	p	P	*VW*	PC	11100	p	P	*VW*	PC
11093	p	P	*VW*	PC	11101	p	P	*VW*	PC

AD1J (FO)   OPEN FIRST

Dia. AD111. Mark 4. Air conditioned. Known as 'Pullman open' by GNER. 46/ – 1T. BT41 bogies. ETH 6.

11264–71 were cancelled.

Lot No. 31046 Metro-Cammell 1989–92. 39.7 t.

11200	**GN**	F	*GN*	BN	11234	**GN**	F	*GN*	BN
11201	p **GN**	F	*GN*	BN	11235	p **GN**	F	*GN*	BN
11202	**GN**	F	*GN*	BN	11236	**GN**	F	*GN*	BN
11203	p **GN**	F	*GN*	BN	11237	p **GN**	F	*GN*	BN
11204	p **GN**	F	*GN*	BN	11238	**GN**	F	*GN*	BN
11205	**GN**	F	*GN*	BN	11239	p **GN**	F	*GN*	BN
11206	**GN**	F	*GN*	BN	11240	**GN**	F	*GN*	BN
11207	p **GN**	F	*GN*	BN	11241	**GN**	F	*GN*	BN
11208	**GN**	F	*GN*	BN	11242	p **GN**	F	*GN*	BN
11209	**GN**	F	*GN*	BN	11243	p **GN**	F	*GN*	BN
11210	**GN**	F	*GN*	BN	11244	**GN**	F	*GN*	BN
11211	p **GN**	F	*GN*	BN	11245	p **GN**	F	*GN*	BN
11212	**GN**	F	*GN*	BN	11246	p	F	*GN*	BN
11213	p **GN**	F	*GN*	BN	11247	p	F	*GN*	BN
11214	p **GN**	F	*GN*	BN	11248	**GN**	F	*GN*	BN
11215	**GN**	F	*GN*	BN	11249	p **GN**	F	*GN*	BN
11216	**GN**	F	*GN*	BN	11250	**GN**	F	*GN*	BN
11217	p **GN**	F	*GN*	BN	11251	p **GN**	F	*GN*	BN
11218	**GN**	F	*GN*	BN	11252	**GN**	F	*GN*	BN
11219	p **GN**	F	*GN*	BN	11253	p **GN**	F	*GN*	BN
11220	**GN**	F	*GN*	BN	11254	**GN**	F	*GN*	BN
11221	p **GN**	F	*GN*	BN	11255	p **GN**	F	*GN*	BN
11222	p **GN**	F	*GN*	BN	11256	**GN**	F	*GN*	BN
11223	**GN**	F	*GN*	BN	11257	p **GN**	F	*GN*	BN
11224	p **GN**	F	*GN*	BN	11258	**GN**	F	*GN*	BN
11225	p **GN**	F	*GN*	BN	11259	p **GN**	F	*GN*	BN
11226	**GN**	F	*GN*	BN	11260	**GN**	F	*GN*	BN
11227	p **GN**	F	*GN*	BN	11261	p **GN**	F	*GN*	BN
11228	p **GN**	F	*GN*	BN	11262	**GN**	F	*GN*	BN
11229	p **GN**	F	*GN*	BN	11263	p **GN**	F	*GN*	BN
11230	**GN**	F	*GN*	BN	11272	**GN**	F	*GN*	BN
11231	p **GN**	F	*GN*	BN	11273	**GN**	F	*GN*	BN
11232	**GN**	F	*GN*	BN	11274	**GN**	F	*GN*	BN
11233	p **GN**	F	*GN*	BN	11275	**GN**	F	*GN*	BN

11276 **GN** F *GN* BN |

AC2G (TSO) OPEN STANDARD

Dia. AC213 (AC220 z). Mark 3A. Air conditioned. All now refurbished with modified seat backs and new layout. 12169–72 have been converted from open composites 11908–10/22, formerly FOs 11008–10/22. –/76 2T (–/74 2T wz) BT10 (BREL T4 §) bogies. Secondary door locks. ETH 6X.

Lot No. 30877 Derby 1975–7. 34.3 t.

r Further refurbished with new light blue seat trim.
* Further refurbished with new light blue seat trim and two wheelchair spaces. –/70 2T.

No.						No.					
12004			P	VW	PC	12044	r	V	P	VW	MA
12005			P	VW	PC	12045	r	V	P	VW	MA
12007	r	V	P	VW	MA	12046			P	VW	PC
12008	r	V	P	VW	MA	12047	z		P	VW	PC
12009			P	VW	PC	12048			P	VW	MA
12010	r	V	P	VW	MA	12049			P	VW	PC
12011			P	VW	PC	12050	w		P	VW	PC
12012			P	VW	MA	12051			P	VW	PC
12013	r		P	VW	MA	12052			P	VW	PC
12014			P	VW	PC	12053	r	V	P	VW	MA
12015			P	VW	PC	12054	*	V	P	VW	MA
12016			P	VW	PC	12055			P	VW	MA
12017	r	V	P	VW	MA	12056			P	VW	PC
12019			P	VW	PC	12057	r	V	P	VW	MA
12020			P	VW	MA	12058			P	VW	MA
12021			P	VW	MA	12059	w		P	VW	PC
12022	r	V	P	VW	MA	12060			P	VW	PC
12023			P	VW	PC	12061	w		P	VW	MA
12024	w		P	VW	PC	12062			P	VW	PC
12025	r		P	VW	MA	12063			P	VW	PC
12026			P	VW	PC	12064			P	VW	PC
12027	r		P	VW	MA	12065			P	VW	PC
12028	r		P	VW	MA	12066	r	V	P	VW	MA
12029			P	VW	PC	12067	r	V	P	VW	MA
12030			P	VW	PC	12068			P	VW	PC
12031			P	VW	PC	12069	r	V	P	VW	MA
12032			P	VW	PC	12070			P	VW	PC
12033	z*	V	P	VW	MA	12071			P	VW	PC
12034			P	VW	PC	12072	r	V	P	VW	MA
12035	r	V	P	VW	PC	12073	r	V	P	VW	MA
12036	w		P	VW	PC	12075	r	V	P	VW	PC
12037			P	VW	PC	12076			P	VW	PC
12038			P	VW	PC	12077			P	VW	PC
12040			P	VW	PC	12078			P	VW	MA
12041			P	VW	PC	12079			P	VW	PC
12042	w		P	VW	PC	12080	r	V	P	VW	PC
12043	r	V	P	VW	MA	12081			P	VW	PC

No.						No.					
12082			P	VW	PC	12127	r	V	P	VW	MA
12083	r	V	P	VW	MA	12128	w		P	VW	PC
12084			P	VW	PC	12129			P	VW	PC
12085	*	V	P	VW	MA	12130	r	V	P	VW	MA
12086	*	V	P	VW	MA	12131			P	VW	PC
12087	*	V	P	VW	PC	12132			P	VW	PC
12088			P	VW	PC	12133			P	VW	PC
12089			P	VW	PC	12134			P	VW	PC
12090			P	VW	PC	12135			P	VW	PC
12091			P	VW	MA	12136	r	V	P	VW	MA
12092			P	VW	PC	12137			P	VW	PC
12093			P	VW	PC	12138			P	VW	PC
12094			P	VW	MA	12139			P	VW	PC
12095	r	V	P	VW	MA	12140	z§		P	VW	PC
12096			P	VW	PC	12141			P	VW	PC
12097			P	VW	PC	12142	z		P	VW	MA
12098			P	VW	PC	12143			P	VW	PC
12099			P	VW	PC	12144	w		P	VW	PC
12100	z		P	VW	PC	12145			P	VW	MA
12101	w		P	VW	PC	12146			P	VW	PC
12102			P	VW	PC	12147			P	VW	PC
12103	w		P	VW	PC	12148			P	VW	PC
12104			P	VW	MA	12149			P	VW	PC
12105			P	VW	PC	12150			P	VW	PC
12106	r	V	P	VW	MA	12151			P	VW	PC
12107			P	VW	PC	12152	r	V	P	VW	MA
12108	w		P	VW	PC	12153			P	VW	PC
12109	w		P	VW	PC	12154			P	VW	PC
12110			P	VW	PC	12155	*	V	P	VW	MA
12111	r	V	P	VW	MA	12156	r	V	P	VW	MA
12112	z*	V	P	VW	MA	12157			P	VW	PC
12113	r	V	P	VW	MA	12158	r	V	P	VW	MA
12114			P	VW	PC	12159			P	VW	PC
12115	w		P	VW	PC	12160	w		P	VW	PC
12116			P	VW	PC	12161	z		P	VW	PC
12117	r	V	P	VW	MA	12163	r	V	P	VW	MA
12118			P	VW	MA	12164	r	V	P	VW	PC
12119			P	VW	PC	12165			P	VW	MA
12120			P	VW	PC	12166			P	VW	PC
12121			P	VW	PC	12167			P	VW	PC
12122	z		P	VW	MA	12168	w		P	VW	PC
12123			P	VW	MA	12169	*	V	P	VW	MA
12124			P	VW	PC	12170	*		P	VW	MA
12125			P	VW	PC	12171	z		P	VW	PC
12126	r	V	P	VW	MA	12172	z		P	VW	PC

AI2J (TSOE) OPEN STANDARD (END)

Dia. AI201. Mark 4. Air conditioned. –/74 2T. BT41 bogies. ETH 6.
Lot No. 31047 Metro-Cammell 1989–91. 39.5 t.

12232 was converted from the original 12405.

12200		F	GN	BN		12216	GN	F	GN	BN
12201	GN	F	GN	BN		12217	GN	F	GN	BN
12202	GN	F	GN	BN		12218	GN	F	GN	BN
12203	GN	F	GN	BN		12219	GN	F	GN	BN
12204	GN	F	GN	BN		12220	GN	F	GN	BN
12205	GN	F	GN	BN		12222	GN	F	GN	BN
12206	GN	F	GN	BN		12223	GN	F	GN	BN
12207	GN	F	GN	BN		12224		F	GN	BN
12208	GN	F	GN	BN		12225		GN	F	GN	BN
12209	GN	F	GN	BN		12226	GN	F	GN	BN
12210	GN	F	GN	BN		12227	GN	F	GN	BN
12211	GN	F	GN	BN		12228	GN	F	GN	BN
12212	GN	F	GN	BN		12229	GN	F	GN	BN
12213	GN	F	GN	BN		12230	GN	F	GN	BN
12214	GN	F	GN	BN		12231	GN	F	GN	BN
12215	GN	F	GN	BN		12232	GN	F	GN	BN

AL2J (TSOD) OPEN STANDARD (DISABLED ACCESS)

Dia. AL201. Mark 4. Air conditioned. –/72 + wheelchair space 1T (suitable for a disabled person). BT41 bogies. p. ETH 6.

Lot No. 31048 Metro-Cammell 1989–91. 39.4 t.

12300	GN	F	GN	BN		12316	GN	F	GN	BN
12301	GN	F	GN	BN		12317	GN	F	GN	BN
12302	GN	F	GN	BN		12318	GN	F	GN	BN
12303	GN	F	GN	BN		12319	GN	F	GN	BN
12304	GN	F	GN	BN		12320	GN	F	GN	BN
12305	GN	F	GN	BN		12321	GN	F	GN	BN
12306	GN	F	GN	BN		12322	GN	F	GN	BN
12307	GN	F	GN	BN		12323		F	GN	BN
12308	GN	F	GN	BN		12324	GN	F	GN	BN
12309	GN	F	GN	BN		12325	GN	F	GN	BN
12310	GN	F	GN	BN		12326	GN	F	GN	BN
12311	GN	F	GN	BN		12327	GN	F	GN	BN
12312	GN	F	GN	BN		12328	GN	F	GN	BN
12313	GN	F	GN	BN		12329	GN	F	GN	BN
12314	GN	F	GN	BN		12330	GN	F	GN	BN
12315	GN	F	GN	BN						

AC2J (TSO)   OPEN STANDARD

Dia. AC214. Mark 4. Air conditioned. –/74 2T. BT41 bogies. ETH 6X.

Lot No. 31049 Metro-Cammell 1989 onwards. 39.9 t.

12405 is the second coach to carry that number. It was built from the bodyshell originally intended for 12221. The original 12405 is now 12232. 12490–12512 were cancelled.

12400	GN	F	GN	BN		12451	GN	F	GN	BN
12401	GN	F	GN	BN		12452	GN	F	GN	BN
12402	GN	F	GN	BN		12453	GN	F	GN	BN
12403	GN	F	GN	BN		12454	GN	F	GN	BN
12404	GN	F	GN	BN		12455	GN	F	GN	BN
12405	GN	F	GN	BN		12456	GN	F	GN	BN
12406	GN	F	GN	BN		12457	GN	F	GN	BN
12407	GN	F	GN	BN		12458	GN	F	GN	BN
12408	GN	F	GN	BN		12459	GN	F	GN	BN
12409	GN	F	GN	BN		12460	GN	F	GN	BN
12410	GN	F	GN	BN		12461	GN	F	GN	BN
12411	GN	F	GN	BN		12462	GN	F	GN	BN
12412	GN	F	GN	BN		12463	GN	F	GN	BN
12413	GN	F	GN	BN		12464	GN	F	GN	BN
12414	GN	F	GN	BN		12465	GN	F	GN	BN
12415	GN	F	GN	BN		12466	GN	F	GN	BN
12416	GN	F	GN	BN		12467		F	GN	BN
12417	GN	F	GN	BN		12468		F	GN	BN
12418	GN	F	GN	BN		12469		F	GN	BN
12419	GN	F	GN	BN		12470	GN	F	GN	BN
12420	GN	F	GN	BN		12471	GN	F	GN	BN
12421	GN	F	GN	BN		12472	GN	F	GN	BN
12422	GN	F	GN	BN		12473	GN	F	GN	BN
12423	GN	F	GN	BN		12474	GN	F	GN	BN
12424	GN	F	GN	BN		12475	GN	F	GN	BN
12425	GN	F	GN	BN		12476	GN	F	GN	BN
12426	GN	F	GN	BN		12477	GN	F	GN	BN
12427	GN	F	GN	BN		12478	GN	F	GN	BN
12428	GN	F	GN	BN		12479	GN	F	GN	BN
12429	GN	F	GN	BN		12480	GN	F	GN	BN
12430	GN	F	GN	BN		12481	GN	F	GN	BN
12431	GN	F	GN	BN		12482	GN	F	GN	BN
12432	GN	F	GN	BN		12483	GN	F	GN	BN
12433	GN	F	GN	BN		12484	GN	F	GN	BN
12434	GN	F	GN	BN		12485	GN	F	GN	BN
12435	GN	F	GN	BN		12486	GN	F	GN	BN
12436	GN	F	GN	BN		12487	GN	F	GN	BN
12437	GN	F	GN	BN		12488	GN	F	GN	BN
12438	GN	F	GN	BN		12489	GN	F	GN	BN
12439	GN	F	GN	BN		12513	GN	F	GN	BN
12440	GN	F	GN	BN		12514	GN	F	GN	BN
12441	GN	F	GN	BN		12515	GN	F	GN	BN
12442	GN	F	GN	BN		12516	GN	F	GN	BN
12443	GN	F	GN	BN		12517	GN	F	GN	BN
12444	GN	F	GN	BN		12518	GN	F	GN	BN
12445	GN	F	GN	BN		12519	GN	F	GN	BN
12446	GN	F	GN	BN		12520	GN	F	GN	BN
12447	GN	F	GN	BN		12521	GN	F	GN	BN
12448	GN	F	GN	BN		12522	GN	F	GN	BN
12449	GN	F	GN	BN		12523	GN	F	GN	BN
12450	GN	F	GN	BN		12524	GN	F	GN	BN

12525	**GN**	F	*GN*	BN		12532	**GN**	F	*GN*	BN
12526	**GN**	F	*GN*	BN		12533	**GN**	F	*GN*	BN
12527	**GN**	F	*GN*	BN		12534	**GN**	F	*GN*	BN
12528	**GN**	F	*GN*	BN		12535	**GN**	F	*GN*	BN
12529	**GN**	F	*GN*	BN		12536	**GN**	F	*GN*	BN
12530		F	*GN*	BN		12537	**GN**	F	*GN*	BN
12531	**GN**	F	*GN*	BN		12538	**GN**	F	*GN*	BN

AA11 (FK)   CORRIDOR FIRST

Dia. AA101. Mark 1. 42/– 2T. ETH 3.

13225–13230. Lot No. 30381 Ashford/Swindon 1959. B4 bogies. 33 t.
13306–13341. Lot No. 30667 Swindon 1962. Commonwealth bogies. 36 t.

f Fitted with fluorescent lighting.

13225	xk **RR**	F	CP		13318			RS *SS*	EC
13227	xk **CH**	RV *SS*	CP		13320	v	**BG**	W	CS
13228	xk **M**	SP	BO		13321	x	**M**	W *SS*	CS
13229	xk **M**	SP *SS*	BO		13323	xf	**M**	W	CS
13230	xk **M**	SP *SS*	BO		13331	vf	**N**	LN	CP
13306	v **BG**	E	KM		13341	f		RS *SS*	EC
13317	x **M**	W	CS	

AA1D (FK)   CORRIDOR FIRST

Dia. AA109. Mark 2D. Air conditioned (Stones). 42/– 2T. B4 bogies. 13585–13607 require at least 800 V train heat supply. ETH 5.

Lot No. 30825 Derby 1971–2. 34.5 t.

13575	**N**	F	OM		13585	CC	KI
13581		RR	WB		13604	CC	BN
13582		E	KI		13607	CC	Hornsey Up CS
13583		RR	WB	

AA31 (CK)   CORRIDOR COMPOSITE

Dia. AA301. Mark 1. 24/18 1T. ETH 2.

Lot No. 30665 Derby 1961. Commonwealth bogies and metal window frames. 37 t. Numbered 7167/87/91 for a time.

16167	v	**N**	VS	SL		16191	x	**CH**	RV	CP
16187	x	**CH**	RV	CP	

AB11 (BFK)   CORRIDOR BRAKE FIRST

Dia. AB101. Mark 1. 24/– 1T. Commonwealth bogies. ETH 2.

17007. Lot No. 30382 Ashford/Swindon 1959. 35 t.
17013–17019. Lot No. 30668 Swindon 1961. 36 t.
17023. Lot No. 30718 Swindon 1963. Metal window frames. 36 t.
Originally numbered 14007/13/15/19/23.

17007	x	**PC**	O	*SU*	SZ		17019	v	**M**	O	*SU*	GT
17013	v	**M**	FS		SZ		17023	x	**G**	RS	*SS*	BN
17015	x	**W**	RS	*SS*	BN							

AB1Z (BFK)   CORRIDOR BRAKE FIRST

Dia. AB102. Mark 2. Pressure ventilated. 24/– 1T. B4 bogies. ETH 4.

Lot No. 30756 Derby 1966. 31.5 t.

Originally numbered 14039/54.

| 17039 | v | **RX** | F | *EW* | CD | | 17054 | v | **BG** | F | | CB |

AB1A (BFK)   CORRIDOR BRAKE FIRST

Dia. AB103. Mark 2A. Pressure ventilated. 24/– 1T. B4 bogies. ETH 4.

17056–17077. Lot No. 30775 Derby 1967–8. 32 t.
17086–17102. Lot No. 30786 Derby 1968. 32 t.

Originally numbered 14056–102. 17090 was renumbered 35503 for a time when declassified.

17096 is leased to the Venice Simplon Orient Express.

17056		**N**	CC		CP		17086		**FD**	F	*SS*	CP
17058		**N**	F		LM		17090	v	**RR**	F		LT
17064	v	**RR**	F		LT		17091	v	**RR**	F		LT
17073		**N**	F		LM		17096		**G**	F	*SS*	SL
17076		**N**	F		Easteigh Down Sdgs		17099	v	**RR**	F		LT
17077		**N**	F		CP		17102		**M**	W	*SS*	CS

AB1D (BFK)   CORRIDOR BRAKE FIRST

Dia. AB106. Mark 2D. Air conditioned (Stones equipment). 24/– 1T. B4 Bogies. 17163–17172 require at least 800 V train heat supply. ETH 5.

Lot No. 30823 Derby 1971–2. 33.5 t.

Non-Standard Livery: 17141 & 17164 are Waterman VIP without lining.

Originally numbered 14141–72.

17141	**0**	CC *CA*	CF		17163		F	KI
17144		CC	DY		17164	**0**	CC *CA*	CF
17146		CC	DY		17165		CC	Ferme Park
17148		F	KI		17166		F	LT
17151		CC	EC		17167		CC	EC
17153	**W**	CC	CS		17168	**M**	W *SS*	CS
17155		F	KI		17169		CC	CS
17156		CC	DY		17170		CC	DY
17159		CC	Hornsey Up CS		17171		E	KM
17161		E	OM		17172		CC	Ferme Park

AE1G (BFO)   OPEN BRAKE FIRST

Dia. AE101. Mark 3B. Air conditioned. Fitted with hydraulic handbrake. 36/–1T (35/–1T w). BT10 bogies. pg. Secondary door locks. ETH 5X.

Lot No. 30990 Derby 1986. 35.81 t.

r Refurbished with table lamps and new burgundy seat trim.

17173		P	*VW*	MA	
17174	r **V**	P	*VW*	PC	

17175	w	P	*VW*	PC	

AB31 (BCK)   CORRIDOR BRAKE COMPOSITE

Dia. AB301 (AB302*). Mark 1. There are two variants depending upon whether the standard, class compartments have armrests. Each vehicle has two first class and three standard class compartments. 12/18 2T (12/24 2T *). ETH 2.

21096. Lot No. 30185. Metro-Cammell 1956. BR Mark 1 bogies. Steam heat only. 32.5 t.
21224. Lot No. 30245. Metro-Cammell 1958. B4 bogies. 33 t.
21236–21246. Lot No. 30669 Swindon 1961–2. Commonwealth bogies. 36 t.
21256. Lot No. 30731 Derby 1963. Commonwealth bogies. 37 t.
21265–21272. Lot No. 30732 Derby 1964. Commonwealth bogies. 37 t.

21096	x	**M**	O	*SU*	BQ
21224	**RC**		RA	*SS*	CP
21236	v	**M**	RV	*SU*	ZG
21241	x		RS	*SS*	EC
21245	x	**CC**	RS	*SS*	BN
21246			RS	*SS*	BN

21256	x	**M**	W	*SS*	CS
21265	*	**BG**	E		KM
21266	*		FS		SZ
21268	*		FS		SZ
21269	*	**WV**	RS	*SS*	BN
21272	x*	**M**	RS	*SS*	EC

AA21 (SK)   CORRIDOR STANDARD

Dia. AA201 (AA202*). Mark 1. There are two variants depending upon whether the standard class compartments have armrests. Each vehicle has eight compartments. All remaining vehicles have metal window frames and melamine interior panelling. Commonwealth bogies. –/48 2T (–/64 2T *). ETH 4.

25729–25893. Lot No. 30685 Derby 1961–2. 36 t.
25955. Lot No. 30686 Derby 1962. 36 t.
26013. Lot No. 30719 Derby 1962. 37 t.

Non-Standard Livery: 25767, 25837, 25893 & 26013 are Pilkington's K (green with white red chevron and light blue block).

f Facelifted with fluorescent lighting.
† Rebuilt internally as TSO using components from 4936. -/64 2T.

These coaches were renumbered 18729–19013 for a time.

25729	x*f	**M**	W	*SS*	CS
25756	x	**M**	W	*SS*	CS
25767	x	**0**	W	*SS*	CS

25806	x†	**M**	W	*SS*	CS
25808	x	**M**	W	*SS*	CS
25837	x	**0**	W	*SS*	CS

25862	x	**M**	W SS	CS		25955	x*f	**M**	W SS	CS
25893	x	**0**	W SS	CS		26013	x	**0**	W SS	CS

AB21 (BSK) CORRIDOR BRAKE STANDARD

Dia. AB201 (AB 202*). Mark 1. There are two variants depending upon whether the standard class compartments have armrests. Each vehicle has four compartments. Lots 30699, 30721 and 30728 have metal window frames and melamine interior panelling. –/24 1T (–/32 1T*). ETH2.

g Converted to e.t.h. generator vehicle.

34525–34556. Lot No. 30095 Wolverton 1955. BR Mark 1 bogies. 34 t. (34525 C 36 t.).
34952–34991. Lot No. 30229 Metro-Cammell 1956–7. BR Mark 1 bogies. 34 t. (34991 C 36 t.).
35073. Lot No. 30233 Gloucester 1956–7. BR Mark 1 bogies. 35 t.
35185–35207. Lot No. 30427 Wolverton 1959. B4 bogies. 33 t.
35290. Lot No. 30573 Gloucester 1960. B4 bogies. 33 t.
35317–35333. Lot No. 30699 Wolverton 1962–3. Commonwealth bogies. 37 t.
35449. Lot No. 30728 Wolverton 1963. Commonwealth bogies. 37 t.
35407, 35452–35486. Lot No. 30721 Wolverton 1963. Commonwealth bogies. 37 t.

Non-Standard Liveries: 35290 is black. 35407 is in London & North Western Railway livery.

34525	g	**M**	GS		CS	35449	v	**CH** O		SU SZ
34556	v	**BG**	VS		SL	35452	x	**RR** F		NW LL
34952	v*	**BG**	VS		SL	35453	x	**CH** RV	SS	CP
34991	*	**PC**	VS	SS	SL	35457	v	**M** O		SU BQ
35073	v	**M**	W		CS	35459	x	**M** W		SS CS
35185	x	**M**	SP		BO	35461	x	**CH** RV		CP
35204	v	**M**	VS		SL	35463	v	**M** W		SU CS
35207	x*	**PC**	VS	SU	SL	35465	x	**WV** LN		SU CQ
35290	v	**0**	CC		CQ	35467	v	**M** RV		SU KR
35317	v	**M**	MH	SS	RL	35468	v	**M** O		SU YM
35329	v	**G**	MH	SS	RL	35469	xg	**CC** RS		SS BN
35333	x	**CH**	O	SU	DI	35486	v	**M** O		SU KR
35407	xg	**0**	RE	SS	CJ					

AB2A/AB2C (BSK) CORRIDOR BRAKE STANDARD

Dia. AB204. Mark 2A (2C*). Pressure ventilated. Renumbered from BFK. –/24 1T. B4 bogies. ETH 4.

35507–9/11. Lot No. 30796 Derby 1969–70. 32.5 t.
35510/12–14. Lot No. 30775 Derby 1967–68. 32 t.
35515–18. Lot No. 30786 Derby 1968. 32 t.

§ Cage removed from brake compartment.

35507 (14123, 17123) * **RR** F LM

35508	(14128, 17128)	*	**RR**	CC		CP
35509	(14138, 17138)	*	**RR**	F		LM
35510	(14075, 17075)		**RR**	F		LM
35511	(14130, 17130)	*	**RR**	F		KI
35512	(14057, 17057)	§	**RR**	F	NW	LL
35513	(14063, 17063)	§	**RR**	F	NW	LL
35514	(14069, 17069)	§	**RR**	F	NW	LL
35515	(14079, 17079)	§	**RR**	F	NW	LL
35516	(14080, 17080)	§	**RR**	F	NW	LL
35517	(14088, 17088)	§	**RR**	F	NW	LL
35518	(14097, 17097)	§	**RR**	F	NW	LL

NAMED COACHES

The following miscellaneous coaches carry names:

1683	CAROL
1953	LANCASTRIAN
3105	Julia
3125	LOCH SHIEL
3130	BERYL
3181	MONARCH
3188	SOVEREIGN
3240	PENDENNIS
3267	TREGENNA
3273	RESTORMEL
5132	CLAN MUNRO
5154	CLAN FRASER
5166	CLAN MACKENZIE
5191	CLAN DONALD

5193	CLAN MACLEOD
5212	CAPERKAILZIE
5275	Wendy
5365	Deborah
5373	Felicity
9385	BALMACARA
9388	BAILECHAUL
9414	BRAHAN SEER
10541	STATE CAR 5
10556	SERVICE CAR
10569	LEVIATHAN
17007	MERCATOR
17086	Georgina
34991	BAGGAGE CAR No.9

2.2. HIGH SPEED TRAIN TRAILER CARS

HSTs run in formations of 7 or 8 trailer cars with a driving motor brake (power car) at each end. All vehicles are classified mark 3. All trailer cars have BT10 bogies with disc brakes. Heating is by a three-phase supply and vehicles have air conditioning. Max. Speed is 125 m.p.h.

All vehicles underwent a mid-life refurbishment in the 1980s, and they are at present undergoing a further refurbishment, each train operating company having a different scheme. GWT vehicles have new green seat covers and extra partitions between seat bays. GNER vehicles have new ceiling lighting panels and brown seat covers. First class vehicles have table lamps and imitation plastic walnut end panels. Virgin Cross-Country vehicles have green seat covers and standard class vehicles have four seats in the centre of each standard class carriage replaced with a luggage stack. Midland Mainline vehicles have grey seat covers, redesigned seat squabs, side carpeting and two seats in the centre of each standard class carriage replaced with a luggage stack. The 'r' symbol is not used in this section, as all re-liveried vehicles have been refurbished.

TOPS CODES

TOPS codes for HST trailer cars are made up as follows:

(1) Two letters denoting the layout of the vehicle as follows:

GH   Open
GJ   Open with Guard's compartment
GK   Buffet
GL   Kitchen
GN   Buffet

(2) A digit for the class of passenger accommodation:

1   First
2   Standard (formerly second)
4   Unclassified

(3) A suffix relating to the build of coach:

G   Mark 3

OPERATOR CODES

The normal operator codes are given in brackets after the TOPS codes. These are as follows:

TCSD   Trailer Conductor Standard
TF   Trailer First
TGS   Trailer Guard's Standard
TRB   Trailer Buffet First
TRFB   Trailer Buffet First
TRFK   Trailer Kitchen First
TRFM   Trailer Modular Buffet First
TRSB   Trailer Buffet Standard
TS   Trailer Standard

GN4G (TRB)   TRAILER BUFFET FIRST

Dia. GN401. Converted from TRSB by fitting first class seats. Renumbered from 404xx series by subtracting 200. Secondary door locks. pq. 23/– (22/– w).

40204–40228. Lot No. 30883 Derby 1976–7. 36.12 t.
40231–40233. Lot No. 30899 Derby 1978–9. 36.12 t.

40204	**GW**	A	*GW*	LA	40211	w	P	*VX*	LA
40205	**GW**	A	*GW*	PM	40212	w	P	*VX*	LA
40206		A	*GW*	PM	40213		A	*GW*	PM
40207		A	*GW*	PM	40221		A	*GW*	PM
40208	**GW**	A	*GW*	LA	40228		A	*GW*	PM
40209	**GW**	A	*GW*	PM	40231	**GW**	A	*GW*	LA
40210	**GW**	A	*GW*	PM	40233	w	P	*VX*	LA

GK2G (TRSB)   TRAILER BUFFET STANDARD

Dia. GK202. Renumbered from 400xx series by adding 400. Secondary door locks. pq. –/33 + tip up seat and wheelchair space.

40401–40427. Lot No. 30883 Derby 1976–7. 36.12 t.
40429–40437. Lot No. 30899 Derby 1978–9. 36.12 t.

40432/4 were numbered 40232/4 for a time when fitted with first class seats.

40401		P	*VX*	EC	40424		P	*VX*	LA
40402		P	*VX*	EC	40425		P	*VX*	LA
40403		P	*VX*	LA	40426		P	*VX*	LA
40414		P	*VX*	LA	40427	**V**	P	*VX*	EC
40415	**V**	P	*VX*	LA	40429		P	*VX*	EC
40416		P	*VX*	EC	40430		P	*VX*	EC
40417		P	*VX*	LA	40432	**V**	P	*VX*	LA
40418	**V**	P	*VX*	LA	40434		P	*VX*	LA
40419		P	*VX*	EC	40435	**V**	P	*VX*	EC
40420	**V**	P	*VX*	EC	40436		P	*VX*	LA
40422	**V**	P	*VX*	LA	40437		P	*VX*	EC
40423		P	*VX*	EC					

GL1G (TRFK)   TRAILER KITCHEN FIRST

Dia. GL101. Reclassified from TRUK. pq. 24/–.

Lot No. 30884 Derby 1976–7. 37 t.

40501	d	P		ZD	40511		A		KI
40505		A		KI	40513	d	P		ZD

GK1G (TRFM)   TRAILER MODULAR BUFFET FIRST

Dia. GK102. Converted to modular catering from 40719. Secondary door locks. pq. 17/–.

Lot No. 30921 Derby 1978–9. 38.16 t.

40619		P		DY

GK1G (TRFB)   TRAILER BUFFET FIRST

Dia. GK101. These vehicles have larger kitchens than the 402xx and 404xx series vehicles, and are used in trains where full meal service is required. They were renumbered from the 403xx series (in which the seats were unclassified) by adding 400 to previous number. Secondary door locks. pq. 17/–.

40700–40721. Lot No. 30921 Derby 1978–9. 38.16 t.
40722–40735. Lot No. 30940 Derby 1979–80. 38.16 t.
40736–40753. Lot No. 30948 Derby 1980–1. 38.16 t.
40754–40757. Lot No. 30966 Derby 1982. 38.16 t.

* Prototype refurbished vehicle for Porterbrook Leasing Company.

40700	MM	P	ML	NL		40730	MM	P	ML	NL
40701		P	ML	NL		40731		A	GW	LA
40702		P	ML	NL		40732		A	VW	MA
40703	GW	A	GW	LA		40733	GW	A	GW	LA
40704	GN	A	GN	EC		40734	GW	A	GW	LA
40705		A	GN	EC		40735	GN	A	GN	EC
40706	GN	A	GN	EC		40736	GW	A	GW	LA
40707	GW	A	GW	LA		40737	GN	A	GN	EC
40708		P	ML	NL		40738	GW	A	GW	LA
40709	GW	A	GW	LA		40739	GW	A	GW	PM
40710		A	GW	LA		40740		A	GN	EC
40711	GN	A	GN	EC		40741	MM	P	ML	NL
40712	GW	A	GW	LA		40742		A	VW	MA
40713	GW	A	GW	LA		40743	GW	A	GW	LA
40714	GW	A	GW	PM		40744	GW	A	GW	PM
40715	GW	A	GW	PM		40745		A	GW	ZD (U)
40716	GW	A	GW	PM		40746	MM	P	ML	NL
40717		A	GW	PM		40747		A	GW	PM
40718	GW	A	GW	LA		40748		A	GN	EC
40720	GN	A	GN	EC		40749	MM	P	ML	NL
40721	GW	A	GW	LA		40750	GN	A	GN	EC
40722	GW	A	GW	LA		40751	MM	P	ML	NL
40723		A	VW	MA		40752	GW	A	GW	PM
40724	GW	A	GW	PM		40753	MM	P	ML	NL
40725		A	GW	LA		40754	*	MM	P	ML	NL
40726		A	GW	LA		40755		A	GW	LA
40727	GW	A	GW	LA		40756		P	ML	NL
40728	MM	P	ML	NL		40757		A	GW	LA
40729	MM	P	ML	NL						

GH1G (TF)   TRAILER FIRST

Dia. GH102. Secondary door locks. 48/– 2T (47/– 2T w).

41003–41056. Lot No. 30881 Derby 1976–7. 33.66 t.
41057–41120. Lot No. 30896 Derby 1977–8. 33.66 t.
41121–41148. Lot No. 30938 Derby 1979–80. 33.66 t.
41149–41166. Lot No. 30947 Derby 1980. 33.66 t.
41167–41169. Lot No. 30963 Derby 1982. 33.66 t.
41170. Lot No. 30967 Derby 1982. Ex prototype vehicle. 33.66 t.
41178. Lot No. 30882 Derby 1976–7. 33.60 t.

§ Fitted with wheelchair space, disabled persons toilet and centre luggage stack. 46/– 2T.

41170 was converted from 41001. 41178 is a prototype refurbished vehicle and has been converted from 42011 which was damaged by fire.

41003	p	**GW**	A	*GW*	PM		41037	p		A	*GW*	LA
41004	w	**GW**	A	*GW*	PM		41038			A	*GW*	LA
41005	p	**GW**	A	*GW*	PM		41039		**GN**	A	*GN*	EC
41006		**GW**	A	*GW*	PM		41040		**GN**	A	*GN*	EC
41007	p		A	*GW*	PM		41041	p§	**MM**	P	*ML*	NL
41008	w		A	*GW*	PM		41042	w		A	*GW*	PM
41009	pw	**GW**	A	*GW*	LA		41043	w		A	*GN*	EC
41010	w	**GW**	A	*GW*	LA		41044		**GN**	A	*GN*	EC
41011	p	**GW**	A	*GW*	PM		41045	w		P	*VX*	LA
41012		**GW**	A	*GW*	PM		41046	§	**MM**	P	*ML*	NL
41013	pw		A	*GW*	PM		41049			A	*GW*	ZD
41014	w		A	*GW*	PM		41051	w		A	*GW*	LA
41015	p		A	*GW*	PM		41052			A	*GW*	LA
41016			A	*GW*	PM		41055	w	**GW**	A	*GW*	LA
41017	pw	**GW**	A	*GW*	LA		41056	w	**GW**	A	*GW*	LA
41018		**GW**	A	*GW*	PM		41057			P	*ML*	NL
41019	p		A	*GW*	PM		41058	w		P	*ML*	NL
41020	w		A	*GW*	PM		41059	w		P	*VX*	EC
41021	pw	**GW**	A	*GW*	PM		41060	w		A	*GW*	LA
41022		**GW**	A	*GW*	PM		41061		**MM**	P	*ML*	NL
41023	p		A	*GW*	LA		41062	w	**MM**	P	*ML*	NL
41024	w		A	*GW*	LA		41063		**MM**	P	*ML*	NL
41025	p		A	*VW*	MA		41064	§	**MM**	P	*ML*	NL
41026			A	*VW*	MA		41065	w	**GW**	A	*GW*	LA
41027	pw		A	*GW*	LA		41066	pw		A	*VW*	MA
41028	w		A	*GW*	LA		41067	§	**MM**	P	*ML*	NL
41029	pw		A	*GW*	LA		41068	§	**MM**	P	*ML*	NL
41030	w		A	*GW*	LA		41069	§	**MM**	P	*ML*	NL
41031	p		A	*GW*	LA		41070	§	**MM**	P	*ML*	NL
41032	w		A	*GW*	LA		41071		**MM**	P	*ML*	NL
41033	p		A	*GW*	LA		41072	§	**MM**	P	*ML*	NL
41034			A	*GW*	I A		41075			P	*ML*	NL
41035	p		A	*VW*	MA		41076	w		P	*ML*	NL
41036			A	*VW*	MA		41077		**MM**	P	*ML*	NL

No.		Bold	P/A	Code	Reg.
41078		**MM**	P	*ML*	NL
41079		**MM**	P	*ML*	NL
41080	§	**MM**	P	*ML*	NL
41081	w	**V**	P	*VX*	EC
41082	w		P	*ML*	NL
41083			P	*ML*	NL
41084	w		P	*ML*	NL
41085			P	*VX*	EC
41086	w	**V**	P	*VX*	EC
41087		**GN**	A	*GN*	EC
41088	w	**GN**	A	*GN*	EC
41089		**GW**	A	*GW*	LA
41090	w	**GN**	A	*GN*	EC
41091		**GN**	A	*GN*	EC
41092	w	**GN**	A	*GN*	EC
41093		**GW**	A	*GW*	LA
41094		**GW**	A	*GW*	LA
41095	w	**V**	P	*VX*	EC
41096	w		P	*VX*	EC
41097		**GN**	A	*GN*	EC
41098		**GN**	A	*GN*	EC
41099		**GN**	A	*GN*	EC
41100	w	**GN**	A	*GN*	EC
41101		**GW**	A	*GW*	LA
41102	w	**GW**	A	*GW*	LA
41103	w	**GW**	A	*GW*	LA
41104	w	**GW**	A	*GW*	LA
41105	w	**GW**	A	*GW*	PM
41106	w	**GW**	A	*GW*	PM
41107	w		P	*VX*	EC
41108	w		P	*VX*	LA
41109	w	**V**	P	*VX*	LA
41110		**GW**	A	*GW*	PM
41111	w	**MM**	P	*ML*	NL
41112		**MM**	P	*ML*	NL
41113	§	**MM**	P	*ML*	NL
41114	w		P	*VX*	EC
41115			P	*ML*	NL
41116		**GW**	A	*GW*	LA
41117		**MM**	P	*ML*	NL
41118	w	**GN**	A	*GN*	EC
41119	w	**V**	P	*VX*	EC
41120			A	*GN*	EC
41121	p	**GW**	A	*GW*	LA
41122	w	**GW**	A	*GW*	LA
41123	p	**GW**	A	*GW*	PM
41124		**GW**	A	*GW*	PM
41125		**GW**	A	*GW*	PM
41126	p	**GW**	A	*GW*	PM
41127	p		A	*GW*	PM
41128	w		A	*GW*	PM
41129	pw	**GW**	A	*GW*	PM
41130	w	**GW**	A	*GW*	PM
41131	p	**GW**	A	*GW*	LA
41132	w	**GW**	A	*GW*	LA
41133	p	**GW**	A	*GW*	LA
41134	w	**GW**	A	*GW*	LA
41135	p	**GW**	A	*GW*	LA
41136	w	**GW**	A	*GW*	PM
41137	pw		A	*GW*	PM
41138			A	*GW*	PM
41139	p	**GW**	A	*GW*	LA
41140	w	**GW**	A	*GW*	LA
41141	p	**GW**	A	*GW*	LA
41142	w	**GW**	A	*GW*	LA
41143	p	**GW**	A	*GW*	LA
41144	w	**GW**	A	*GW*	LA
41145	p	**GW**	A	*GW*	PM
41146	w	**GW**	A	*GW*	PM
41147	w		P	*VX*	EC
41148	w		P	*VX*	EC
41149	w	**V**	P	*VX*	EC
41150	w		A	*GN*	EC
41151			A	*GN*	EC
41152	w		A	*GN*	EC
41153		**MM**	P	*ML*	NL
41154	§	**MM**	P	*ML*	NL
41155		**MM**	P	*ML*	NL
41156		**MM**	P	*ML*	NL
41157		**GW**	A	*GW*	LA
41158	w	**GW**	A	*GW*	LA
41159	w		P	*VX*	LA
41160	w		P	*VX*	LA
41161	w		P	*VX*	LA
41162	w		P	*VX*	LA
41163	w		P	*VX*	LA
41164	p		A	*VW*	MA
41165	w		P	*VX*	LA
41166	w		P	*VX*	LA
41167	w		P	*VX*	LA
41168	w	**V**	P	*VX*	LA
41169	w		P	*VX*	LA
41170		**GN**	A	*GN*	EC
41178			A	*GW*	PM

GH2G (TS)   TRAILER STANDARD

Dia. GH203. Secondary door locks. –/76 2T.

42003–42090. Lot No. 30882 Derby 1976–7. 33.60 t.
42091–42250. Lot No. 30897 Derby 1977–9. 33.60 t.
42251–42305. Lot No. 30939 Derby 1979–80. 33.60 t.
42306–42322. Lot No. 30969 Derby 1982. 33.60 t.
42323–42341. Lot No. 30983 Derby 1984–5. 33.60 t.
42342. Lot No. 30949 Derby 1982. 33.47 t. Converted from TGS.
42343/5. Lot No. 30970 Derby 1982. 33.47 t. Converted from TGS.
42344. Lot No. 30964 Derby 1982. 33.47 t. Converted from TGS.
42346/7/50/1. Lot No. 30881 Derby 1976–7. 33.66 t. Converted from TF.
42348/9. Lot No. 30896 Derby 1977–8. 33.66 t. Converted from TF.
42353/5–7. Lot No. 30967 Derby 1982. Ex prototype vehicles. 33.66 t.
42352/4. Lot No. 30897 Derby 1977. Were TF from 1983 to 1992. 33.66 t.

§ Centre luggage stack –/74 2T (–72 2T w).
† Centre luggage stack –/72 2T.
• Centre luggage stack –/72 2T. Fitted with pt.
* disabled persons toilet and 5 tip-up seats. –/65 2T.

42158 was also numbered 41177 for a time when fitted with first class seats.

42003		A	GW	PM	42032	GW	A	GW	PM
42004	GW	A	GW	LA	42033		A	GW	LA
42005	GW	A	GW	PM	42034		A	GW	LA
42006		A	GW	PM	42035		A	GW	LA
42007		A	GW	PM	42036		A	VW	MA
42008	GW	A	GW	LA	42037		A	VW	MA
42009		A	GW	PM	42038		A	VW	MA
42010	GW	A	GW	PM	42039		A	GW	LA
42012	GW	A	GW	LA	42040		A	GW	LA
42013	GW	A	GW	LA	42041		A	GW	LA
42014	GW	A	GW	LA	42042		A	GW	LA
42015	GW	A	GW	PM	42043		A	GW	LA
42016	GW	A	GW	PM	42044		A	GW	LA
42017	GW	A	GW	PM	42045	GW	A	GW	LA
42018		A	GW	PM	42046	GW	A	GW	LA
42019		A	GW	PM	42047		A	GW	LA
42020		A	GW	PM	42048		A	GW	LA
42021		A	GW	PM	42049	GW	A	GW	LA
42022		A	GW	PM	42050	GW	A	GW	LA
42023		A	GW	PM	42051		A	VW	MA
42024	GW	A	GW	LA	42052		A	VW	MA
42025	GW	A	GW	LA	42053		A	VW	MA
42026	GW	A	GW	LA	42054		A	GW	LA
42027		A	GW	PM	42055		A	GW	LA
42028		A	GW	PM	42056		A	GW	LA
42029		A	GW	PM	42057	GN	A	GN	EC
42030	GW	A	GW	PM	42058	GN	A	GN	EC
42031	GW	A	GW	PM	42059	GN	A	GN	EC

No.						No.					
42060			A	*GW*	PM	42111			P	*ML*	NL
42061			A	*GW*	PM	42112			P	*ML*	NL
42062		**GW**	A	*GW*	LA	42113			P	*ML*	NL
42063			A	*GN*	EC	42115			P	*VX*	EC
42064			A	*GN*	EC	42116			P	*VX*	EC
42065			A	*GN*	EC	42117			P	*VX*	EC
42066		**GW**	A	*GW*	LA	42118		**GW**	A	*GW*	PM
42067		**GW**	A	*GW*	LA	42119	§	**MM**	P	*ML*	NL
42068		**GW**	A	*GW*	LA	42120	§	**MM**	P	*ML*	NL
42069		**GW**	A	*GW*	PM	42121	§	**MM**	P	*ML*	NL
42070		**GW**	A	*GW*	PM	42122			A	*VW*	MA
42071		**GW**	A	*GW*	PM	42123	§	**MM**	P	*ML*	NL
42072			A	*GW*	PM	42124	§	**MM**	P	*ML*	NL
42073			A	*GW*	PM	42125	§	**MM**	P	*ML*	NL
42074			A	*GW*	PM	42126		**GW**	A	*GW*	LA
42075			A	*GW*	LA	42127			P	*VX*	EC
42076			A	*GW*	LA	42128			P	*VX*	EC
42077			A	*GW*	LA	42129		**GW**	A	*GW*	LA
42078			A	*GW*	LA	42130			P	*VX*	LA
42079			A	*GW*	PM	42131	§	**MM**	P	*ML*	NL
42080			A	*GW*	PM	42132	§	**MM**	P	*ML*	NL
42081		**GW**	A	*GW*	LA	42133	§	**MM**	P	*ML*	NL
42082		**GW**	A	*GW*	LA	42134			A	*VW*	MA
42083		**GW**	A	*GW*	LA	42135	§	**MM**	P	*ML*	NL
42084			P	*VX*	LA	42136	§	**MM**	P	*ML*	NL
42085			P	*VX*	LA	42137	§	**MM**	P	*ML*	NL
42086			P	*VX*	LA	42138	*	**GW**	A	*GW*	PM
42087			P	*VX*	LA	42139	§	**MM**	P	*ML*	NL
42088			P	*VX*	LA	42140	§	**MM**	P	*ML*	NL
42089		**GW**	A	*GW*	PM	42141	§	**MM**	P	*ML*	NL
42090			P	*VX*	LA	42143		**GW**	A	*GW*	PM
42091			P	*VX*	LA	42144		**GW**	A	*GW*	PM
42092			P	*VX*	LA	42145		**GW**	A	*GW*	PM
42093			P	*VX*	LA	42146			A	*GN*	EC
42094			P	*VX*	LA	42147	§	**MM**	P	*ML*	NL
42095			P	*VX*	LA	42148			P	*ML*	NL
42096			A	*GW*	LA	42149			P	*ML*	NL
42097			A	*VW*	MA	42150			A	*GN*	EC
42098			A	*GW*	LA	42151	w§	**MM**	P	*ML*	NL
42099			A	*GW*	LA	42152	§	**MM**	P	*ML*	NL
42100	§	**MM**	P	*ML*	NL	42153	§	**MM**	P	*ML*	NL
42101			P	*ML*	NL	42154			A	*GN*	EC
42102			P	*ML*	NL	42155	w§	**MM**	P	*ML*	NL
42103			P	*VX*	EC	42156	§	**MM**	P	*ML*	NL
42104		**GN**	A	*GN*	EC	42157	§	**MM**	P	*ML*	NL
42105			P	*VX*	LA	42158		**GN**	A	*GN*	EC
42106		**GN**	A	*GN*	EC	42159			P	*ML*	NL
42107		**GW**	A	*GW*	LA	42160			P	*ML*	NL
42108			P	*VX*	LA	42161			P	*ML*	NL
42109			P	*VX*	LA	42162	†	**V**	P	*VX*	EC
42110			P	*VX*	LA	42163			P	*ML*	NL

42164			P	*ML*	NL
42165			P	*ML*	NL
42166	•	V	P	*VX*	EC
42167			P	*VX*	EC
42168			P	*VX*	EC
42169			P	*VX*	EC
42170	†	V	P	*VX*	EC
42171		GN	A	*GN*	EC
42172		GN	A	*GN*	EC
42173			P	*VX*	EC
42174	†	V	P	*VX*	EC
42175			P	*VX*	LA
42176			P	*VX*	LA
42177			P	*VX*	LA
42178			P	*VX*	EC
42179		GN	A	*GN*	EC
42180		GN	A	*GN*	EC
42181		GN	A	*GN*	EC
42182			A	*GN*	EC
42183		GW	A	*GW*	LA
42184		GW	A	*GW*	LA
42185		GW	A	*GW*	LA
42186			A	*GN*	EC
42187	†	V	P	*VX*	EC
42188	†	V	P	*VX*	EC
42189	†	V	P	*VX*	EC
42190			A	*GN*	EC
42191		GN	A	*GN*	EC
42192		GN	A	*GN*	EC
42193		GN	A	*GN*	EC
42194	w§	MM	P	*ML*	NL
42195			P	*VX*	EC
42196			A	*GW*	PM
42197			A	*GW*	PM
42198		GN	A	*GN*	EC
42199		GN	A	*GN*	EC
42200		GW	A	*GW*	LA
42201		GW	A	*GW*	LA
42202		GW	A	*GW*	LA
42203		GW	A	*GW*	LA
42204		GW	A	*GW*	LA
42205	§	MM	P	*ML*	NL
42206		GW	A	*GW*	LA
42207		GW	A	*GW*	LA
42208		GW	A	*GW*	LA
42209		GW	A	*GW*	LA
42210	§	MM	P	*ML*	NL
42211		GW	A	*GW*	PM
42212		GW	A	*GW*	PM
42213		GW	A	*GW*	PM
42214		GW	A	*GW*	PM
42215			A	*GN*	EC
42216		GW	A	*GW*	LA
42217			P	*VX*	EC
42218			P	*VX*	EC
42219		GN	A	*GN*	EC
42220	w§	MM	P	*ML*	NL
42221			A	*GW*	LA
42222	†	V	P	*VX*	LA
42223	†	V	P	*VX*	LA
42224	†	V	P	*VX*	LA
42225	§	MM	P	*ML*	NL
42226			A	*GN*	EC
42227	§	MM	P	*ML*	NL
42228	§	MM	P	*ML*	NL
42229	§	MM	P	*ML*	NL
42230	§	MM	P	*ML*	NL
42231			P	*VX*	EC
42232			P	*VX*	EC
42233			P	*VX*	EC
42234			P	*VX*	EC
42235		GN	A	*GN*	EC
42236			A	*GW*	PM
42237		V	P	*VX*	EC
42238		V	P	*VX*	EC
42239		V	P	*VX*	EC
42240		GN	A	*GN*	EC
42241		GN	A	*GN*	EC
42242		GN	A	*GN*	EC
42243		GN	A	*GN*	EC
42244		GN	A	*GN*	EC
42245		GW	A	*GW*	LA
42246	†	V	P	*VX*	EC
42247	†	V	P	*VX*	EC
42248	†	V	P	*VX*	EC
42249	†	V	P	*VX*	EC
42250		GW	A	*GW*	LA
42251		GW	A	*GW*	LA
42252		GW	A	*GW*	LA
42253		GW	A	*GW*	LA
42254			P	*VX*	EC
42255		GW	A	*GW*	PM
42256		GW	A	*GW*	PM
42257		GW	A	*GW*	PM
42258			P	*VX*	EC
42259		GW	A	*GW*	PM
42260		GW	A	*GW*	PM
42261		GW	A	*GW*	PM
42262			P	*VX*	EC
42263		GW	A	*GW*	PM
42264		GW	A	*GW*	LA
42265		GW	A	*GW*	LA

No.				
42266		P	*VX*	EC
42267	**GW**	A	*GW*	PM
42268	**GW**	A	*GW*	PM
42269	**GW**	A	*GW*	PM
42270		P	*VX*	EC
42271	**GW**	A	*GW*	LA
42272	**GW**	A	*GW*	LA
42273	**GW**	A	*GW*	LA
42274		P	*VX*	EC
42275	**GW**	A	*GW*	LA
42276	**GW**	A	*GW*	LA
42277	**GW**	A	*GW*	LA
42278		P	*VX*	EC
42279	**GW**	A	*GW*	LA
42280	**GW**	A	*GW*	LA
42281	**GW**	A	*GW*	LA
42282		P	*VX*	EC
42283		A	*GW*	PM
42284		A	*GW*	PM
42285		A	*GW*	PM
42286		P	*VX*	LA
42287	**GW**	A	*GW*	LA
42288	**GW**	A	*GW*	LA
42289	**GW**	A	*GW*	LA
42290		P	*VX*	LA
42291	**GW**	A	*GW*	LA
42292	**GW**	A	*GW*	LA
42293	**GW**	A	*GW*	LA
42294		P	*VX*	LA
42295	**GW**	A	*GW*	LA
42296	**GW**	A	*GW*	LA
42297	**GW**	A	*GW*	LA
42298		P	*VX*	LA
42299	**GW**	A	*GW*	PM
42300	**GW**	A	*GW*	PM
42301	**GW**	A	*GW*	PM
42302		P	*VX*	LA
42303		P	*VX*	LA

No.				
42304		P	*VX*	LA
42305		P	*VX*	LA
42306		P	*VX*	LA
42307		P	*VX*	LA
42308		P	*VX*	LA
42309		P	*VX*	LA
42310		P	*VX*	LA
42311		P	*VX*	LA
42312		P	*VX*	LA
42313		P	*VX*	LA
42314	† **V**	P	*VX*	LA
42315	† **V**	P	*VX*	LA
42316	† **V**	P	*VX*	LA
42317	† **V**	P	*VX*	LA
42318		P	*VX*	LA
42319		P	*VX*	LA
42320		P	*VX*	LA
42321		P	*VX*	LA
42322		P	*VX*	LA
42323	**GN**	A	*GN*	EC
42324	w§ **MM**	P	*ML*	NL
42325	**GW**	A	*GW*	PM
42326	† **V**	P	*VX*	EC
42327	w§ **MM**	P	*ML*	NL
42328		P	*ML*	NL
42329	w§ **MM**	P	*ML*	NL
42330	**V**	P	*VX*	EC
42331	w§ **MM**	P	*ML*	NL
42332		A	*GW*	PM
42333	**GW**	A	*GW*	PM
42334	† **V**	P	*VX*	LA
42335		P	*ML*	NL
42336		P	*VX*	EC
42337	w§ **MM**	P	*ML*	NL
42338		P	*VX*	EC
42339	w§ **MM**	P	*ML*	NL
42340		A	*GN*	EC
42341		P	*ML*	NL

No.					
42342	(44082)		A	*VW*	MA
42343	(44095)		A	*GW*	LA
42344	(44092)	**GW**	A	*GW*	PM
42345	(44096)	**GW**	A	*GW*	LA
42346	(41053)	**GW**	A	*GW*	PM
42347	(41054)	**GW**	A	*GW*	LA
42348	(41073)	**GW**	A	*GW*	LA
42349	(41074)	**GW**	A	*GW*	PM
42350	(41047)	**GW**	A	*GW*	LA
42351	(41048)	**GW**	A	*GW*	PM
42352	(42142, 41176)		P	*ML*	NL
42353	(42001, 41171)	† **V**	P	*VX*	EC

42354	(42114, 41175)		A	*GN*	EC
42355	(42000, 41172)		A	*VW*	MA
42356	(42002, 41173)	**GW**	A	*GW*	LA
42357	(41002, 41174)		A	*VW*	MA

GJ2G (TGS) TRAILER GUARD'S STANDARD

Dia. GJ205. Secondary door locks. pg. –/65 1T (–/63 1T + one tip-up seat and wheelchair space w).

44000. Lot No. 30953 Derby 1980. 33.47 t.
44001–44090. Lot No. 30949 Derby 1980–2. 33.47 t.
44091–44094. Lot No. 30964 Derby 1982. 33.47 t.
44097–44101. Lot No. 30970 Derby 1982. 33.47 t.

§ Fitted with centre luggage stack –/63 1T.
† Fitted with centre luggage stack –/61 1T.

No.						No.					
44000			P	*VX*	EC	44034	w	**GW**	A	*GW*	LA
44001	w	**GW**	A	*GW*	LA	44035	w	**GW**	A	*GW*	LA
44002	w	**GW**	A	*GW*	PM	44036	w		A	*GW*	PM
44003	w	**GW**	A	*GW*	PM	44037	w	**GW**	A	*GW*	LA
44004	w	**GW**	A	*GW*	LA	44038	w	**GW**	A	*GW*	LA
44005	w	**GW**	A	*GW*	PM	44039	w	**GW**	A	*GW*	LA
44006	w		A	*GW*	PM	44040	w	**GW**	A	*GW*	PM
44007	w		A	*GW*	PM	44041	w		P	*ML*	NL
44008	w	**GW**	A	*GW*	LA	44042			P	*VX*	EC
44009	w		A	*GW*	PM	44043	w	**GW**	A	*GW*	LA
44010	w	**GW**	A	*GW*	PM	44044	§	**MM**	P	*ML*	NL
44011	w		A	*GW*	LA	44045	w		A	*GN*	EC
44012	w		A	*VW*	MA	44046	§	**MM**	P	*ML*	NL
44013	w		A	*GW*	LA	44047	§	**MM**	P	*ML*	NL
44014	w		A	*GW*	LA	44048	§	**MM**	P	*ML*	NL
44015	w		A	*GW*	LA	44049	w	**GW**	A	*GW*	PM
44016	w	**GW**	A	*GW*	LA	44050	w		P	*ML*	NL
44017	w		A	*VW*	MA	44051	§	**MM**	P	*ML*	NL
44018	w		A	*GW*	LA	44052	§	**MM**	P	*ML*	NL
44019	w	**GN**	A	*GN*	EC	44053			P	*ML*	NL
44020	w		A	*GW*	PM	44054	w		P	*ML*	NL
44021			P	*VX*	EC	44055			P	*VX*	EC
44022	w	**GW**	A	*GW*	LA	44056	w	**GN**	A	*GN*	EC
44023	w	**GW**	A	*GW*	PM	44057	w		P	*VX*	LA
44024	w		A	*GW*	PM	44058	w	**GN**	A	*GN*	EC
44025	w		A	*GW*	LA	44059	w		A	*GW*	LA
44026	w		A	*GW*	PM	44060		**V**	P	*VX*	EC
44027	§	**MM**	P	*ML*	NL	44061	w	**GN**	A	*GN*	EC
44028	w	**GW**	A	*GW*	LA	44062			P	*VX*	EC
44029	w	**GW**	A	*GW*	PM	44063	w	**GN**	A	*GN*	EC
44030	w	**GW**	A	*GW*	PM	44064	w	**GW**	A	*GW*	LA
44031	w		A	*VW*	MA	44065			P	*VX*	LA
44032	w	**GW**	A	*GW*	PM	44066	w	**GW**	A	*GW*	LA
44033	w	**GW**	A	*GW*	LA	44067	w	**GW**	A	*GW*	PM

44068			P	VX	LA	44085	§	MM	P	ML	NL
44069			P	VX	LA	44086	w	GW	A	GW	LA
44070	§	MM	P	ML	NL	44087			P	VX	LA
44071	w		P	ML	NL	44088			P	VX	LA
44072			P	VX	LA	44089			P	VX	LA
44073	§	MM	P	ML	NL	44090	†	V	P	VX	LA
44074			P	VX	EC	44091			P	VX	LA
44075	†	V	P	VX	EC	44093	w		A	GW	LA
44076			P	VX	LA	44094	w		A	GN	EC
44077	w		A	GN	EC	44097			P	VX	EC
44078	†	V	P	VX	EC	44098	w	GN	A	GN	EC
44079	†	V	P	VX	EC	44099			A	GW	PM
44080	w		A	GN	EC	44100			P	VX	EC
44081			P	VX	LA	44101	†	V	P	VX	LA
44083	§	MM	P	ML	NL						

GH2G (TCSD)   TRAILER CONDUCTOR STANDARD

Dia. GH201. Converted from 44084. Guard's compartment converted to walk-through conductor's compartment with a disabled persons toilet also provided. The car is marshalled adjacent to the buffet. Secondary door locks.

45084. Lot No. 30949 Derby 1982. 33.47 t.

45084		A	GW	LA	

2.3. NIGHTSTAR STOCK

These coaches are designed for use on new 'Nightstar' services between Britain and Continental Europe via the Channel Tunnel. The new generation of overnight trains offer high quality accommodation to both business and leisure customers.

This innovative venture is being developed by European Night Services Limited (ENS), a joint company of Eurostar (UK) Ltd., SNCF, DB and NS. It was originally intended that the trains would operate on the following routes:

London Waterloo–Amsterdam CS.
London Waterloo–Dortmund Hbf./Frankfurt Hbf.
Glasgow/Manchester–Paris Nord.
Plymouth/Swansea–Paris Nord.

All but the first of these have now been cancelled, and it is not known when or whether the first will operate, as the project has been bedevilled with both technical and commercial problems, e.g. there is no locomotive in Belgium which has enough power for the train heating and air conditioning!

Both sleeping cars and reclining seat coaches have been built. Each train was due to be formed of two half-sets, London services having two reclining seat coaches, a service vehicle and five sleeping cars in each half-set to form a sixteen coach train, whilst services from the Provinces to Paris were to be fourteen coaches long with each portion consisting of three sleeping cars, a service vehicle and three reclining seat coaches. The "regional" half-sets are numbered 1–9, whilst the "London" half-sets are numbered 10–18.

In the following lists, the UIC number for each vehicle is followed by the set number to which it belongs.

RECLINING SEAT CARS   SO End

Each car has 50 seats which are fully reclining, with generous leg space. A table is provided at each seat, and footrests will offer extra comfort. The seats are mounted on plinths, which enhance the customer's sense of personal space. Main luggage is stored beneath the seat, while hand baggage is stored in overhead lockers. Individually controlled reading lights are provided, with different levels of ambient lighting for sleeping and non-sleeping hours. Each car has three toilet compartments with washing facilities. These include facilities such as shaver sockets and hot-air hand dryers.

61 19 20-90 001-0 *1*	KI	61 19 20-90 010-1 *10*
61 19 20-90 002-8 *2*	KI	61 19 20-90 011-9 *11*
61 19 20-90 003-6 *3*	KI	61 19 20-90 012-7 *12*
61 19 20-90 004-4 *4*	KI	61 19 20-90 013-5 *13*
61 19 20-90 005-1 *5*	KI	61 19 20-90 014-3 *14*
61 19 20-90 006-9 *6*	KI	61 19 20-90 015-0 *15*
61 19 20-90 007-7 *7*		61 19 20-90 016-8 *16*
61 19 20-90 008-5 *8*		61 19 20-90 017-6 *17*
61 19 20-90 009-3 *9*		61 19 20-90 018-4 *18*

RECLINING SEAT CAR   SO

Details as above, but no coupling for locomotive.

61 19 20-90 019-2 *1*	KI	61 19 20-90 034-1 *8*	
61 19 20-90 020-0 *1*	KI	61 19 20-90 035-8 *9*	
61 19 20-90 021-8 *2*	KI	61 19 20-90 036-6 *9*	
61 19 20-90 022-6 *2*	KI	61 19 20-90 037-4 *10*	
61 19 20-90 023-4 *3*	KI	61 19 20-90 038-2 *11*	
61 19 20-90 024-2 *3*	KI	61 19 20-90 039-0 *12*	
61 19 20-90 025-9 *4*	KI	61 19 20-90 040-8 *13*	
61 19 20-90 026-7 *4*	KI	61 19 20-90 041-6 *14*	
61 19 20-90 027-5 *5*	KI	61 19 20-90 042-4 *15*	
61 19 20-90 028-3 *5*	KI	61 19 20-90 043-2 *16*	
61 19 20-90 029-1 *6*	KI	61 19 20-90 044-0 *17*	
61 19 20-90 030-9 *6*	KI	61 19 20-90 045-7 *18*	
61 19 20-90 031-7 *7*		61 19 20-90 046-5 *S*	KI
61 19 20-90 032-5 *7*		61 19 20-90 047-3 *S*	KI
61 19 20-90 033-3 *8*			

SLEEPING CARS   SLF End

ENS sleeping cars will set high standards, with service quality and facilities like those found in a good hotel. Main users are expected to be business travellers and comfort-seeking leisure travellers Each sleeping car will have 10 cabins. Six of these will have a compact en-suite shower room, with a washbasin, toilet and hairdryers. The remaining four cabins will include en-suite toilet and washing facilities, but without the shower.

All cabins will be convertible so that when the bunks are folded away by the attendant after passengers have got up, two comfortable armchairs with fold-out tables are revealed. The bunks themselves are generously sized one above the other and will already be made up with duvets, sheets and pillows When passengers arrive. Each cabin will have a fitted wardrobe and cupboard, together with facilities for making hot drinks. Cabin telephones are provided for room service.

61 19 70-90 001-9 *1*	KI	61 19 70-90 010-0 *10*	
61 19 70-90 002-7 *2*	KI	61 19 70-90 011-8 *11*	
61 19 70-90 003-5 *3*	KI	61 19 70-90 012-6 *12*	
61 19 70-90 004-3 *4*	KI	61 19 70-90 013-4 *13*	
61 19 70-90 005-0 *5*	KI	61 19 70-90 014-2 *14*	
61 19 70-90 006-8 *6*	KI	61 19 70-90 015-9 *15*	
61 19 70-90 007-6 *7*		61 19 70-90 016-7 *16*	
61 19 70-90 008-4 *8*		61 19 70-90 017-5 *17*	
61 19 70-90 009-2 *9*		61 19 70-90 018-3 *18*	

SLEEPING CARS   SLF

Details as above, but no coupling for locomotive.

61 19 70-90 019-1 *1*	KI	61 19 70-90 046-4 *12*	
61 19 70-90 020-9 *1*	KI	61 19 70-90 047-2 *12*	
61 19 70-90 021-7 *2*	KI	61 19 70-90 048-0 *12*	
61 19 70-90 022-5 *2*	KI	61 19 70-90 049-8 *13*	
61 19 70-90 023-3 *3*	KI	61 19 70-90 050-6 *13*	
61 19 70-90 024-1 *3*	KI	61 19 70-90 051-4 *13*	
61 19 70-90 025-8 *4*	KI	61 19 70-90 052-2 *13*	
61 19 70-90 026-6 *4*	KI	61 19 70-90 053-0 *14*	
61 19 70-90 027-4 *5*	KI	61 19 70-90 054-8 *14*	
61 19 70-90 028-2 *5*	KI	61 19 70-90 055-5 *14*	
61 19 70-90 029-0 *6*	KI	61 19 70-90 056-3 *14*	
61 19 70-90 030-8 *6*	KI	61 19 70-90 057-1 *15*	
61 19 70-90 031-6 *7*		61 19 70-90 058-9 *15*	
61 19 70-90 032-4 *7*		61 19 70-90 059-7 *15*	
61 19 70-90 033-2 *8*		61 19 70-90 060-5 *15*	
61 19 70-90 034-0 *8*		61 19 70-90 061-3 *16*	
61 19 70-90 035-7 *9*		61 19 70-90 062-1 *16*	
61 19 70-90 036-5 *9*		61 19 70-90 063-9 *16*	
61 19 70-90 037-3 *10*	KI	61 19 70-90 064-7 *16*	
61 19 70-90 038-1 *10*		61 19 70-90 065-4 *17*	
61 19 70-90 039-9 *10*		61 19 70-90 066-2 *17*	
61 19 70-90 040-7 *10*		61 19 70-90 067-0 *17*	
61 19 70-90 041-5 *11*		61 19 70-90 068-8 *17*	
61 19 70-90 042-3 *11*		61 19 70-90 069-6 *18*	
61 19 70-90 043-1 *11*		61 19 70-90 070-4 *18*	
61 19 70-90 044-9 *11*		61 19 70-90 071-2 *18*	
61 19 70-90 045-6 *12*		61 19 70-90 072-0 *18*	

SERVICE VEHICLE/LOUNGE CAR   SV

Lounge cars are positioned in each half of the train, between the sleeping cars and the seated accommodation. These vehicles consist of of a sleeping cabin for a disabled passenger and companion with en-suite washroom, a parcels room, offices for train manager and control authority, a lounge with bar for sleeping car passengers and public telephone and a bar for seated passengers.

The vehicle also acts as a base for the sleeping car attendants and for the trolley service which will be provided for the seated passengers in the evening. There is also a seated passengers' counter so that snacks and drinks can be obtained during sleeping hours.

61 19 89-90 001-8 *1*	KI	61 19 89-90 011-7 *11*	
61 19 89-90 002-6 *2*	KI	61 19 89-90 012-5 *12*	
61 19 89-90 003-4 *3*	KI	61 19 89-90 013-3 *13*	
61 19 89-90 004-2 *4*	KI	61 19 89-90 014-1 *14*	
61 19 89-90 005-9 *5*	KI	61 19 89-90 015-8 *15*	
61 19 89-90 006-7 *6*	KI	61 19 89-90 016-6 *16*	
61 19 89-90 007-5 *7*		61 19 89-90 017-4 *17*	
61 19 89-90 008-3 *8*		61 19 89-90 018-2 *18*	
61 19 89-90 009-1 *9*		61 19 89-90 019-0 *S*	
61 19 89-90 010-9 *10*		61 19 89-90 020-8 *S*	

2.4. SALOONS

Several specialist passenger carrying vehicles, normally referred to as saloons are permitted to run on the Railtrack system. Many of these are to pre-nationalisation designs.

LNER GENERAL MANAGERS SALOON

Built 1945 by LNER, York. Gangwayed at one end with a verandah at the other. The interior has a dining saloon seating twelve, kitchen, toilet, office and nine seat lounge. B4 bogies.

| 1999 | (902260) | **M** | GS | *SS* | EN |

GNR FIRST CLASS SALOON

Built 1912 by GNR, Doncaster. Contains entrance vestibule, lavatory, two seperate saloons and luggage space. Gresley bogies.

Non-Standard Livery: Teak.

| 4807 | (807) | x | **0** | RE | *SS* | CJ |

L&NWR DINING SALOON

Built 1891 by L&NWR, Wolverton. Mounted on the underframe of LMS GUV 37908 in 198x. Gresley bogies.

Non-Standard Livery: London & North Western Railway.

| 5159 | (159) | x | **0** | RE | *SS* | CJ |

LMR GENERAL MANAGERS SALOON

Dia. AZ501. Renumbered 1989 from LMR departmental series. Formerly the LMR General Manager's saloon. Rebuilt from LMS period 1 BFK M 5033 M to dia. 1654 and mounted on the underframe of BR suburban BS M 43232. B4 bogies. This vehicle has a maximum speed of 100 m.p.h. Screw coupling has been removed.

LMS Lot No. 326 Derby 1927.   t.

Non-Standard Livery: Aircraft blue with gold lining.

| 6320 | (5033, DM 395707) | x | **0** | RV | *SS* | CP |

GWR FIRST CLASS SALOON

Built 1930 by GWR, Swindon. Contains saloons at either end with body end observation windows. A seperate first class compartment and a central kitchen/ Pantry and lavatory. B5 bogies. Numbered DE321011 when in departmental service.

| 9004 | x | **CH** | RA | *SS* | CP |

WCJS OBSERVATION SALOON

Built 1897 by L&NWR, Wolverton. Originally dining saloon mounted on six-wheel bogies. Rebuilt with new underframe with four-wheel bogies in 1927. Rebuilt 1960 as observation saloon with DMU end. Gangwayed at other end. The interior has a saloon, kitchen, guards vestibule and observation lounge. Gresley bogies.

Non-Standard Livery: London & North Western Railway.

45018 (484, 15555)   x **0**   RE *SS*   CJ

LMSR INSPECTION SALOON

Built as engineers inspection saloons non-gangwayed. Observation windows at each end. The interior layout consists of two saloons interspersed by a central lavatory/kitchen/guards section. BR1 bogies.

45020–45026. Lot No. LMS 1356 Wolverton 1944.
45029. Lot No. LMS 1327 Wolverton 1942.
999503–999504. Lot No. BR Wagon Lot. 3093 Wolverton 1957.

Non-Standard Livery: Mainline Freight blue with five white stripes on lower bodyside.

45026 & 999503 are currently hired to Racal-BRT who have contracted maintenance to the Severn Valley Railway.

45020		**M**	E	*SS*	ML
45026	v	**M**	E	*SS*	KR
45029	v	**M**	E	*SS*	CD
999503	v	**BG**	E	*SS*	KR
999504	v	**0**	E	*SS*	TO

RAILFILMS 'LMS CLUB CAR'

Converted from BR Mark 1 TSO at Carnforth Railway Restoration and Engineering Services in 1994. Commonwealth bogies. ETH 4.

Lot. No. 30724 York 1963. 37 t.

99993 (5067)   **M**   RA *SS*   CP

BR INSPECTION SALOON

Mark 1. Short frames. Non-gangwayed. Observation windows at each end. The interior layout consists of two saloons interspersed by a central lavatory/kitchen/guards/luggage section. BR1 bogies.

Lot No. BR Wagon Lot. 3379 Swindon 1960.

999509   **BG**   E   *SS*   CF

2.5. PULLMAN CAR COMPANY STOCK

Pullman cars have never generally been numbered as such, although many have carried numbers, instead they have carried titles. However, a scheme of schedule numbers exists which generally lists cars in chronological order. In this section those numbers are shown followed by the cars title. Cars described as 'kitchen' contain a kitchen in addition to passenger accomodation. Cars described as 'parlour' consist entirely of passenger accomodation.

PULLMAN PARLOUR FIRST

Built 1927 by Midland Carriage and Wagon Company. Gresley bogies. 26/–. ETH 2.

213 MINERVA   **PC** VS *SS* SL

Built 1928 by Metropolitan Carriage and Wagon Company, Birmingham. Gresley bogies. 24/–. ETH 4.

239 AGATHA   **PC** VS   SL
243 LUCILLE   **PC** VS *SS* SL

PULLMAN KITCHEN FIRST

Built 1925 by Birmingham Railway Carriage and Wagon Company, Birmingham. Rebuilt by Midland Carriage & Wagon Company in 1928. Gresley bogies. 20/–. ETH 4.

245 IBIS   **PC** VS *SS* SL

PULLMAN PARLOUR FIRST

Built 1928 by Metropolitan Carriage and Wagon Company, Birmingham. Gresley bogies. 24/–. ETH 4.

254 ZENA   **PC** VS *SS* SL

PULLMAN KITCHEN FIRST

Built 1928 by Metropolitan Carriage and Wagon Company, Birmingham. Gresley bogies. 20/–. ETH 4.

255 IONE   **PC** VS *SS* SL

PULLMAN PARLOUR THIRD

Built 1931 by Birmingham Railway Carriage and Wagon Company, Birmingham, Gresley bogies. –/42.

261 CAR No. 83   **PC** VS   SL

PULLMAN KITCHEN FIRST

Built 1932 by Metropolitan Carriage and Wagon Company, Birmingham. Originally included in 'Brighton Belle' EMUs but now used as hauled stock. B5S
bogies (§ EMU bogies). 20/–. ETH 2.

280	AUDREY		**PC**	VS	*SS*	SL
281	GWEN	§	**PC**	VS		SL
284	VERA		**PC**	VS	*SS*	SL

PULLMAN PARLOUR THIRD

Built 1932 by Metropolitan Carriage and Wagon Company, Birmingham. Originally included in 'Brighton Belle' EMU but now used as hauled stock. EMU
bogies. –/56.

286	CAR No. 86	**PC**	VS	SL

PULLMAN PARLOUR FIRST

Built 1951 by Birmingham Railway Carriage & Wagon Company, Birmingham.
Gresley bogies. 32/–. ETH 3.

301	PERSEUS	**PC**	VS	*SS*	SL

Built 1952 by Pullman Car Company, Preston Park using underframe and bogies from 176 RAINBOW, the body of which had been destroyed by fire. Gresley
bogies. 26/–. ETH 4.

302	PHOENIX	**PC**	VS	*SS*	SL

PULLMAN KITCHEN FIRST

Built 1951 by Birmingham Railway Carriage & Wagon Company, Birmingham.
Gresley bogies. 22/–.

307	CARINA	**PC**	VS	SL

PULLMAN PARLOUR FIRST

Built 1951 by Birmingham Railway Carriage & Wagon Company, Birmingham.
Gresley bogies. 32/–. ETH 3.

308	CYGNUS	**PC**	VS	*SS*	SL

PULLMAN KITCHEN FIRST

Built by Metro-Cammell 1960/1 for East Coast Main-line services. Common-wealth bogies. 20/– 2T. 40 t. Used in the 'Royal Scotsman' charter train set, some of these vehicles have been modified. Names not carried. Current title and use is shown.

313	FINCH	x	**M**	GS	*SS*	EN	STATE CAR 4	Sleeping Car
317	RAVEN	x	**M**	GS	*SS*	EN	DINING CAR 1	Kitchen & Dining Car
319	SNIPE	x	**M**	GS	*SS*	EN	OBSERVATION CAR	Observation Car

PULLMAN PARLOUR FIRST

Built by Metro-Cammell 1960/1 for East Coast Main-line services. Common-wealth bogies. 29/– 2T. 38.5 t. Used in the 'Royal Scotsman' charter train set, some of these vehicles have been modified. Names not carried. Current title and use is shown.

324	AMBER	x	**M**	GS	*SS*	EN	STATE CAR 1	Sleeping Car
329	PEARL	x	**M**	GS	*SS*	EN	STATE CAR 2	Sleeping Car
331	TOPAZ	x	**M**	GS	*SS*	EN	STATE CAR 3	Sleeping Car

PULLMAN KITCHEN SECOND

Built by Metro-Cammell 1960/1 for East Coast Main-line services. Common-wealth bogies. –/30 1T. 40 t.

| 335 | CAR No. 335 | x | **PC** | FS | On loan to Swanage Railway |

PULLMAN PARLOUR SECOND

Built by Metro-Cammell 1960/1 for East Coast Main-line services. Common-wealth bogies. –/42 2T. 38.5 t.

347	CAR No. 347	x	**PC**	FS	Crewe Carriage Shed
348	CAR No. 348	x	**PC**	FS	On loan to Swanage Railway
349	CAR No. 349	x	**PC**	FS	On loan to Kent & East Sussex Railway
350	CAR No. 350	x	**PC**	FS	Crewe Carriage Shed
351	CAR No. 347	x	**PC**	FS	Crewe Carriage Shed
352	CAR No. 348	x	**PC**	FS	Crewe Carriage Shed
353	CAR No. 349	x	**PC**	FS	On loan to Swanage Railway

THE HADRIAN BAR

Built by Metro-Cammell 1961 for East Coast Main-line services. Common-wealth bogies. 24/– + bar seating 1T. 38.5 t.

| 354 | THE HADRIAN BAR | x | **PC** | FS | Crewe Carriage Shed |

2.6. COACHING STOCK AWAITING DISPOSAL

This list contains the last known locations of coaching stock awaiting disposal. The definition of which vehicles are "awaiting disposal" is somewhat vague, but generally speaking these are vehicles of types not now in normal service or vehicles which have been damaged by fire, vandalism or collision.

6339	EC
6345	EC
6900	Cambridge Station Yard
6901	Cambridge Station Yard
7183	Crewe Brook Sidings
7213	OM
9533	MA
18416	Crewe Brook Sidings
18750	Crewe Brook Sidings
19500	Crewe Brook Sidings

2.7. ADDITIONAL INFORMATION

The following table is presented to help readers identify vehicles which may still have their former private owner numbers painted on them. The private owner number is shown in column 1 and the number by which the vehicle is now identified is shown in column 2.

99000	4946	99325	5727	99568	3068	99822	1859
99001	4996	99326	4954	99670	546	99823	4832
99002	5008	99327	5044	99671	548	99824	4831
99052	45018	99328	5033	99672	549	99826	13229
99053	9004	99329	4931	99673	550	99827	3096
99080	21096	99356	21245	99674	551	99828	13230
99121	3105	99371	3128	99675	552	99829	4856
99125	3113	99405	35486	99676	553	99830	5028
99127	3117	99530	301	99677	586	99831	4836
99129	21272	99531	302	99678	504	99880	5159
99131	1999	99532	308	99679	506	99881	4807
99241	35449	99534	245	99680	17102	99886	35407
99304	21256	99535	213	99710	25767	99887	2127
99311	1882	99536	254	99712	25893	99953	35468
99312	35463	99537	280	99713	26013	99961	324
99314	25729	99538	34991	99716	25808	99962	329
99315	25955	99539	255	99717	25837	99963	331
99316	13321	99540	3069	99718	25862	99964	313
99317	3766	99541	243	99721	25756	99965	319
99318	4912	99542	889202	99722	25806	99966	34525
99319	14168	99543	284	99723	35459	99967	317
99321	5299	99544	35207	99782	17007	99968	10541
99322	5600	99545	80207	99792	17019	99969	10556
99323	5704	99554	92904	99818	1730	99995	35457
99324	5714	99566	3066	99821	9227		

The following table lists support coaches and the locomotives which they normally support at present.

17007	35028	35333	6024	35465	D 172	35486	KR locos
17019	60532	35449	34027	35467	KR locos	80217	75014
21096	4498	35457	44767	35468	NRM locos	80220	31625
21236	30828	35463	48151				

PROTECT YOUR BOOKS!

Don't let your books become tatty and dog-eared. Keep them in first class condition with hardwearing Platform 5 PVC book covers. Available in three different sizes to fit British Railways Pocket Books, BR Locomotives & Coaching Stock and a wide range of A5 size books. Each cover is finished in an attractive synthetic leather effect and comes complete with Platform 5 logo foil embossed on the front. Cover sleeves can be used for storing handy notes and papers.

All sizes now available in four different colours: Blue, Red, Green or Grey.

A6 Size (to fit British Railways Pocket Books) £1.00
BR Locomotives & Coaching Stock Size £1.20
A5 Size (to fit European Handbooks etc.) £1.50

Available from the Platform 5 Mail Order Department. To place an order, please follow the instructions on page 383 of this book.

DIESEL MULTIPLE UNITS - INTRODUCTION

Diesel Multiple Unit operation on Britain's main line railways has increased enormously since the end of the steam era but there have been many changes in recent years. Electrification has meant the replacement of DMUs with EMUs on many routes, whilst on other services, DMUs have replaced loco-hauled trains. Very few first generation DMUs remain and most DMU services are now operated by "Pacer", "Sprinter" or other modern air-braked Express units. A few DEMUs will be found operating on the former Southern Region, but some of these now have ex-EMU centre cars. One vehicle (71634) started life as loco-hauled coach No. 4059, was converted to an EMU trailer and is now part of DEMU 205 205!

NUMBERING

Diesel mechanical and diesel hydraulic multiple unit vehicles are numbered in the series 51000-59999. All vehicles numbered in the 53000-53999 series were originally numbered in the series 50000-50999, and were renumbered by having 3000 added to their original numbers. All vehicles in the series 54000-54504 were originally numbered in the series 56000-56504, and were renumbered by having 2000 subtracted from their original numbers.

Diesel electric multiple unit vehicles are numbered in the series 60000-60918. A number of vehicles which were numbered in the series 60001-60100 were renumbered in 1989 to avoid conflicting with Class 60 locomotives.

DESIGN CONSIDERATIONS

Unless stated otherwise, all diesel multiple unit vehicles are of BR design, or designed by contractors for BR and have buckeye couplings and tread brakes. Seating is 3+2 or 2+2 in standard class open vehicles, 2+2 or 2+1 in first class open vehicles, 8 to a corridor standard class compartment and 6 to a corridor first class compartment.

VEHICLE CODES

The codes used by the Operating Department to describe the various different types of DMU vehicles and quoted in the class headings are as follows:

Diesel Mechanical & Diesel Hydraulic Units.

DMBS	Driving Motor Brake Standard
DMC	Driving Motor Composite
DMS	Driving Motor Standard
DMPMV	Driving Motor Parcels & Miscellaneous Van
DTPMV	Driving Trailer Parcels & Miscellaneous Van
DTC	Driving Trailer Composite
DTS	Driving Trailer Standard
MS	Motor Standard
TS	Trailer Standard

It should be noted that as all vehicles are of an open configuration the letter 'O' is omitted for all vehicles. An 'L' suffix denotes that the vehicle is fitted with a lavatory compartment. The letters (A) and (B) may be added to the above codes to differentiate between two cars of the same operating type which have differences between them. Note that a consistent system is used here, rather than the official operator codes which are sometimes inconsistent.

A composite is a vehicle containing both first and standard class accommodation, and vehicles are described as such even though most first class accommodation has now been declassified on most vehicles. This is done so as to differentiate between the different styles of seat provided in standard and erstwhile first class areas of a vehicle. At the time of writing no heritage units retained first class accommodation in use as such.

A brake vehicle is a vehicle containing seperate specific accommodation for the guard (as opposed to the use of spare driving cabs on second generation units).

Diesel Electric Units.

DMBSO Driving Motor Brake Standard (Open).
DTCsoL Driving Trailer Composite with Lavatory (Semi-Open).
DTSOL Driving Trailer Standard with Lavatory (Open).
DTSO Driving Trailer Standard (Open).
TSO Trailer Standard (Open).
TCsoL Trailer Composite with Lavatory (Semi-open).
TSOL Trailer Standard with Lavatory (Open).

The notes as above apply regarding composite and brake vehicles. A semi-open composite vehicle has first class accommodation in compartments with a side corridor and standard class accommodation provided in an open saloon.

WEIGHTS & DIMENSIONS

Approximate weights in working order are given in tonnes for all vehicle types in the class headings and sub headings as appropriate.

The dimensions of each type of vehicle are given in metric units, with length followed by width. All lengths quoted are over buffers (1st generation vehicles) or couplings (2nd generation vehicles). All widths quoted are maxima.

DIAGRAMS AND DESIGN CODES

For each type of vehicle, the official design code consists of a seven character code of two letters, four numbers and another letter, e.g. DP2010A. The first five characters of this are the diagram code and are given in the class heading or sub heading. These are explained as follows:

1st Letter

This is always 'D' for a diesel multiple unit vehicle, although certain DEMUs contain EMU trailers which have an 'E' as the first letter.

2nd Letter

as follows for various vehicle types (DMMU or DHMU unless otherwise stated):

B	Driving motor passenger vehicles with a brake compartment (DEMU).
E	Driving trailer passenger vehicles (DEMU).
H	Trailer passenger vehicles without a brake compartment (DEMU).
P	Driving motor passenger vehicles without a brake compartment.
Q	Driving motor passenger vehicles with a brake compartment.
R	Non-Driving motor passenger vehicles.
S	Driving trailer passenger vehicles.
T	Trailer passenger vehicles without a brake compartment.
X	Parcels and Mails vehicles and single unit railcars.

1st Figure

This denotes the class of accommodation as follows:

2	Standard class accommodation (incl. declassified seats).
3	Composite accommodation.
5	No passenger accommodation.

2nd & 3rd Figures

These distinguish between the different designs of vehicle, each different design being allocated a unique two digit number.

Special Note

Where vehicles have been declassified the correct design code for a declassified vehicle is given, even though this may be at variance with official records which do not show the reality of the current position. A declassified composite is still referred to as a composite if it still retains the first class style seats in the erstwhile first class section of the vehicle. Its declassification is denoted by the fact that the first figure of the design code is a '2'.

ACCOMMODATION

This information is given in class headings and sub headings in the form F/S nT, where F & S denote the number of first class nd standard class seats followed by n which denotes the number of toilets. (e.g. 12/54 1T denotes 12 first class seats, 54 standard class seats and one toilet). In declassified vehicles, the capacity is still shown in terms of first and standard class seats to differentiate between the two physically different seat types available, although all seats are officially standard class in such instances. TD denotes a toilet suitable for a disabled person.

BUILD DETAILS

LOT NUMBERS

Each batch of vehicles is allocated a Lot (or batch) number when ordered and these are quoted in class headings and sub headings.

BUILDERS

These are shown in class headings. A list of builders is given on page 361.

ADDITIONAL FEATURE CODES

r   Fitted with radio electronic token block apparatus.

LAYOUT

The layout in this section is as follows:

(1) Unit number.	(5) Operator code.
(2) Notes (if any).	(6) Depot code.
(3) Livery code.	(7) Individual car numbers.
(4) Owner code.	(8) Name (if any).

Thus an example of the layout is as follows:

No.	Notes	Livery	Owner	Operation	Depot	Car 1	Car 2	Name
150 257	†	**RR**	P	*AR*	NC	52257	57257	QUEEN BOADICEA

For off-loan or stored vehicles, the last storage location is given where known.

3.1. 'HERITAGE' DIESEL MULTIPLE UNITS

Very few first generation diesel multiple units remain. These are now referred to as 'heritage' units. Standard features are as follows:

Brakes:

All units are vacuum braked.

Lighting:

All cars are now fitted with fluorescent lighting.

Couplings:

Screw couplings are used on all vehicles. All remaining first generation vehicles may be coupled together to work in multiple up to a maximum of 6 motor cars or 12 cars in total in a formation. First generation vehicles may not be coupled in multiple with second generation vehicles.

CLASS 101   METRO-CAMMELL

Engines: Two Leyland of 112 kW (150 h.p.) per power car.
Transmission: Mechanical. Cardan shaft and freewheel to a four-speed epicyclic gearbox with a further cardan shaft to the final drive, each engine driving the inner axle of one bogie.
Gangways: Midland scissors type. Within unit only.
Doors: Slam.
Bogies: DD15 (motor) and DT11 (trailer).
Dimensions: 18.49 x 2.82 m.
Seats: 3+2 facing (2+2 in first class).

51175–51253. DMBS. Dia. DQ202. Lot No. 30467 1958–59. –/52. 32.5 t.
51426–51463. DMBS. Dia. DQ202. Lot No. 30500 1959. –/52. (–/49 with additional luggage rack for Gatwick sets – Dia. DQ232) 32.5 t.
53164. DMBS. Dia. DQ202. Lot No. 30254 1956. –/52. 32.5 t.
53198–53204. DMBS. Dia. DQ202. Lot No. 30259 1957. –/52. 32.5 t.
53211–53228. DMBS. Dia. DQ202. Lot No. 30261 1957. –/52. 32.5 t.
53253–53256. DMBS. Dia. DQ202. Lot No. 30266 1957. –/52. 32.5 t.
53311–53314. DMBS. Dia. DQ202. Lot No. 30275 1958. –/52. (–/49 with additional luggage rack for Gatwick sets – Dia. DQ232) 32.5 t.
51496–51533. DMCL or DMSL. Dia. DP317 or DP210. Lot No. 30501 1959. 12/46 1T with additional luggage racks. 32.5t.
51800. DMBS. Dia. DQ202. Lot No. 30587 1956. –/52. 32.5 t.
51803. DMSL. Dia. DP210. Lot No. 30588 1959. –/72 1T. 32.5 t.
53160–53163. DMSL. Dia. DP214. Lot No. 30253 1956. –/72 1T. 32.5 t.
53170–53171. DMSL. Dia. DP214. Lot No. 30255 1957. –/72 1T. 32.5 t.
53177. DMSL. Dia. DP214. Lot No. 30256 1957. –/72 1T. 32.5 t.
53266–53269. DMSL. Dia. DP210. Lot No. 30267 1957. –/72 1T. 32.5 t.
53322–53327. DMCL. Dia. DP317. Lot No. 30276 1958. 12/46 1T with additional luggage racks. 32.5 t.

53746. DMSL. Dia. DP210. Lot No. 30271 1957. –/72 1T. 32.5 t.
54055–54061. DTSL. DS206. Lot No. 30260 1957. –/72 1T. 25.5 t.
54062–54091. DTSL. Dia. DS206. Lot No. 30262 1957. –/72 1T. 25.5 t.
54343–54408. DTSL. Dia. DS206. Lot No. 30468 1958. –/72 1T. 25.5 t.
59303. TSL. Dia. DT202. Lot No. 30273 1957. –/71 1T. 25.5 t.
59539. TSL. Dia. DT228. Lot No. 30502 1959. –/72 1T. 25.5 t.

Refurbished 2-car Sets. DMBS–DTSL.

101 651	**RR**	A		ZD	53201	54379
101 652	**RR**	A		ZH	53198	54346
101 653	**RR**	A	*NW*	LO	51426	54358
101 654	**RR**	A	*NW*	LO	51800	54408
101 655	**RR**	A	*NW*	LO	51428	54062
101 656	**RR**	A	*NW*	LO	51230	54056
101 657	**RR**	A	*NW*	LO	53211	54085
101 658	**RR**	A	*NW*	LO	51175	54091
101 659	**RR**	A	*NW*	LO	51213	54352
101 660	**RR**	A	*NW*	LO	51189	54343
101 661	**RR**	A	*NW*	LO	51463	54365
101 662	**RR**	A	*NW*	LO	53228	54055
101 663	**RR**	A	*NW*	LO	51201	54347
101 664	**RR**	A	*NW*	LO	51442	54061
101 665	**RR**	A	*NW*	LO	51429	54393

Refurbished Twin Power Car and 3-Car Sets. DMBS–DMSL or DMBS–TSL–DMSL.

Non-Standard livery: Caledonian Blue.

101 676	**RR**	A	*NW*	LO	51205		51803
101 677	**RR**	A	*NW*	LO	51179		51496
101 678	**RR**	A	*NW*	LO	51210		53746
101 679	**RR**	A	*NW*	LO	51224		51533
101 680	**RR**	A	*NW*	LO	53204		53163
101 681	**RR**	A	*NW*	LO	51228		51506
101 682	**RR**	A	*NW*	LO	53256		51505
101 683	**RR**	A	*NW*	LO	51177	59303	53269
101 684	**S**	A	*SR*	CK	51187		51509
101 685	**G**	A	*NW*	LO	53164	59539	53160
101 686	**S**	A	*SR*	CK	51231		51500
101 687	**S**	A	*SR*	CK	51247		51512
101 688	**S**	A	*SR*	CK	51431		51501
101 689	**S**	A	*SR*	CK	51185		51511
101 690	**S**	A	*SR*	CK	51435		53177
101 691	**S**	A	*SR*	CK	51253		53171
101 692	**0**	A	*SR*	CK	53253		53170
101 693	**S**	A	*SR*	CK	51192		53266
101 694	**S**	A	*SR*	CK	51188		53268
101 695	**S**	A	*SR*	CK	51226		51499

Note: The trailers of units 101 683 and 101 685 are normally removed for the winter period and stored at Chester CSD.

Unrefurbished Twin Power Car Sets. DMBS–DMCL. These sets have seats removed and additional luggage racks. These modifications were carried out when they were used on Reading–Gatwick Airport services.

Note: Some units show 'L' instead of the official class prefix.

101 835	**RR**	A	*NW*	LO	51432	51498
101 840	**N**	A	*NW*	LO	53311	53322
101 842	**N**	A		KI	53314	53327

CLASS 117   PRESSED STEEL SUBURBAN

DMBS–TSL–DMS (refurbished) or DMBS–DMS (facelifted).
Engines: Two Leyland 680/1 of 112 kW (150 h.p.) per power car.
Transmission: Mechanical. Cardan shaft and freewheel to a four-speed epicyclic gearbox with a further cardan shaft to the final drive, each engine driving the inner axle of one bogie.
Gangways: GWR suspension type. Within unit only.
Bogies: DD10 (motor) and DT9 (trailer).
Dimensions: 20.45 x 2.82 m.
Seats: 3+2 facing.

DMBS. Dia. DQ220. Lot No. 30546 1959–60. –/65. 36.5 t.
TSL. Dia. DT230. Lot No. 30547 1959–60. –78 2T. 30.5 t.
DMS. Dia. DP221. Lot No. 30548 1959–60. –/89. 36.5 t.

Notes: Some units show 'L' instead of the official class prefix.

117 301	**RR**	A	*SR*	HA	51353	59505	51395
117 306	**RR**	A	*SR*	HA	51369	59521	51411
117 308	**RR**	A	*SR*	HA	51371	59509	51413
117 310	**RR**	A	*SR*	HA	51373	59486	51381
117 311	**RR**	A	*SR*	HA	51334	59500	51376
117 313	**RR**	A	*SR*	HA	51339	59492	51382
117 314	**RR**	A	*SR*	HA (S)	51352	59489	51394
117 700	**N**	A	*SL*	BY	51332		51374
117 701	**N**	A	*SL*	BY	51350		51392
117 702	**N**	A	*SL*	BY	51356		51398
117 704	**N**	A	*SL*	BY	51341		51383
117 705	**N**	A	*SL*	BY	51358		51400
117 706	**N**	A	*SL*	BY	51366		51408
117 707	**N**	A	*SL*	BY	51335		51377
117 708	**N**	A		KI	51336		51378
117 709	**N**	A		KI	51344		51386
117 720	**N**	A	*SL*	BY	51354		51396
117 721	**N**	A	*SL*	BY	51363		51405
117 724	**N**	A	*SL*	BY	51333		51375

Names:

51332 of 117 700 is named 'Marston Vale'.
51358 of 117 705 is named 'LESLIE CRABBE'.

CLASS 121   PRESSED STEEL SUBURBAN

DMBS.
Engines: Two Leyland 1595 of 112 kW (150 h.p.).
Transmission: Mechanical. Cardan shaft and freewheel to a four-speed epicylic gearbox with a further cardan shaft to the final drive, each engine driving the inner axle of one bogie.
Gangways: Non gangwayed single cars with cabs at each end.
Bogies: DD10 (motor) and DT9 (trailer).
Dimensions: 20.45 x 2.82 m.
Seats: 3+2 facing.

DMBS. Dia. DX201. Lot No. 30518 1960. –/65. 38.0 t.

Note: Some of the cars show 'L' instead of the official class prefix.

121 127	**N**	A *GE*	IL	55027
121 129	**N**	A *GE*	IL	55029
121 131	**N**	A *GE*	IL	55031

CLASS 122   GLOUCESTER SUBURBAN

DMBS.
Engines: Two AEC 220 of 112 kW (150 h.p.).
Transmission: Mechanical. Cardan shaft and freewheel to a four-speed epicylic gearbox with a further cardan shaft to the final drive, each engine driving the inner axle of one bogie.
Gangways: Non gangwayed single car with cabs at each end.
Bogies: DD10 (motor) and DT9 (trailer).
Dimensions: 20.45 x 2.82 m.
Seats: 3+2 facing.

DMBS. Dia. DX202. Lot No. 30419 1958. 36.5 t.

Note: This unit is used as a crew-training vehicle and does not carry its unit number.

122 012	**LH**	E *CR*	TE	55012

3.2. SECOND GENERATION DMUS

Unit Types :

There are six basic types of second generation vehicle as referred to in the class headings as follows:

Pacers (Railbuses). Folding power operated exterior doors. Bus-type 3+2 (2+2 on class 141) largely unidirectional seating. Limited luggage space. Four wheel chassis. 75 m.p.h.

Sprinter. Sliding power operated exterior double doors to large entrance vestibules. High backed 3+2 seating. Limited luggage space. 75 m.p.h.

Super Sprinter. Sliding/sliding plug power-operated exterior doors. High backed 2+2 largely unidirectional seating with some tables. 75 m.p.h.

Express. Sliding plug power-operated exterior doors. Air conditioned. High backed 2+2 half-facing and half-unidirectional seating with some tables. 90 m.p.h.

Network Turbo. Sliding power operated exterior double doors to large entrance vestibules. 3+2 seating. Limited luggage space. 75 or 90 m.p.h.

Turbostar. Sliding power operated exterior double doors to large entrance vestibules. 2+2 seating. Limited luggage space. 100 m.p.h.

Public Address System:

All vehicles are equipped with public address, with transmission equipment on driving vehicles.

Gangways:

Unless stated otherwise, all vehicles have flexible diaphragm gangways.

Couplings:

Unless otherwise stated all vehicles are fitted with BSI automatic couplings at their outer ends. Railbus types are fitted with bar couplings at their inner ends, but all other types have BSI couplings at their inner ends unless otherwise stated.

Brakes:

All vehicles are fitted with electro-pneumatic and air brakes.

Multiple Working:

Classes 141–158/170 can work in multiple with one another as can Classes 165/6/8.

CLASS 141   LEYLAND BUS/BREL RAILBUS

DMS–DMSL. Built from Leyland National bus parts on four-wheeled underframes.

Engines: One Leyland TL11 152 kW (205 h.p.) (* Cummins LT10-R) per car.
Transmission: Hydraulic. Voith T211r with Gmeinder final drive.
Gangways: Within unit only.
Doors: Folding.
Dimensions: 15.45 x 2.50 m.
Accommodation: 2+2 bus style.
Maximum Speed: 75 m.p.h.

DMS. Dia. DP228 Lot No. 30977 Derby 1984. Modified by Barclay 1988–89. –/50. 26.0 t.
DMSL. Dia. DP229 Lot No. 30978 Derby 1984. Modified by Barclay 1988–89. –/44 1T. 26.5 t.

141 101		**Y**	P	*NE*	NL (S)	55521	55541
141 102		**Y**	P	*NE*	HT (S)	55502	55522
141 103		**Y**	P		ZB	55503	55523
141 105		**Y**	P		ZB	55505	55525
141 106		**Y**	P		ZB	55506	55526
141 107		**Y**	P		ZB	55507	55527
141 108		**Y**	P		ZB	55508	55528
141 109		**Y**	P	*NE*	HT (S)	55509	55529
141 110		**Y**	P		ZB	55510	55530
141 111		**Y**	P	*NE*	HT (S)	55511	55531
141 112		**Y**	P		ZB	55512	55532
141 113	*	**Y**	P	*NE*	NL	55513	55533
141 114		**Y**	P	*NE*	NL (S)	55514	55534
141 115		**Y**	P	*NE*	HT (S)	55515	55535
141 116		**Y**	P		ZB	55516	55536
141 117		**Y**	P	*NE*	HT (S)	55517	55537
141 118		**Y**	P		ZB	55518	55538
141 119		**Y**	P	*NE*	NL	55519	55539
141 120		**Y**	P		ZB	55520	55540

CLASS 142   LEYLAND BUS/BREL RAILBUS

DMS–DMSL. Development of Class 141 with wider body and improved appearance.

Engines: One Cummins LTA10-R of 170 kW (225 h.p.) per car.
Transmission: Hydraulic. Voith T211r with Gmeinder final drive.
Gangways: Within unit only.
Doors: Folding.
Dimensions: 15.55 x 2.80 m.
Accommodation: 2+3 bus style.
Maximum Speed: 75 m.p.h.

Non-Standard Livery: Chocolate & Cream.

55542–55591. DMS. Dia. DP234 Lot No. 31003 Derby 1985–6. –/62. 24.5 t.
55592–55641. DMSL. Dia. DP235 Lot No. 31004 Derby 1985–6. –/59 1T. 25.0 t.

55701–55746. DMS. Dia. DP234 Lot No. 31013 Derby 1986–7. –/62. 24.5 t.
55747–55792. DMSL. Dia. DP235 Lot No. 31014 Derby 1986–7. –/59 1T. 25.0 t.

142 001	**GM**	A	*NW*	NH	55542	55592
142 002	**GM**	A	*NW*	NH	55543	55593
142 003	**GM**	A	*NW*	NH	55544	55594
142 004	**GM**	A	*NW*	NH	55545	55595
142 005	**GM**	A	*NW*	NH	55546	55596
142 006	**GM**	A	*NW*	NH	55547	55597
142 007	**GM**	A	*NW*	NH	55548	55598
142 008	**GM**	A	*NW*	NH	55549	55599
142 009	**GM**	A	*NW*	NH	55550	55600
142 010	**GM**	A	*NW*	NH	55551	55601
142 011	**GM**	A	*NW*	NH	55552	55602
142 012	**GM**	A	*NW*	NH	55553	55603
142 013	**GM**	A	*NW*	NH	55554	55604
142 014	**GM**	A	*NW*	NH	55555	55605
142 015	**RR**	A	*NE*	HT	55556	55606
142 016	**RR**	A	*NE*	HT	55557	55607
142 017	**T**	A	*NE*	HT	55558	55608
142 018	**T**	A	*NE*	HT	55559	55609
142 019	**T**	A	*NE*	HT	55560	55610
142 020	**T**	A	*NE*	HT	55561	55611
142 021	**T**	A	*NE*	HT	55562	55612
142 022	**T**	A	*NE*	HT	55563	55613
142 023	**RR**	A	*NW*	NH	55564	55614
142 024	**RR**	A	*NE*	HT	55565	55615
142 025	**0**	A	*NE*	HT	55566	55616
142 026	**0**	A	*NE*	HT	55567	55617
142 027	**GM**	A	*NW*	NH	55568	55618
142 028	**GM**	A	*NW*	NH	55569	55619
142 029	**GM**	A	*NW*	NH	55570	55620
142 030	**GM**	A	*NW*	NH	55571	55621
142 031	**GM**	A	*NW*	NH	55572	55622
142 032	**GM**	A	*NW*	NH	55573	55623
142 033	**RR**	A	*NW*	NH	55574	55624
142 034	**GM**	A	*NW*	NH	55575	55625
142 035	**GM**	A	*NW*	NH	55576	55626
142 036	**RR**	A	*NW*	NH	55577	55627
142 037	**GM**	A	*NW*	NH	55578	55628
142 038	**GM**	A	*NW*	NH	55579	55629
142 039	**GM**	A	*NW*	NH	55580	55630
142 040	**GM**	A	*NW*	NH	55581	55631
142 041	**GM**	A	*NW*	NH	55582	55632
142 012	**GM**	A	*NW*	NH	55583	55633
142 043	**GM**	A	*NW*	NH	55584	55634
142 044	**RR**	A	*NW*	NH	55585	55635

142 045	**GM**	A	*NW*	NH	55586	55636
142 046	**GM**	A	*NW*	NH	55587	55637
142 047	**RR**	A	*NW*	NH	55588	55638
142 048	**RR**	A	*NW*	NH	55589	55639
142 049	**GM**	A	*NW*	NH	55590	55640
142 050	**PR**	A	*NE*	HT	55591	55641
142 051	**MT**	A	*NW*	NH	55701	55747
142 052	**MT**	A	*NW*	NH	55702	55748
142 053	**MT**	A	*NW*	NH	55703	55749
142 054	**MT**	A	*NW*	NH	55704	55750
142 055	**MT**	A	*NW*	NH	55705	55751
142 056	**MT**	A	*NW*	NH	55706	55752
142 057	**MT**	A	*NW*	NH	55707	55753
142 058	**MT**	A	*NW*	NH	55708	55754
142 060	**GM**	A	*NW*	NH	55710	55756
142 061	**GM**	A	*NW*	NH	55711	55757
142 062	**GM**	A	*NW*	NH	55712	55758
142 063	**GM**	A	*NW*	NH	55713	55759
142 064	**GM**	A	*NW*	NH	55714	55760
142 065	**PR**	A	*NE*	HT	55715	55761
142 066	**PR**	A	*NE*	HT	55716	55762
142 067	**GM**	A	*NW*	NH	55717	55763
142 068	**GM**	A	*NW*	NH	55718	55764
142 069	**GM**	A	*NW*	NH	55719	55765
142 070	**GM**	A	*NW*	NH	55720	55766
142 071	**RR**	A	*NE*	NL	55721	55767
142 072	**RR**	A	*NE*	NL	55722	55768
142 073	**RR**	A	*NE*	NL	55723	55769
142 074	**RR**	A	*NE*	NL	55724	55770
142 075	**RR**	A	*NE*	NL	55725	55771
142 076	**RR**	A	*NE*	NL	55726	55772
142 077	**RR**	A	*NE*	NL	55727	55773
142 078	**RR**	A	*NE*	NL	55728	55774
142 079	**RR**	A	*NE*	NL	55729	55775
142 080	**RR**	A	*NE*	NL	55730	55776
142 081	**RR**	A	*NE*	NL	55731	55777
142 082	**RR**	A	*NE*	NL	55732	55778
142 083	**RR**	A	*NE*	NL	55733	55779
142 084	**RR**	A	*NE*	NL	55734	55780
142 085	**RR**	A	*NE*	NL	55735	55781
142 086	**RR**	A	*NE*	NL	55736	55782
142 087	**RR**	A	*NE*	NL	55737	55783
142 088	**RR**	A	*NE*	NL	55738	55784
142 089	**RR**	A	*NE*	NL	55739	55785
142 090	**RR**	A	*NE*	NL	55740	55786
142 091	**RR**	A	*NE*	NL	55741	55787
142 092	**RR**	A	*NE*	NL	55742	55788
142 093	**RR**	A	*NE*	NL	55743	55789
142 094	**RR**	A	*NE*	NL	55744	55790
142 095	**RR**	A	*NE*	NL	55745	55791
142 096	**RR**	A	*NE*	NL	55746	55792

CLASS 143   ALEXANDER/BARCLAY RAILBUS

DMS–DMSL. Similar design to Class 142, but bodies built by W. Alexander with Barclay underframes.

Engines: One Cummins LTA10-R of 170 kW (225 h.p.) per car.
Transmission: Hydraulic. Voith T211r with Gmeinder final drive.
Gangways: Within unit only.
Doors: Folding.
Dimensions: 15.55 x 2.70 m.
Accommodation: 2+3 bus style.
Maximum Speed: 75 m.p.h.

DMS. Dia. DP236 Lot No. 31005 Andrew Barclay 1985–6. –/62. 24.5 t.
DMSL. Dia. DP237 Lot No. 31006 Andrew Barclay 1985–6. –/60 1T. 25.0 t.

Note: 143 601/10/4 are owned by Mid-Glamorgan County Council, 143 609 is owned by South Glamorgan County Council and 143 617–9 are owned by West Glamorgan County Council although managed by Porterbrook Leasing Company.

g Fitted with global positioning system.

143 601		**RR**	P	*WW*	CF	55642	55667
143 602	g	**RR**	P	*CA*	CF	55651	55668
143 603	g	**RR**	P	*CA*	CF	55658	55669
143 604	g	**RR**	P	*CA*	CF	55645	55670
143 605	g	**RR**	P	*CA*	CF	55646	55671
143 606	g	**RR**	P	*CA*	CF	55647	55672
143 607	g	**RR**	P	*CA*	CF	55648	55673
143 608	g	**RR**	P	*CA*	CF	55649	55674
143 609	g	**RR**	P	*CA*	CF	55650	55675
143 610		**RR**	P	*WW*	CF	55643	55676
143 611	g	**RR**	P	*CA*	CF	55652	55677
143 612		**RR**	P	*WW*	CF	55653	55678
143 613	g	**RR**	P	*CA*	CF	55654	55679
143 614		**RR**	P	*WW*	CF	55655	55680
143 615	g	**RR**	P	*CA*	CF	55656	55681
143 616	g	**RR**	P	*CA*	CF	55657	55682
143 617		**RR**	P	*WW*	CF	55644	55683	Bewick's Swan
143 618		**RR**	P	*WW*	CF	55659	55684	Mute Swan
143 619		**RR**	P	*WW*	CF	55660	55685	Whooper Swan
143 620		**RR**	P	*WW*	CF	55661	55686	
143 621		**RR**	P	*WW*	CF	55662	55687	
143 622		**RR**	P	*WW*	CF	55663	55688	
143 623		**RR**	P	*WW*	CF	55664	55689	
143 624		**RR**	P	*WW*	CF	55665	55690	
143 625		**RR**	P	*WW*	CF	55666	55691	

CLASS 144   ALEXANDER/BREL RAILBUS

DMS–DMSL or DMS–MS–DMSL. Similar design to Class 143, but under-frames built by BREL as subcontractor to W. Alexander.

Engines: One Cummins LTA10-R of 170 kW (225 h.p.) per car.
Transmission: Hydraulic. Voith T211r with Gmeinder final drive.
Gangways: Within unit only.
Doors: Folding.
Dimensions: 15.25 x 2.70 m.
Accommodation: 2+3 bus style.
Maximum Speed: 75 m.p.h.

DMS. Dia. DP240 Lot No. 31015 Derby 1986–7. –/62 and wheelchair space. 24.2 t.
MS. Dia. DR205 Lot No. Derby 31037 1987. –/73. 22.6 t.
DMSL. Dia. DP241 Lot No. Derby 31016 1986–7. –/60 1T. 25.0 t.

Note: The centre cars of the three-car units are owned by West Yorkshire PTE, although managed by Porterbrook Leasing Company.

144 001	**Y**	P	*NE*	NL	55801		55824
144 002	**Y**	P	*NE*	NL	55802		55825
144 003	**Y**	P	*NE*	NL	55803		55826
144 004	**Y**	P	*NE*	NL	55804		55827
144 005	**Y**	P	*NE*	NL	55805		55828
144 006	**Y**	P	*NE*	NL	55806		55829
144 007	**Y**	P	*NE*	NL	55807		55830
144 008	**Y**	P	*NE*	NL	55808		55831
144 009	**Y**	P	*NE*	NL	55809		55832
144 010	**Y**	P	*NE*	NL	55810		55833
144 011	**RR**	P	*NE*	NL	55811		55834
144 012	**RR**	P	*NE*	NL	55812		55835
144 013	**RR**	P	*NE*	NL	55813		55836
144 014	**Y**	P	*NE*	NL	55814	55850	55837
144 015	**Y**	P	*NE*	NL	55815	55851	55838
144 016	**Y**	P	*NE*	NL	55816	55852	55839
144 017	**Y**	P	*NE*	NL	55817	55853	55840
144 018	**Y**	P	*NE*	NL	55818	55854	55841
144 019	**Y**	P	*NE*	NL	55819	55855	55842
144 020	**Y**	P	*NE*	NL	55820	55856	55843
144 021	**Y**	P	*NE*	NL	55821	55857	55844
144 022	**Y**	P	*NE*	NL	55822	55858	55845
144 023	**Y**	P	*NE*	NL	55823	55859	55846

CLASS 150/0   BREL PROTOTYPE SPRINTER

DMSL–MS–DMS. Prototype Sprinter.

Engines: One Cummins NT855R5 of 210 kW (285 h.p.) per car.
Transmission: Hydraulic. Voith T211r with Gmeinder final drive.
Bogies: One BX8P and one BX8T.
Couplings: BSI at outer end of driving vehicles, bar non-driving ends.

Gangways: Within unit only.
Doors: Sliding.
Accommodation: 2+3 (mainly unidirectional).
Dimensions: 20.06 x 2.82 m (outer cars), 20.18 x 2.82 m (inner car).
Maximum Speed: 75 m.p.h.

DMSL. Dia. DP230. Lot No. 30984 York 1984. –/72 1T. 35.8 t.
MS. Dia. DR202. Lot No. 30986 York 1984. –/92. 34.4 t.
DMS. Dia. DP231. Lot No. 30985 York 1984. –/76. 35.6 t.

Note: 150 002 was converted to 154 002 at RTC Derby in 1986, but was later converted back to a Class 150.

150 001	**CE**	A	*CT*	TS	55200	55400	55300
150 002	**CE**	A	*CT*	TS	55201	55401	55301

CLASS 150/1   BREL SPRINTER

DMSL–DMS or DMSL–DMSL–DMS or DMSL–DMS–DMS.

Engines: One Cummins NT855R5 of 210 kW (285 h.p.) per car.
Transmission: Hydraulic. Voith T211r with Gmeinder final drive.
Bogies: One BP38 and one BT38.
Gangways: Within unit only.
Doors: Sliding.
Accommodation: 2+3 facing. († Reseated with part unidirectional seating and part facing).
Dimensions: 20.06 x 2.82 m.
Maximum Speed: 75 m.p.h.

DMSL. Dia. DP238. Lot No. 31011 York 1985–6. –/68 1T (–/64 1T*, –/72 1T†). 36.5 t.
DMS. Dia. DP239. Lot No. 31012 York 1985–6. –/70 (–/76†, –/66§). 38.45 t.

Note: The centre cars of three-car units are Class 150/2 vehicles. For details see next Class.

150 010	r†	**CE**	A	*CT*	TS	52110	57226	57110
150 011	r†	**CE**	A	*CT*	TS	52111	57206	57111
150 012	r†	**CE**	A	*CT*	TS	52112	52204	57112
150 013	r†	**CE**	A	*CT*	TS	52113	52226	57113
150 014	r†	**CE**	A	*CT*	TS	52114	57204	57114
150 015	r†	**CE**	A	*CT*	TS	52115	52206	57115
150 016	r†	**CE**	A	*CT*	TS	52116	57212	57116
150 017	r†	**CE**	A	*CT*	TS	52117	57209	57117
150 101	r†	**CE**	A	*CT*	TS	52101		57101
150 102	r†	**CE**	A	*CT*	TS	52102		57102
150 103	r†	**CE**	A	*CT*	TS	52103		57103
150 104	r†	**CE**	A	*CT*	TS	52104		57104
150 105	r†	**CE**	A	*CT*	TS	52105		57105
150 106	r†	**CE**	A	*CT*	TS	52106		57106
150 107	r†	**CE**	A	*CT*	TS	52107		57107
150 108	r†	**CE**	A	*CT*	TS	52108		57108

▲ 3-car Class 101 No. 101 683 waits at Blaneau Ffestiniog on 15th September 1997 before working the 17.27 to Llandudno. Regional Railways livery is carried. **Peter Fox**

▼ Network SouthEast liveried Class 117 No. 117 706 pauses at Ridgemont, Bedfordshire on 1st October 1996 with the 12.40 Bedford–Bletchley. **Kevin Conkey**

▲ Class 121 'Bubble Cars' Nos. L127 and L129 at Colchester depot on 21st September 1997. These units are now based here for use on the Sudbury branch.

Michael J. Collins

▼ The only Class 122 still numbered in the capital stock series is No. 55012 and is in fact used as a crew-training vehicle. The unit, which carries Loadhaul livery, is pictured here with an Immingham to Landor Street run on 2nd May 1997.

Bob Sweet

▲ There are now only two Class 141 units in traffic. One of them No. 141 113 has Voith transmission and is pictured here near Doncaster with the 13.44 Scunthorpe–Doncaster on 20th January 1997. The unit is in West Yorkshire PTE livery.   **Gary Pierrepont**

▼ Merseytravel liveried Class 142 No. 142 058 at Siddick on the Solway coast on 8th March 1997 with the 10.15 Whitehaven–Carlisle.   **Dave McAlone**

▲ Class 143 No. 143 608 passes Burton-on-Trent running ECS to Derby on 12th June 1996. **Hugh Ballantyne**

▼ Class 144 No. 144 008 pauses at Milford Sidings with a crew-training run on 10th October 1996. **Nic Joynson**

Centro Trains liveried Class 150/1 No. 150 016 departs from Hatton with the 11.00 Leamington Spa–Great Malvern service on 29th February 1996.

Hugh Ballantyne

▲ Greater Manchester PTE liveried Class 150/2 No. 150 224 en-route between Manchester Oxford Road and Piccadilly on 28th May 1997. **Hugh Ballantyne**

▼ Class 153 No. 153 324 pauses at Bamber Bridge on 17th June 1997 with the 19.09 Colne–Blackpool South service. **Martyn Hilbert**

Class 155 No. 155 342 passes the Rochdale Canal near Castleton with the 07.40 Selby–Manchester Victoria service during May 1997.

Vincent Eastwood

▲ Class 156 No. 156 425 pauses at Lostock Hall with a Sundays only 08.39 Blackpool North–Carlisle 'Dalesrail' service. The unit is painted in unbranded North West Regional Railways livery. **Martyn Hilbert**

▼ Class 158/0 No. 158 765 passes Heeley, Sheffield on 1st August 1997 with the 16.12 Cleethorpes–Manchester Airport service. The unit carries Regional Railways Express livery. **P. Renard**

West Yorkshire PTE liveried Class 158/9 No. 158 902 passes New Barnetby with the 14.28 Cleethorpes–Manchester Airport on 30th May 1997.

Brian Denton

A six-car Class 159 formation led by unit No. 159 017 passes Micheldever with the diverted 08.13 Basingstoke–Paignton service on 14th September 1997. The units are in Network SouthEast livery with South West Trains branding.

Nic Joynson

▲ Class 165/0 No. 165 003 at Reading on 26th August 1997.　**Dave McAlone**

▼ Class 166 No. 166 213 passes Buckland with the 14.03 Gatwick Airport–Reading on 4th October 1997.　**Alex Dasi-Sutton**

▲ The first car to be completed of the new Class 168 'Turbostar' units.
ADtranz

▼ Hastings Diesels owned Class 201 No. 1001 approaches Edenbridge Town with the 15.30 Uckfield–Oxted on the occasion of the Uckfield line gala weekend. The centre car is from a Class 411/5 EMU.
David Brown

▲ Connex liveried Class 205 No. 205 028 works the 17.00 Uckfield–Oxted near Edenbridge Town on 23rd August 1997. **Chris Wilson**

▼ Class 207/2 No. 207 201 'Ashford Fayre' departs from Ashford with the 08.54 service to Eastbourne on 10th April 1997. Note the centre car from a Class 411/5 EMU. **David Brown**

Eurotunnel shuttle loco No. 9002 'STUART BURROWS' arrives at Cheriton terminal with a car carrying train from France. The date is 6th September 1996.

Hugh Ballantyne

▲ Docklands Light Railway Class B92 cars, Nos. 80 and 53 enter Crossharbour station with an Island Gardens service on 24th May 1997. **Peter Fox**

▼ Tyne & Wear Metro car No. 4034 at West Jesmond with an Airport–South Shields service on 25th October 1997. **Peter Fox**

South Yorkshire Supertram car No. 16 at Birley Moor Road on 1st January 1997. Trams and trains were the only public transport operating in Sheffield on New Years Day, as no buses ran.   **Peter Fox**

150 109	r†	**CE**	A	*CT*	TS	52109	57109
150 118	r†	**CE**	A	*CT*	TS	52118	57118
150 119	r†	**CE**	A	*CT*	TS	52119	57119
150 120	r†	**CE**	A	*CT*	TS	52120	57120
150 121	r†	**CE**	A	*CT*	TS	52121	57121
150 122	r†	**CE**	A	*CT*	TS	52122	57122
150 123	r†	**CE**	A	*CT*	TS	52123	57123
150 124	r†	**CE**	A	*CT*	TS	52124	57124
150 125	r†	**CE**	A	*CT*	TS	52125	57125
150 126	r†	**CE**	A	*CT*	TS	52126	57126
150 127	r†	**CE**	A	*CT*	TS	52127	57127
150 128	r†	**CE**	A	*CT*	TS	52128	57128
150 129	r†	**CE**	A	*CT*	TS	52129	57129
150 130	r†	**CE**	A	*CT*	TS	52130	57130
150 131	r†	**CE**	A	*CT*	TS	52131	57131
150 132	r†	**CE**	A	*CT*	TS	52132	57132
150 133	r*	**GM**	A	*NW*	NH	52133	57133
150 134	r*	**GM**	A	*NW*	NH	52134	57134
150 135	r*	**GM**	A	*NW*	NH	52135	57135
150 136	r*	**GM**	A	*NW*	NH	52136	57136
150 137	r§	**GM**	A	*NW*	NH	52137	57137
150 138	r*	**GM**	A	*NW*	NH	52138	57138
150 139	r*	**GM**	A	*NW*	NH	52139	57139
150 140	r*	**GM**	A	*NW*	NH	52140	57140
150 141	r*	**GM**	A	*NW*	NH	52141	57141
150 142	r§	**GM**	A	*NW*	NH	52142	57142
150 143	r*	**P**	A	*NW*	NH	52143	57143
150 144	r§	**P**	A	*NW*	NH	52144	57144
150 145	r§	**P**	A	*NW*	NH	52145	57145
150 146	r§	**RR**	A	*NW*	NH	52146	57146
150 147	r§	**P**	A	*NW*	NH	52147	57147
150 148	r§	**P**	A	*NW*	NH	52148	57148
150 149	r§	**P**	A	*NW*	NH	52149	57149
150 150	r§	**P**	A	*NW*	NH	52150	57150

CLASS 150/2   BREL SPRINTER

DMSL–DMS.

Engines: One Cummins NT855R5 of 210 kW (285 h.p.) per car.
Transmission: Hydraulic. Voith T211r with Gmeinder final drive.
Bogies: One BP38 and one BT38.
Gangways: Throughout.
Doors: Sliding.
Accommodation: 2+3 mainly unidirectional.
Dimensions: 20.06 x 2.82 m.
Maximum Speed: 75 m.p.h.

DMSL. Dia. DP242. Lot No. 31017 York 1986–87. –/73 1T (–70 1T*). 35.8 t.
DMS. Dia. DP243. Lot No. 31018 York 1986–7. –/76 (–/73§, –/68†) and luggage space. 34.90 t.

g Fitted with global positioning system.

150 201	*	**MT**	A	*NW*	NH	52201	57201	
150 202		**CE**	A	*CT*	TS	52202	57202	
150 203	*	**MT**	A	*NW*	NH	52203	57203	
150 205	*	**MT**	A	*NW*	NH	52205	57205	
150 207	§	**MT**	A	*NW*	NH	52207	57207	
150 208		**RR**	P	*SR*	HA	52208	57208	
150 210		**CE**	A	*CT*	TS	52210	57210	
150 211	§	**MT**	A	*NW*	NH	52211	57211	
150 213	†	**RR**	P	*AR*	NC	52213	57213	LORD NELSON
150 214		**CE**	A	*CT*	TS	52214	57214	
150 215		**GM**	A	*NW*	NH	52215	57215	
150 216		**CE**	A	*CT*	TS	52216	57216	
150 217	†	**RR**	P	*AR*	NC	52217	57217	OLIVER CROMWELL
150 218	*§	**GM**	A	*NW*	NH	52218	57218	
150 219		**RR**	P	*WW*	CF	52219	57219	
150 220		**CE**	A	*CT*	TS	52220	57220	
150 221		**RR**	P	*WW*	CF	52221	57221	
150 222	*§	**GM**	A	*NW*	NH	52222	57222	
150 223	*	**GM**	A	*NW*	NH	52223	57223	
150 224	*	**GM**	A	*NW*	NH	52224	57224	
150 225	§	**GM**	A	*NW*	NH	52225	57225	
150 227	†	**RR**	P	*AR*	NC	52227	57227	SIR ALF RAMSEY
150 228		**RR**	P	*SR*	HA	52228	57228	
150 229	†	**RR**	P	*AR*	NC	52229	57229	GEORGE BORROW
150 230		**RR**	P	*WW*	CF	52230	57230	
150 231	†	**RR**	P	*AR*	NC	52231	57231	KING EDMUND
150 232		**RR**	P	*WW*	CF	52232	57232	
150 233		**RR**	P	*WW*	CF	52233	57233	
150 234		**RR**	P	*WW*	CF	52234	57234	
150 235	†	**RR**	P	*AR*	NC	52235	57235	CARDINAL WOLSEY
150 236		**RR**	P	*WW*	CF	52236	57236	
150 237	†	**RR**	P	*AR*	NC	52237	57237	HEREWARD THE WAKE
150 238		**RR**	P	*WW*	CF	52238	57238	
150 239		**RR**	P	*WW*	CF	52239	57239	
150 240		**RR**	P	*WW*	CF	52240	57240	
150 241		**RR**	P	*WW*	CF	52241	57241	
150 242		**RR**	P	*WW*	CF	52242	57242	
150 243		**RR**	P	*WW*	CF	52243	57243	
150 244		**RR**	P	*WW*	CF	52244	57244	
150 245		**RR**	P	*SR*	HA	52245	57245	
150 246		**RR**	P	*WW*	CF	52246	57246	
150 247		**RR**	P	*WW*	CF	52247	57247	
150 248		**RR**	P	*WW*	CF	52248	57248	
150 249		**RR**	P	*WW*	CF	52249	57249	
150 250		**RR**	P	*SR*	HA	52250	57250	
150 251		**RR**	P	*WW*	CF	52251	57251	
150 252		**RR**	P	*SR*	HA	52252	57252	
150 253		**RR**	P	*WW*	CF	52253	57253	
150 254		**RR**	P	*WW*	CF	52254	57254	

150 255	†	**RR**	P	*AR*	NC	52255	57255	HENRY BLOGG
150 256		**RR**	P	*SR*	HA	52256	57256	
150 257	†	**RR**	P	*AR*	NC	52257	57257	QUEEN BOADICEA
150 258		**RR**	P	*SR*	HA	52258	57258	
150 259		**RR**	P	*SR*	HA	52259	57259	
150 260		**RR**	P	*SR*	HA	52260	57260	
150 261		**RR**	P	*WW*	CF	52261	57261	
150 262		**RR**	P	*SR*	HA	52262	57262	
150 263		**RR**	P	*WW*	CF	52263	57263	
150 264		**RR**	P	*SR*	HA	52264	57264	
150 265	g	**RR**	P	*WW*	CF	52265	57265	
150 266	g	**RR**	P	*CA*	CF	52266	57266	
150 267	g	**RR**	P	*CA*	CF	52267	57267	
150 268	g	**RR**	P	*CA*	CF	52268	57268	
150 269	g	**RR**	P	*CA*	CF	52269	57269	
150 270	g	**RR**	P	*CA*	CF	52270	57270	
150 271	g	**RR**	P	*CA*	CF	52271	57271	
150 272	g	**RR**	P	*CA*	CF	52272	57272	
150 273	g	**RR**	P	*CA*	CF	52273	57273	
150 274	g	**RR**	P	*CA*	CF	52274	57274	
150 275	g	**RR**	P	*CA*	CF	52275	57275	
150 276	g	**RR**	P	*CA*	CF	52276	57276	
150 277	g	**RR**	P	*CA*	CF	52277	57277	
150 278	g	**RR**	P	*CA*	CF	52278	57278	
150 279	g	**RR**	P	*CA*	CF	52279	57279	
150 280	g	**RR**	P	*CA*	CF	52280	57280	
150 281	g	**RR**	P	*CA*	CF	52281	57281	
150 282	g	**RR**	P	*CA*	CF	52282	57282	
150 283		**RR**	P	*SR*	HA	52283	57283	
150 284		**RR**	P	*SR*	HA	52284	57284	
150 285		**RR**	P	*SR*	HA	52285	57285	EDINBURGH–BATHGATE 1986–1996

CLASS 153   LEYLAND BUS SUPER SPRINTER

DMSL. Converted by Hunslet-Barclay, Kilmarnock from Class 155 two-car units.

Engines: One Cummins NT855R5 of 213 kW (285 h.p.) per car.
Transmission: Hydraulic. Voith T211r with Gmeinder final drive.
Bogies: One P3-10 and one BT38.
Gangways: Throughout.
Doors: Sliding plug.
Accommodation: 2+2 facing/unidirectional with wheelchair space.
Dimensions: 23.21 x 2.70 m.
Maximum Speed: 75 m.p.h.

52301–52335. DMSL. Dia. DX203. Lot No. 31026 1987–8. Converted under Lot No. 31115 1991–2. –/72 1TD (–/66 1TD†) + 3 tip-up seats. 41.2 t.
57301–57335. DMSL. Dia. DX203. Lot No. 31027 1987–8. Converted under Lot No. 31115 1991–2. –/72 1TD + 3 tip-up seats. 41.2 t.

Notes:

Cars numbered in the 573XX series have been renumbered by adding 50 to the
number so that the last two digits correspond with the set number.
Certain Central Trains units have been fitted with new-style seating and certain
Wales & West units have been fitted with Class 158-style seating.

153 301		**RR**	A	*NE*	HT	52301
153 302		**RR**	A	*WW*	CF	52302
153 303		**RR**	A	*WW*	CF	52303
153 304		**RR**	A	*NE*	HT	52304
153 305		**RR**	A	*WW*	CF	52305
153 306	†	**RR**	P	*AR*	NC	52306	EDITH CAVELL
153 307		**RR**	A	*NE*	HT	52307
153 308		**RR**	A	*WW*	CF	52308
153 309	†	**RR**	P	*AR*	NC	52309	GERARD FIENNES
153 310		**RR**	P	*NW*	NH	52310
153 311	†	**RR**	P	*AR*	NC	52311	JOHN CONSTABLE
153 312		**RR**	A	*WW*	CF	52312
153 313		**RR**	P	*NW*	NH	52313
153 314	†	**RR**	P	*AR*	NC	52314	DELIA SMITH
153 315		**RR**	A	*NE*	HT	52315
153 316		**RR**	P	*NW*	NH	52316
153 317		**RR**	A	*NE*	HT	52317
153 318		**RR**	A	*WW*	CF	52318
153 319		**RR**	A	*NE*	HT	52319
153 320		**RR**	P	*CT*	TS	52320
153 321		**RR**	P	*CT*	TS	52321
153 322	†	**RR**	P	*AR*	NC	52322	BENJAMIN BRITTEN
153 323		**RR**	P	*CT*	TS	52323
153 324		**RR**	P	*NW*	NH	52324
153 325		**RR**	P	*CT*	TS	52325
153 326	†	**RR**	P	*AR*	NC	52326	TED ELLIS
153 327		**RR**	A	*WW*	CF	52327
153 328		**RR**	A	*NE*	HT	52328
153 329		**RR**	P	*CT*	TS	52329
153 330		**RR**	P	*NW*	NH	52330
153 331		**RR**	A	*NE*	HT	52331
153 332		**RR**	P	*NW*	NH	52332
153 333		**RR**	P	*CT*	TS	52333
153 334		**RR**	P	*CT*	TS	52334
153 335	†	**RR**	P	*AR*	NC	52335	MICHAEL PALIN
153 351		**RR**	A	*NE*	HT	57351
153 352		**RR**	A	*NE*	HT	57352
153 353		**RR**	A	*WW*	CF	57353
153 354		**RR**	P	*CT*	TS	57354
153 355		**RR**	A	*WW*	CF	57355
153 356		**RR**	P	*CT*	TS	57356
153 357		**RR**	A	*NE*	HT	57357
153 358		**RR**	P	*NW*	NH	57358
153 359		**RR**	P	*NW*	NH	57359
153 360		**RR**	P	*NW*	NH	57360

153 361	**RR**	P	*NW*	NH	57361
153 362	**RR**	A	*WW*	CF	57362
153 363	**RR**	P	*NW*	NH	57363
153 364	**RR**	P	*CT*	TS	57364
153 365	**RR**	P	*CT*	TS	57365
153 366	**RR**	P	*CT*	TS	57366
153 367	**RR**	P	*NW*	NH	57367
153 368	**RR**	A	*WW*	CF	57368
153 369	**RR**	P	*CT*	TS	57369
153 370	**RR**	A	*WW*	CF	57370
153 371	**RR**	P	*CT*	TS	57371
153 372	**RR**	A	*WW*	CF	57372
153 373	**RR**	A	*WW*	CF	57373
153 374	**RR**	A	*WW*	CF	57374
153 375	**RR**	P	*CT*	TS	57375
153 376	**RR**	P	*CT*	TS	57376
153 377	**RR**	A	*WW*	CF	57377
153 378	**RR**	A	*NE*	HT	57378
153 379	**RR**	P	*CT*	TS	57379
153 380	**RR**	A	*WW*	CF	57380
153 381	**RR**	P	*CT*	TS	57381
153 382	**RR**	A	*WW*	CF	57382
153 383	**RR**	P	*CT*	TS	57383
153 384	**RR**	P	*CT*	TS	57384
153 385	**RR**	P	*CT*	TS	57385

CLASS 155   LEYLAND BUS SUPER SPRINTER

DMSL–DMS.

Engines: One Cummins NT855R5 of 213 kW (285 h.p.) per car.
Transmission: Hydraulic. Voith T211r with Gmeinder final drive.
Bogies: One P3-10 and one BT38.
Gangways: Throughout.
Doors: Sliding plug.
Accommodation: 2+2 facing/unidirectional with wheelchair space in DMSL.
Dimensions: 23.21 x 2.70 m.
Maximum Speed: 75 m.p.h.

DMSL. Dia. DP248. Lot No. 31057 1988. –/80 1TD. 39.0 t.
DMS. Dia. DP249. Lot No. 31058 1988. –/80 and parcels area. 38.7 t.

Note: These units are owned by West Yorkshire PTE, although managed by Porterbrook Leasing Company.

155 341	**Y**	P	*NE*	NL	52341	57341
155 342	**Y**	P	*NE*	NL	52342	57342
155 343	**Y**	P	*NE*	NL	52343	57343
155 344	**Y**	P	*NE*	NL	52344	57344
155 345	**Y**	P	*NE*	NL	52345	57345
155 346	**Y**	P	*NE*	NL	52346	57346
155 347	**Y**	P	*NE*	NL	52347	57347

CLASS 156   METRO-CAMMELL SUPER SPRINTER

DMSL–DMS.

Engines: One Cummins NT855R5 of 210 kW (285 h.p.) per car.
Transmission: Hydraulic. Voith T211r with Gmeinder final drive.
Bogies: One P3-10 and one BT38.
Gangways: Throughout.
Doors: Sliding.
Accommodation: 2+2 facing/unidirectional with wheelchair space in DMSL.
Dimensions: 23.03 x 2.73 m.
Maximum Speed: 75 m.p.h.

DMSL. Dia. DP244. Lot No. 31028 1988–9. –/74 (–/72 †, –/70 §•) 1TD. 36.1 t.
DMS. Dia. DP245. Lot No. 31029 1987–9. 35.5 t. –/76 (72 •) + parcels area.

Notes: 156 500–514 are owned by Strathclyde PTE, although managed by Angel Trains Contracts. Units reliveried in **RE** or **RN** livery have been fitted with new-style seats.

156 401	†	**RE**	P	*CT*	TS	52401	57401
156 402	†	**RE**	P	*CT*	TS	52402	57402
156 403	†	**RE**	P	*CT*	TS	52403	57403
156 404	†	**RE**	P	*CT*	TS	52404	57404
156 405	†	**RE**	P	*CT*	TS	52405	57405
156 406	†	**RE**	P	*CT*	TS	52406	57406
156 407	†	**RE**	P	*CT*	TS	52407	57407
156 408	†	**RE**	P	*CT*	TS	52408	57408
156 409	†	**RE**	P	*CT*	TS	52409	57409
156 410	†	**RE**	P	*CT*	TS	52410	57410
156 411	†	**RE**	P	*CT*	TS	52411	57411
156 412	†	**RE**	P	*CT*	TS	52412	57412
156 413	†	**RE**	P	*CT*	TS	52413	57413
156 414	†	**RE**	P	*CT*	TS	52414	57414
156 415	†	**RE**	P	*CT*	TS	52415	57415
156 416	†	**RE**	P	*CT*	TS	52416	57416
156 417	†	**RE**	P	*CT*	TS	52417	57417
156 418	†	**RE**	P	*CT*	TS	52418	57418
156 419	†	**RE**	P	*CT*	TS	52419	57419
156 420	§	**RN**	P	*NW*	NH	52420	57420
156 421	§	**RN**	P	*NW*	NH	52421	57421
156 422	†	**RE**	P	*CT*	TS	52422	57422
156 423	§	**RN**	P	*NW*	NH	52423	57423
156 424	§	**RN**	P	*NW*	NH	52424	57424
156 425	§	**RN**	P	*NW*	NH	52425	57425
156 426	§	**RN**	P	*NW*	NH	52426	57426
156 427	§	**RN**	P	*NW*	NH	52427	57427
156 428	§	**RN**	P	*NW*	NH	52428	57428
156 429	§	**RN**	P	*NW*	NH	52429	57429
156 430		**P**	A	*SR*	CK	52430	57430
156 431	r•	**P**	A	*SR*	CK	52431	57431
156 432	r•	**P**	A	*SR*	CK	52432	57432

156 433		**CC**	A	*SR*	CK	52433	57433	The Kilmarnock Edition
156 434	r•	**P**	A	*SR*	CK	52434	57434	
156 435	r•	**P**	A	*SR*	CK	52435	57435	
156 436	r•	**P**	A	*SR*	CK	52436	57436	
156 437		**P**	A	*SR*	CK	52437	57437	
156 438		**P**	A	*NE*	NL	52438	57438	
156 439		**P**	A	*SR*	CK	52439	57439	
156 440	§	**RN**	P	*NW*	NH	52440	57440	
156 441	§	**RN**	P	*NW*	NH	52441	57441	
156 442		**P**	A	*SR*	CK	52442	57442	
156 443		**P**	A	*NE*	HT	52443	57443	
156 444		**P**	A	*NE*	HT	52444	57444	
156 445	r•	**P**	A	*SR*	CK	52445	57445	
156 446	r•	**P**	A	*SR*	IS	52446	57446	
156 447	r•	**P**	A	*SR*	CK	52447	57447	
156 448		**P**	A	*NE*	HT	52448	57448	
156 449	r•	**P**	A	*SR*	CK	52449	57449	Saint Columba
156 450	r•	**P**	A	*SR*	CK	52450	57450	
156 451		**P**	A	*NE*	HT	52451	57451	
156 452	§	**RN**	P	*NW*	NH	52452	57452	
156 453	r•	**P**	A	*SR*	CK	52453	57453	
156 454		**P**	A	*NE*	HT	52454	57454	
156 455	†	**RN**	P	*NW*	NH	52455	57455	
156 456	r•	**P**	A	*SR*	CK	52456	57456	
156 457	r•	**P**	A	*SR*	IS	52457	57457	
156 458	r•	**P**	A	*SR*	IS	52458	57458	
156 459	†	**RN**	P	*NW*	NH	52459	57459	
156 460	†	**RN**	P	*NW*	NH	52460	57460	
156 461	†	**RN**	P	*NW*	NH	52461	57461	
156 462	r	**P**	A	*SR*	CK	52462	57462	
156 463		**P**	A	*NE*	HT	52463	57463	
156 464	§	**RN**	P	*NW*	NH	52464	57464	
156 465	r•	**P**	A	*SR*	CK	52465	57465	Bonnie Prince Charlie
156 466	§	**RN**	P	*NW*	NH	52466	57466	
156 467	r•	**P**	A	*SR*	CK	52467	57467	
156 468		**P**	A	*NE*	NL	52468	57468	
156 469		**P**	A	*NE*	HT	52469	57469	
156 470		**P**	A	*NE*	NL	52470	57470	
156 471		**P**	A	*NE*	NL	52471	57471	
156 472		**P**	A	*NE*	NL	52472	57472	
156 473		**P**	A	*NE*	NL	52473	57473	
156 474	r•	**P**	A	*SR*	IS	52474	57474	
156 475		**P**	A	*NE*	NL	52475	57475	
156 476		**P**	A	*SR*	CK	52476	57476	
156 477	r•	**P**	A	*SR*	IS	52477	57477	HIGHLAND FESTIVAL
156 478	r•	**P**	A	*SR*	IS	52478	57478	
156 479		**P**	A	*NE*	NL	52479	57479	
156 480		**P**	A	*NE*	NL	52480	57480	
156 481		**P**	A	*NE*	NL	52481	57481	
156 482		**P**	A	*NE*	NL	52482	57482	
156 483		**P**	A	*NE*	NL	52483	57483	

156 484		P	A	NE	NL	52484	57484
156 485	r•	P	A	SR	CK	52485	57485
156 486		P	A	NE	NL	52486	57486
156 487		P	A	NE	NL	52487	57487
156 488		P	A	NE	NL	52488	57488
156 489		P	A	NE	NL	52489	57489
156 490		P	A	NE	NL	52490	57490
156 491		P	A	NE	NL	52491	57491
156 492	r•	P	A	SR	CK	52492	57492
156 493	r•	P	A	SR	CK	52493	57493
156 494	r•	P	A	SR	CK	52494	57494
156 495	r•	P	A	SR	CK	52495	57495
156 496	r•	P	A	SR	CK	52496	57496
156 497		P	A	NE	NL	52497	57497
156 498		P	A	NE	NL	52498	57498
156 499	r•	P	A	SR	IS	52499	57499
156 500	r•	P	A	SR	CK	52500	57500
156 501		S	A	SR	CK	52501	57501
156 502		S	A	SR	CK	52502	57502
156 503		S	A	SR	CK	52503	57503
156 504	r•	S	A	SR	CK	52504	57504
156 505	r•	S	A	SR	CK	52505	57505
156 506		S	A	SR	CK	52506	57506
156 507		S	A	SR	CK	52507	57507
156 508		S	A	SR	CK	52508	57508
156 509		S	A	SR	CK	52509	57509
156 510		S	A	SR	CK	52510	57510
156 511		S	A	SR	CK	52511	57511
156 512		S	A	SR	CK	52512	57512
156 513		S	A	SR	CK	52513	57513
156 514		S	A	SR	CK	52514	57514

CLASS 158/0 — BREL EXPRESS

DMSL (B)–DMSL (A) or DMCL–DMSL*§ or DMSL (B)–MSL–DMSL (A).

Engines: One Cummins NTA855R of 260 kW (350 h.p.) or 300 kW (400 h.p.) • One Perkins 2006-TWH of 260 kW (350 h.p.)†) per car.
Transmission: Hydraulic. Voith T211r with Gmeinder final drive.
Bogies: One BREL P4 and one BREL T4 per car.
Gangways: Throughout.
Doors: Sliding plug.
Accommodation: 2+2 facing/unidirectional (first & standard classes).
Dimensions: 23.21 x 2.70 m.
Maximum Speed: 90 m.p.h.

DMSL (B).. Dia. DP252. Lot No. 31051 Derby 1990–2. –/68 + wheelchair space 1TD. Public telephone and trolley space. 38.5 t.
DMCL.. Dia. DP252. Lot No. 31051 Derby 1989–90. 15/51*, 9/51§ + wheelchair space 1TD. Public telephone and trolley space. 38.5 t.
MSL. Dia. DR207. Lot No. 31050 Derby 1991. 38 t. –/70 2T.

DMSL (A). Dia. DP251. Lot No. 31052 Derby 1990–92. –/70 1T and parcels area. 37.8 t.

158 701	*	**RE**	P	*SR*	HA	52701	57701	The Scottish Claymores
158 702	*	**RE**	P	*SR*	HA	52702	57702
158 703	*	**RE**	P	*SR*	HA	52703	57703
158 704	*	**RE**	P	*SR*	HA	52704	57704
158 705	*	**RE**	P	*SR*	HA	52705	57705
158 706	*	**RE**	P	*SR*	HA	52706	57706
158 707	*	**RE**	P	*SR*	HA	52707	57707
158 708	*	**RE**	P	*SR*	HA	52708	57708
158 709	*	**RE**	P	*SR*	HA	52709	57709
158 710	*	**RE**	P	*SR*	HA	52710	57710
158 711	*	**RE**	P	*SR*	HA	52711	57711
158 712	*	**RE**	P	*SR*	HA	52712	57712
158 713	*	**RE**	P	*SR*	HA	52713	57713
158 714	*	**RE**	P	*SR*	HA	52714	57714
158 715	*	**RE**	P	*SR*	HA	52715	57715	Haymarket
158 716	*	**RE**	P	*SR*	HA	52716	57716
158 717	*	**RE**	P	*SR*	HA	52717	57717
158 718	*	**RE**	P	*SR*	HA	52718	57718
158 719	*	**RE**	P	*SR*	HA	52719	57719
158 720	*	**RE**	P	*SR*	HA	52720	57720
158 721	*	**RE**	P	*SR*	HA	52721	57721
158 722	*	**RE**	P	*SR*	HA	52722	57722
158 723	*	**RE**	P	*SR*	HA	52723	57723
158 724	*	**RE**	P	*SR*	HA	52724	57724
158 725	*	**RE**	P	*SR*	HA	52725	57725
158 726	*	**RE**	P	*SR*	HA	52726	57726
158 727	*	**RE**	P	*SR*	HA	52727	57727
158 728	*	**RE**	P	*SR*	HA	52728	57728
158 729	*	**RE**	P	*SR*	HA	52729	57729
158 730	*	**RE**	P	*SR*	HA	52730	57730
158 731	*	**RE**	P	*SR*	HA	52731	57731
158 732	*	**RE**	P	*SR*	HA	52732	57732
158 733	*	**RE**	P	*SR*	HA	52733	57733
158 734	*	**RE**	P	*SR*	HA	52734	57734
158 735	*	**RE**	P	*SR*	HA	52735	57735
158 736	*	**RE**	P	*SR*	HA	52736	57736
158 737	*	**RE**	P	*SR*	HA	52737	57737
158 738	*	**RE**	P	*SR*	HA	52738	57738
158 739	*	**RE**	P	*SR*	HA	52739	57739
158 740	*	**RE**	P	*SR*	HA	52740	57740
158 741	*	**RE**	P	*SR*	HA	52741	57741
158 742	*	**RE**	P	*SR*	HA	52742	57742
158 743	*	**RE**	P	*SR*	HA	52743	57743
158 744	*	**RE**	P	*SR*	HA	52744	57744
158 745	*	**RE**	P	*SR*	HA	52745	57745
158 746	*	**RE**	P	*SR*	HA	52746	57746
158 747	§ •	**RE**	P	*VX*	NH	52747	57747

158 748	§ •	**RE**	P	*VX*	NH	52748		57748
158 749	§ •	**RE**	P	*VX*	NH	52749		57749
158 750	§ •	**RE**	P	*VX*	NH	52750		57750
158 751	§ •	**RE**	P	*VX*	NH	52751		57751
158 752		**RE**	P	*NW*	NH	52752		57752
158 753		**RE**	P	*NW*	NH	52753		57753
158 754		**RE**	P	*NW*	NH	52754		57754
158 755		**RE**	P	*NW*	NH	52755		57755
158 756		**RE**	P	*NW*	NH	52756		57756
158 757		**RE**	P	*NW*	NH	52757		57757
158 758		**RE**	P	*NW*	NH	52758		57758
158 759		**RE**	P	*NW*	NH	52759		57759
158 760		**RE**	P	*NE*	NL	52760		57760
158 761		**RE**	P	*NE*	NL	52761		57761
158 762		**RE**	P	*NE*	NL	52762		57762
158 763		**RE**	P	*NE*	NL	52763		57763
158 764		**RE**	P	*NE*	NL	52764		57764
158 765		**RE**	P	*NE*	NL	52765		57765
158 766		**RE**	P	*NE*	NL	52766		57766
158 767		**RE**	P	*NE*	NL	52767		57767
158 768		**RE**	P	*NE*	NL	52768		57768
158 769		**RE**	P	*NE*	NL	52769		57769
158 770		**RE**	P	*NE*	NL	52770		57770
158 771		**RE**	P	*NE*	HT	52771		57771
158 772		**RE**	P	*NE*	NL	52772		57772
158 773		**RE**	P	*NE*	NL	52773		57773
158 774		**RE**	P	*NE*	HT	52774		57774
158 775		**RE**	P	*NE*	HT	52775		57775
158 776		**RE**	P	*NE*	HT	52776		57776
158 777		**RE**	P	*NE*	HT	52777		57777
158 778		**RE**	P	*NE*	HT	52778		57778
158 779		**RE**	P	*NE*	HT	52779		57779
158 780	r	**RE**	A	*CT*	NC	52780		57780
158 781	r	**RE**	P	*NE*	HT	52781		57781
158 782	r	**RE**	A	*CT*	NC	52782		57782
158 783	r	**RE**	A	*CT*	NC	52783		57783
158 784	r	**RE**	A	*CT*	NC	52784		57784
158 785	r	**RE**	A	*CT*	NC	52785		57785
158 786	r	**RE**	A	*CT*	NC	52786		57786
158 787	r	**RE**	A	*CT*	NC	52787		57787
158 788	r	**RE**	A	*CT*	NC	52788		57788
158 789	r	**RE**	A	*CT*	NC	52789		57789
158 790	r	**RE**	A	*CT*	NC	52790		57790
158 791	r	**RE**	A	*CT*	NC	52791		57791
158 792	r	**RE**	A	*CT*	NC	52792		57792
158 793	r	**RE**	A	*CT*	NC	52793		57793
158 794	r	**RE**	A	*CT*	NC	52794		57794
158 795	r	**RE**	A	*CT*	NC	52795		57795
158 706	r	**RE**	A	*CT*	NC	52796		57796
158 797	r	**RE**	A	*CT*	NC	52797		57797
158 798		**RE**	P	*NE*	HT	52798	58715	57798

158 799		**RE**	P	*NE*	HT	52799	58716	57799
158 800		**RE**	P	*NE*	HT	52800	58717	57800
158 801		**RE**	P	*NE*	HT	52801	58701	57801
158 802		**RE**	P	*NE*	HT	52802	58702	57802
158 803		**RE**	P	*NE*	HT	52803	58703	57803
158 804		**RE**	P	*NE*	HT	52804	58704	57804
158 805		**RE**	P	*NE*	HT	52805	58705	57805
158 806		**RE**	P	*NE*	HT	52806	58706	57806
158 807		**RE**	P	*NE*	HT	52807	58707	57807
158 808		**RE**	P	*NE*	HT	52808	58708	57808
158 809		**RE**	P	*NE*	HT	52809	58709	57809
158 810		**RE**	P	*NE*	HT	52810	58710	57810
158 811		**RE**	P	*NE*	HT	52811	58711	57811
158 812		**RE**	P	*NE*	HT	52812	58712	57812
158 813		**RE**	P	*NE*	HT	52813	58713	57813
158 814		**RE**	P	*NE*	HT	52814	58714	57814
158 815	†	**RE**	A	*WW*	CF	52815		57815
158 816	†	**RE**	A	*WW*	CF	52816		57816
158 817	†	**RE**	A	*WW*	CF	52817		57817
158 818	†	**RE**	A	*WW*	CF	52818		57818
158 819	†	**RE**	A	*WW*	CF	52819		57819
158 820	†	**RE**	A	*WW*	CF	52820		57820
158 821	†	**RE**	A	*WW*	CF	52821		57821
158 822	†	**RE**	A	*WW*	CF	52822		57822
158 823	†	**RE**	A	*WW*	CF	52823		57823
158 824	†	**RE**	A	*WW*	CF	52824		57824
158 825	†	**RE**	A	*WW*	CF	52825		57825
158 826	†	**RE**	A	*WW*	CF	52826		57826
158 827	†	**RE**	A	*WW*	CF	52827		57827
158 828	†	**RE**	A	*WW*	CF	52828		57828
158 829	†	**RE**	A	*WW*	CF	52829		57829
158 830	†	**RE**	A	*WW*	CF	52830		57830
158 831	†	**RE**	A	*WW*	CF	52831		57831
158 832	†	**RE**	A	*WW*	CF	52832		57832
158 833	†	**RE**	A	*WW*	CF	52833		57833
158 834	†	**RE**	A	*WW*	CF	52834		57834
158 835	†	**RE**	A	*WW*	CF	52835		57835
158 836	†	**RE**	A	*WW*	CF	52836		57836
158 837	†	**RE**	A	*WW*	CF	52837		57837
158 838	†	**RE**	A	*WW*	CF	52838		57838
158 839	†	**RE**	A	*WW*	CF	52839		57839
158 840	†	**RE**	A	*WW*	CF	52840		57840
158 841	†	**RE**	A	*WW*	CF	52841		57841
158 842	†r	**RE**	A	*WW*	CF	52842		57842
158 843	†r	**RE**	A	*WW*	CF	52843		57843
158 844	†r	**RE**	A	*CT*	NC	52844		57844
158 845	†r	**RE**	A	*CT*	NC	52845		57845
158 846	†r	**RE**	A	*CT*	NC	52846		57846
158 847	†r	**RE**	A	*CT*	NC	52847		57847
158 848	†r	**RE**	A	*CT*	NC	52848		57848
158 849	†r	**RE**	A	*CT*	NC	52849		57849

158 850	†r	**RE**	A	*CT*	NC	52850	57850
158 851	†r	**RE**	A	*CT*	NC	52851	57851
158 852	†r	**RE**	A	*CT*	NC	52852	57852
158 853	†r	**RE**	A	*CT*	NC	52853	57853
158 854	†r	**RE**	A	*CT*	NC	52854	57854
158 855	†r	**RE**	A	*CT*	NC	52855	57855
158 856	†r	**RE**	A	*CT*	NC	52856	57856
158 857	†r	**RE**	A	*CT*	NC	52857	57857
158 858	†r	**RE**	A	*CT*	NC	52858	57858
158 859	†r	**RE**	A	*CT*	NC	52859	57859
158 860	†r	**RE**	A	*CT*	NC	52860	57860
158 861	†r	**RE**	A	*CT*	NC	52861	57861
158 862	†r	**RE**	A	*CT*	NC	52862	57862
158 863	•	**RE**	A	*WW*	CF	52863	57863
158 864	•	**RE**	A	*WW*	CF	52864	57864
158 865	•	**RE**	A	*WW*	CF	52865	57865
158 866	•	**RE**	A	*WW*	CF	52866	57866
158 867	•	**RE**	A	*WW*	CF	52867	57867
158 868	•	**RE**	A	*WW*	CF	52868	57868
158 869	•	**RE**	A	*WW*	CF	52869	57869
158 870	•	**RE**	A	*WW*	CF	52870	57870
158 871	•	**RE**	A	*WW*	CF	52871	57871
158 872	•	**RE**	A	*WW*	CF	52872	57872

CLASS 158/9   BREL EXPRESS

DMSL–DMS. Units leased by West Yorkshire PTE. Details as for Class 158/0
except for seating layout and toilets.

DMSL.. Dia. DP252. Lot No. 31051 Derby 1990–2. –/70 + wheelchair space 1TD.
Public telephone and trolley space. 38.1 t.
DMS. Dia. DP251. Lot No. 31052 Derby 1990–92. –/72 and parcels area. 37.8 t.

Note: Although these units are leased by West Yorkshire PTE, they are
managed by Porterbrook Leasing Company.

158 901	**Y**	P	*NE*	NL	52901	57901
158 902	**Y**	P	*NE*	NL	52902	57902
158 903	**Y**	P	*NE*	NL	52903	57903
158 904	**Y**	P	*NE*	NL	52904	57904
158 905	**Y**	P	*NE*	NL	52905	57905
158 906	**Y**	P	*NE*	NL	52906	57906
158 907	**Y**	P	*NE*	NL	52907	57907
158 908	**Y**	P	*NE*	NL	52908	57908
158 909	**Y**	P	*NE*	NL	52909	57909
158 910	**Y**	P	*NE*	NL	52910	57910

CLASS 159   BREL EXPRESS

DMCL–MSL–DMSL. Built as Class 158 by BREL. Converted before entering
passenger service to Class 159 by Rosyth Dockyard.

Engines: One Cummins NTA855R of 300 kW (400 h.p.) per car.
Transmission: Hydraulic. Voith T211r with Gmeinder final drive.
Bogies: One BREL P4 and one BREL T4 per car.
Gangways: Throughout.
Doors: Sliding plug.
Accommodation: 2+2 facing/unidirectional (standard class), 2+1 facing (first class).
Dimensions: 23.21 x 2.82 m.
Maximum Speed: 90 m.p.h.

DMCL.. Dia. DP322. Lot No. 31051 Derby 1992. 24/28 1TD. 38.5 t.
MSL. Dia. DR209. Lot No. 31050 Derby 1992. 38 t. –/72 2T.
DMSL. Dia. DP260. Lot No. 31052 Derby 1992. –/72 1T and parcels area. 37.8 t.

159 001	**NW**	P	*SW*	SA	52873	58718	57873	CITY OF EXETER
159 002	**NW**	P	*SW*	SA	52874	58719	57874	CITY OF SALISBURY
159 003	**NW**	P	*SW*	SA	52875	58720	57875	TEMPLECOMBE
159 004	**NW**	P	*SW*	SA	52876	58721	57876	BASINGSTOKE AND DEANE
159 005	**NW**	P	*SW*	SA	52877	58722	57877	
159 006	**NW**	P	*SW*	SA	52878	58723	57878	
159 007	**NW**	P	*SW*	SA	52879	58724	57879	
159 008	**NW**	P	*SW*	SA	52880	58725	57880	
159 009	**NW**	P	*SW*	SA	52881	58726	57881	
159 010	**NW**	P	*SW*	SA	52882	58727	57882	
159 011	**NW**	P	*SW*	SA	52883	58728	57883	
159 012	**NW**	P	*SW*	SA	52884	58729	57884	
159 013	**NW**	P	*SW*	SA	52885	58730	57885	
159 014	**NW**	P	*SW*	SA	52886	58731	57886	
159 015	**NW**	P	*SW*	SA	52887	58732	57887	
159 016	**NW**	P	*SW*	SA	52888	58733	57888	
159 017	**NW**	P	*SW*	SA	52889	58734	57889	
159 018	**NW**	P	*SW*	SA	52890	58735	57890	
159 019	**NW**	P	*SW*	SA	52891	58736	57891	
159 020	**NW**	P	*SW*	SA	52892	58737	57892	
159 021	**NW**	P	*SW*	SA	52893	58738	57893	
159 022	**NW**	P	*SW*	SA	52894	58739	57894	

CLASS 165/0   BREL NETWORK TURBO

DMCL–DMS or DMCL–MS–DMS. Built for Chiltern Line services.

Engines: One Perkins 2006-TWH of 260 kW (350 h.p.) per car.
Transmission: Hydraulic. Voith T211r with Gmeinder final drive.
Bogies: One BREL P3 and one BREL T3 per car.
Gangways: Within unit only.
Doors: Sliding plug.
Accommodation: 2+3 facing/unidirectional (standard class), 2+2 facing (first class).
Dimensions: 23.50 x 2.85 m.
Maximum Speed: 75 m.p.h.

58801–58822. 58873–58878. DMCL. Dia. DP319. Lot No. 31087 York 1990. 16/72 1T. 37.0 t.
58823–58833. DMCL. Dia. DP320. Lot No. 31089 York 1991–1992. 24/60 1T. 37.0 t.
MS. Dia. DR208. Lot No. 31090 York 1991–1992. –/106. 37.0 t.
DMS. Dia. DP253. Lot No. 31088 York 1991–1992. –/98. 37.0 t.

165 001	**NW**	A	*TT*	RG	58801		58834
165 002	**NW**	A	*TT*	RG	58802		58835
165 003	**NW**	A	*TT*	RG	58803		58836
165 004	**NW**	A	*TT*	RG	58804		58837
165 005	**NW**	A	*TT*	RG	58805		58838
165 006	**NW**	A	*CH*	AL	58806		58839
165 007	**NW**	A	*CH*	AL	58807		58840
165 008	**NW**	A	*CH*	AL	58808		58841
165 009	**NW**	A	*CH*	AL	58809		58842
165 010	**NW**	A	*CH*	AL	58810		58843
165 011	**NW**	A	*CH*	AL	58811		58844
165 012	**NW**	A	*CH*	AL	58812		58845
165 013	**NW**	A	*CH*	AL	58813		58846
165 014	**NW**	A	*CH*	AL	58814		58847
165 015	**NW**	A	*CH*	AL	58815		58848
165 016	**NW**	A	*CH*	AL	58816		58849
165 017	**NW**	A	*CH*	AL	58817		58850
165 018	**NW**	A	*CH*	AL	58818		58851
165 019	**NW**	A	*CH*	AL	58819		58852
165 020	**NW**	A	*CH*	AL	58820		58853
165 021	**NW**	A	*CH*	AL	58821		58854
165 022	**NW**	A	*CH*	AL	58822		58855
165 023	**NW**	A	*CH*	AL	58873		58867
165 024	**NW**	A	*CH*	AL	58874		58868
165 025	**NW**	A	*CH*	AL	58875		58869
165 026	**NW**	A	*CH*	AL	58876		58870
165 027	**NW**	A	*CH*	AL	58877		58871
165 028	**NW**	A	*CH*	AL	58878		58872
165 029	**NW**	A	*CH*	AL	58823	55404	58856
165 030	**NW**	A	*CH*	AL	58824	55405	58857
165 031	**NW**	A	*CH*	AL	58825	55406	58858
165 032	**NW**	A	*CH*	AL	58826	55407	58859
165 033	**NW**	A	*CH*	AL	58827	55408	58860
165 034	**NW**	A	*CH*	AL	58828	55409	58861
165 035	**NW**	A	*CH*	AL	58829	55410	58862
165 036	**NW**	A	*CH*	AL	58830	55411	58863
165 037	**NW**	A	*CH*	AL	58831	55412	58864
165 038	**NW**	A	*CH*	AL	58832	55413	58865
165 039	**NW**	A	*CH*	AL	58833	55414	58866

CLASS 165/1   BREL NETWORK TURBO

DMCL–DMS or DMCL–MS–DMS. Built for Thames Trains services.

Engines: One Perkins 2006-TWH of 260 kW (350 h.p.) per car.
Bogies: One BREL P3 and one BREL T3 per car.
Transmission: Hydraulic. Voith T211r with Gmeinder final drive.
Gangways: Within unit only.
Doors: Sliding plug.
Accommodation: 2+3 facing/unidirectional (standard class), 2+2 facing (first class).
Dimensions: 23.50 x 2.85 m.
Maximum Speed: 90 m.p.h.

58953–58969. DMCL. Dia. DP320. Lot No. 31098 York 1992. 24/60 1T. 37.0 t.
58879–58898. DMCL. Dia. DP319. Lot No. 31096 York 1992. 16/72 1T. 37.0 t.
MS. Dia. DR208. Lot No. 31099 York 1992. –/106. 37.0 t.
DMS. Dia. DP253. Lot No. 31097 York 1992. –/98. 37.0 t.

165 101	**NW**	A	*TT*	RG	58916	55415	58953
165 102	**NW**	A	*TT*	RG	58917	55416	58954
165 103	**NW**	A	*TT*	RG	58918	55417	58955
165 104	**NW**	A	*TT*	RG	58919	55418	58956
165 105	**NW**	A	*TT*	RG	58920	55419	58957
165 106	**NW**	A	*TT*	RG	58921	55420	58958
165 107	**NW**	A	*TT*	RG	58922	55421	58959
165 108	**NW**	A	*TT*	RG	58923	55422	58960
165 109	**NW**	A	*TT*	RG	58924	55423	58961
165 110	**NW**	A	*TT*	RG	58925	55424	58962
165 111	**NW**	A	*TT*	RG	58926	55425	58963
165 112	**NW**	A	*TT*	RG	58927	55426	58964
165 113	**NW**	A	*TT*	RG	58928	55427	58965
165 114	**NW**	A	*TT*	RG	58929	55428	58966
165 115	**NW**	A	*TT*	RG	58930	55429	58967
165 116	**NW**	A	*TT*	RG	58931	55430	58968
165 117	**NW**	A	*TT*	RG	58932	55431	58969
165 118	**NW**	A	*TT*	RG	58879		58933
165 119	**NW**	A	*TT*	RG	58880		58934
165 120	**NW**	A	*TT*	RG	58881		58935
165 121	**NW**	A	*TT*	RG	58882		58936
165 122	**NW**	A	*TT*	RG	58883		58937
165 123	**NW**	A	*TT*	RG	58884		58938
165 124	**NW**	A	*TT*	RG	58885		58939
165 125	**NW**	A	*TT*	RG	58886		58940
165 126	**NW**	A	*TT*	RG	58887		58941
165 127	**NW**	A	*TT*	RG	58888		58942
165 128	**NW**	A	*TT*	RG	58889		58943
165 129	**NW**	A	*TT*	RG	58890		58944
165 130	**NW**	A	*TT*	RG	58891		58945
165 131	**NW**	A	*TT*	RG	58892		58946
165 132	**NW**	A	*TT*	RG	58893		58947
165 133	**NW**	A	*TT*	RG	58894		58948
165 134	**NW**	A	*TT*	RG	58895		58949
165 135	**NW**	A	*TT*	RG	58896		58950
165 136	**NW**	A	*TT*	RG	58897		58951
165 137	**NW**	A	*TT*	RG	58898		58952

CLASS 166   ABB NETWORK EXPRESS TURBO

DMCL (A)–MS–DMCL (B). Built for Paddington–Oxford/Newbury services. Air conditioned.

Engines: One Perkins 2006-TWH of 260 kW (350 h.p.) per car.
Bogies: One BREL P3 and one BREL T3 per car.
Transmission: Hydraulic. Voith T211r with Gmeinder final drive.
Gangways: Within unit only.
Doors: Sliding plug.
Accommodation: 2+3 facing/unidirectional (standard class) with 20 standard class seats in 2+2 format in DMCL(B), 2+2 facing (first class).
Dimensions: 22.91 x 2.81 m (DMCL), 22.72 x 2.81 m (MS).
Maximum Speed: 90 m.p.h.

DMCL (A). Dia. DP321. Lot No. 31116 York 1992–3. 16/75 1T. 40.62 t.
MS. Dia. DR209. Lot No. 31117 York 1992–3. –/96. 38.04 t.
DMCL (B). Dia. DP321. Lot No. 31116 York 1992–3. 16/72 1T. 40.64 t.

166 201	**NW**	A	*TT*	RG	58101	58601	58122
166 202	**NW**	A	*TT*	RG	58102	58602	58123
166 203	**NW**	A	*TT*	RG	58103	58603	58124
166 204	**NW**	A	*TT*	RG	58104	58604	58125
166 205	**NW**	A	*TT*	RG	58105	58605	58126
166 206	**NW**	A	*TT*	RG	58106	58606	58127
166 207	**NW**	A	*TT*	RG	58107	58607	58128
166 208	**NW**	A	*TT*	RG	58108	58608	58129
166 209	**NW**	A	*TT*	RG	58109	58609	58130
166 210	**NW**	A	*TT*	RG	58110	58610	58131
166 211	**NW**	A	*TT*	RG	58111	58611	58132
166 212	**NW**	A	*TT*	RG	58112	58612	58133
166 213	**NW**	A	*TT*	RG	58113	58613	58134
166 214	**NW**	A	*TT*	RG	58114	58614	58135
166 215	**NW**	A	*TT*	RG	58115	58615	58136
166 216	**NW**	A	*TT*	RG	58116	58616	58137
166 217	**NW**	A	*TT*	RG	58117	58617	58138
166 218	**NW**	A	*TT*	RG	58118	58618	58139
166 219	**NW**	A	*TT*	RG	58119	58619	58140
166 220	**NW**	A	*TT*	RG	58120	58620	58141
166 221	**NW**	A	*TT*	RG	58121	58621	58142

CLASS 168   ADTRANZ TURBOSTAR

DMSL (A)–MSL–MS–DMSL (B). New units under construction for Chiltern Railways. Aluminium bodies. Air conditioned.
Engines: One MTU 6R183TD13H of 315 kW (422 h.p.) at 1900 r.p.m. per car.
Bogies: One ADtranz P3–23 and one BREL T3–23 per car.
Transmission: Hydraulic. Voith T211rzze to ZF final drive.
Gangways: Within unit only.
Doors: Swing plug.

Accommodation: 2+2 facing/unidirectional.
Dimensions: 23.00 x 2.7 m (DMSL), 22.8 x 2.7 m (MS).
Maximum Speed: 100 m.p.h.

DMSL (A). Dia. DP2 . ADtranz Derby 1998. –/56 + 4 tip-up 1TD.
MSL. Dia. DR2 . ADtranz Derby 1998. –/72 1T.
MS. Dia. DR2 . ADtranz Derby 1998. –/76 1T.
DMSL (B). Dia. DP2 . ADtranz Derby 1998. –/72 1T. Catering point.

168 001	**CI**	P	*CH*	58151	58651	58451	58251
168 002	**CI**	P	*CH*	58152	58652	58452	58252
168 003	**CI**	P	*CH*	58153	58653	58453	58253
168 004	**CI**	P	*CH*	58154	58654	58454	58254
168 005	**CI**	P	*CH*	58155	58655	58455	58255

CLASS 170   ADTRANZ TURBOSTAR

Various formations. New units under construction. Aluminium bodies. Air
conditioned.

Engines: One MTU 6R183TD13H of 315 kW (422 h.p.) at 1900 r.p.m. per car.
Bogies: One ADtranz P3–23 and one BREL T3–23 per car.
Transmission: Hydraulic. Voith T211rzze to ZF final drive.
Gangways: Within unit only.
Doors: Swing plug.
Accommodation: 2+2 facing/unidirectional.
Dimensions: 23.00 x 2.7 m (DMSL), 22.8 x 2.7 m (MS).
Maximum Speed: 100 m.p.h.

Class 170/1. DMCL(A)–DMCL(B). Vehicles for Midland Main Line.

DMCL (A). Dia. DP3 . ADtranz Derby 1998. 12/45 1TD.
DMCL (B). Dia. DP3 . ADtranz Derby 1998. 12/52 1T. Catering point.

170 101	**MM**	P	*ML*	50101	79101
170 102	**MM**	P	*ML*	50102	79102
170 103	**MM**	P	*ML*	50103	79103
170 104	**MM**	P	*ML*	50104	79104
170 105	**MM**	P	*ML*	50105	79105
170 106	**MM**	P	*ML*	50106	79106
170 107	**MM**	P	*ML*	50107	79107
170 108	**MM**	P	*ML*	50108	79108
170 109	**MM**	P	*ML*	50109	79109
170 110	**MM**	P	*ML*	50110	79110
170 111	**MM**	P	*ML*	50111	79111
170 112	**MM**	P	*ML*	50112	79112
170 113	**MM**	P	*ML*	50113	79113
170 114	**MM**	P	*ML*	50114	79114
170 115	**MM**	P	*ML*	50115	79115
170 116	**MM**	P	*ML*	50116	79116
170 117	**MM**	P	*ML*	50117	79117

Class 170/2. DMCL-MS-DMSL. Vehicles for Anglia Railways.

DMCL. Dia. DP3 . ADtranz Derby 1998. 30/5 1TD.
MSL. Dia. DR2 . ADtranz Derby 1998. -/65 1T. Catering point.
DMSL. Dia. DP2 . ADtranz Derby 1998. -/70 1T.

170 201	P	*AR*	50201	56201	79201
170 202	P	*AR*	50202	56202	79202
170 203	P	*AR*	50203	56203	79203
170 204	P	*AR*	50204	56204	79204
170 205	P	*AR*	50205	56205	79205
170 206	P	*AR*	50206	56206	79206
170 207	P	*AR*	50207	56207	79207
170 208	P	*AR*	50208	56208	79208

Class 170/3. Vehicles for Porterbrook. Details to be specified by customers.

170 301	P	*SCOTRAIL*	50301	56301	79301
170 302	P		50302	56302	79302
170 303	P		50303	56303	79303
170 304	P		50304	56304	79304
170 305	P		50305	56305	79305
170 306	P		50306	56306	79306
170 307	P		50307	56307	79307
170 308	P		50308	56308	79308
170 309	P		50309	56309	79309
170 310	P		50310		79310
170 311	P		50311		79311
170 312	P		50312		79312
170 313	P		50313		79313
170 314	P		50314		79314
170 315	P		50315		79315
170 316	P		50316		79316
170 317	P		50317		79317
170 318	P		50318		79318
170 319	P		50319		79319
170 320	P		50320		79320
170 321	P		50321		79321
170 322	P		50322		79322

Class 170/4. DMCL-MS-DMSL. Vehicles for a TOC (not yet confirmed).

170 401	P	*CENTRAL*	50401	56401	79401
170 402	P		50402	56402	79402
170 403	P		50403	56403	79403
170 404	P		50404	56404	79404
170 405	P		50405	56405	79405
170 406	P		50406	56406	79406
170 407	P		50407	56407	79407
170 408	P		50408	56408	79408
170 409	P		50409	56409	79409

VALLEY LINES - THE PEOPLE'S RAILWAY

by John Davies & Rhodri Clark

Foreword by the Rt. Hon. Neil Kinnock, European Union Commissioner for Transport

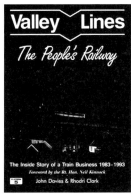

As Britsh Rail's South Wales Passenger Manager between 1983 and 1993, John Davies presided over a most turbulent era in local railway management. In his own words and those of award-winning transport journalist Rhodri Clark, this book reveals the inside story of John's efforts to revitalise train services in the Cardiff Valleys. In the face of substantial opposition from all sides, he transformed the Valley Lines operation from a railway in decline, to a successful well used network. Using unique first-hand knowledge and personal experience, Rhodri Clark and John Davies have produced a remarkable and informative work – essential reading for anyone interested or involved in railway operations. A4 size. 96 pages. Softback. Illustrated in colour and black & white. £9.95.

Available from the Platform 5 Mail Order Department. To place an order, please follow the instructions on page 383 of this book.

3.3. DIESEL ELECTRIC MULTIPLE UNITS

All ex BR Southern Region diesel-electric multiple unit power cars have above-floor-mounted engines and all vehicles are equipped with buckeye couplings and were built at Eastleigh with frames laid at Ashford.

CLASS 201/202   PRESERVED 'HASTINGS' UNIT

DMBSO–3TSOL–DMBSO.
Preserved unit made up from 2 Class 201 short-frame cars and 2 Class 202 long-frame cars. The 'Hastings' units were made with narrow body-profiles for use on the section between Tonbridge and Battle which had tunnels of restricted loading gauge. These tunnels were converted to single track operation in the 1980s thus allowing standard loading gauge stock to be used. The set also contains a Class 411 EMU trailer (not Hastings line gauge).

Engine: English Electric 4SRKT engines of 370 kW (500 h.p.).
Transmission: Two EE 507 traction motors on the inner bogie.
Gangways: Within unit only.
Dimensions: 17.68 x 2.50 m (60000/60501), 19.66 x 2.50 m. (60118/60529). 19.75 x 2.82 m (70262).
Maximum Speed: 75 m.p.h.

60000. DMBSO. Dia DB203. Lot No. 30329 1957. –/22. 54 t.
60118. DMBSO. Dia DB203. Lot No. 30395 1957. –/30. 55 t. Renumbered from 60018.
60501. TSOL. Dia DB204. Lot No. 30331 1957. –/52 2T. 29 t.
60529. TSOL. Dia DH203. Lot No. 30397 1957. –/60 2T. 30 t.
70262. TSOL (ex Class 411/5 EMU). Dia. EH282. Lot No. 30455 1958–9. –/64 2T. 33.78 t.

201 001   **SG**   HD   *SS*   SE   60000 60501 70262 60529 60118

60000 is named 'HASTINGS'.

CLASS 205/0   3H

DMBSO–TSO–DTCsoL or DMBSO–DTCsoL.

Engine: English Electric 4SRKT engines of 450 kW (600 h.p.).
Transmission: Two EE 507 traction motors on the inner bogie.
Gangways: Non-gangwayed.
Dimensions: 20.28 x 2.82 m.
Maximum Speed: 75 m.p.h.

60108–117/154. DMBSO. Dia DB203. Lot No. 30332 1957. –/52. 56 t.
60122–124. DMBSO. Dia DB203. Lot No. 30540 1958–59. –/52. 56 t.
60146–151. DMBSO. Dia DB204. Lot No. 30671 1960–62. –/42. 56 t.
60650–670. TSO. Dia DH203. Lot No. 30542 1958–59. –/104. 30 t.
60673–678. TSO. Dia DH203. Lot No. 30672 1960–62. –/104. 30 t.

60800–811. DTCsoL. Dia DE302. Lot No. 30333 1956–57. 19/50 2T. 32 t.
60822–824. DTCsoL. Dia DE302. Lot No. 30541 1958–59. 19/50 2T. 32 t.
60827–832. DTCsoL. Dia DE303. Lot No. 30673 1960–62. 13/62 2T. 32 t.

§ One compartment of DTCsoL converted to luggage compartment. 13/50 2T.
Dia. DE301.

Note: 60154 was renumbered from 60100.

205 001	§	**N**	P	*SC*	SU	60154	60650	60800
205 009		**N**	P	*SC*	SU	60108	60658	60808
205 012		**N**	P	*SC*	SU	60111	60661	60811
205 018		**N**	P	*SC*	SU	60117	60674	60828
205 023		**N**	P		ZG	60122		60822
205 024	§	**N**	P	*SC*	SU	60123	60669	60823
205 025	§	**N**	P	*SC*	SU	60124	60670	60824
205 028		**CX**	P	*SC*	SU	60146	60673	60827
205 032		**N**	P	*SC*	SU	60150	60677	60831
205 033		**CX**	P	*SC*	SU	60151	60678	60832
Spare		**N**	P		SE		60664	
Spare		**N**	P		SE		60665	
Spare		**N**	P		SE		60668	

CLASS 205/1   3H

DMBSO–TSOL–DTSOL. Refurbished 1980. Fluorescent lighting. PA.

Engine: English Electric 4SRKT engines of 450 kW (600 h.p.).
Transmission: Two EE 507 traction motors on the inner bogie.
Gangways: Within unit only.
Dimensions: 20.28 x 2.82 m.
Maximum Speed: 75 m.p.h.

DMBSO. Dia DB203. Lot No. 30332 1957. –/39. 57 t.
TSOL (ex Class 411/5 EMU). Dia. EH282. Converted from loco-hauled TSO 4059 Lot No. 30149 Ashford/Swindon 1955–7. –/64 2T. 33.78 t.
DTSOL. Dia DE204. Lot No. 30333 1957. –/76 2T. 32 t.

205 205		**N**	P	*SC*	SU	60110	71634	60810

CLASS 207/0   2D

DMBSO–DTSO (formerly DMBSO–TCsoL–DTSO).
These units were built for the Oxted line and therefore referred to as 'Oxted' units. They were made with a narrower body-profile which also allowed them to be used through the restricted loading-gauge Somerhill Tunnel between Tonbridge and Grove Junction (Tunbridge Wells). This tunnel was converted to single track operation in the 1980s thus allowing standard loading gauge stock to be used.

Engine: English Electric 4SRKT engines of 450 kW (600 h.p.).
Transmission: Two EE 507 traction motors on the inner bogie.
Gangways: Non-gangwayed.
Dimensions: 20.34 x 2.74 m. (DMBSO), 20.32 x 2.74 m. (DTSO), 20.34 x 2.74 m. (TCsoL).
Maximum Speed: 75 m.p.h.

DMBSO. Dia DB205. Lot No. 30625 1962. –/42. 56 t.
TCsoL. Dia DH301. Lot No. 30626 1962. 24/42 1T. 31 t.
DTSO. Dia DE201. Lot No. 30627 1962. –/76. 32 t.

207 017	**N**	P	*SC*	SU	60142		60916
Spare	**N**	P		SE	60135	60616	
Spare	**N**	P		ZG	60138		

CLASS 207/1   3D

DMBSO–TSOL–DTSO.
Gangwayed sets with a Class 411 EMU trailer in the centre.

Engine: English Electric 4SRKT engines of 450 kW (600 h.p.).
Transmission: Two EE 507 traction motors on the inner bogie.
Gangways: Fitted with gangways within unit.
Dimensions: 20.34 x 2.74 m. (DMBSO), 20.32 x 2.74 m. (DTSO).
Maximum Speed: 75 m.p.h.

DMBSO. Dia DB205. Lot No. 30625 1962. –/40. 56 t.
70286. TSOL (ex Class 411/5 EMU). Dia. EH282. Lot No. 30455 1958–9. –/64 2T. 33.78 t.
70547/9. TSOL (ex Class 411/5 EMU). Dia. EH282. Lot No. 30620 1960–61 –/64 2T. 33.78 t.
DTSO. Dia DE201. Lot No. 30627 1962. –/75. 32 t.

207 201	**N**	P	*SC*	SU	60129	70286	60903	Ashford Fayre
207 202	**N**	P	*SC*	SU	60130	70549	60904	Brighton Royal Pavilion
207 203	**N**	P	*SC*	SU	60127	70547	60901	

3.4. SERVICE DMUs

This section contains service vehicles, i.e. vehicles not used for the carrying of passengers which are numbered or renumbered in the special service stock number series or in the internal user series. The last capital stock number carried is shown in parentheses.

DERBY LIGHTWEIGHT   LABORATORY COACH

Converted from a Derby Lightweight single unit. DMBS. Laboratory Coach 19.

Engines: Two BUT of 112 kW (150 hp).
Transmission: Mechanical. Cardan shaft and freewheel to a four-speed epicyclic gearbox with a further cardan shaft to the final drive, each engine driving the inner axle of one bogie.
Gangways: Non gangwayed single car with cabs at each end.
Bogies:
Dimensions: 17.53 x 2.82 m.

DMBS. Lot No. 30380 Derby 1957.

–	**G**	SO	*RS*	ZA	975010	(79900)	Iris

CLASS 101   SANDITE UNITS

Converted 1993 from Class 101. DMBS–DMCL or DMBS–DMBS. For details of Class 101 see page 182.

51427. DMBS. Lot No. 30500 Metro-Cammell 1959. 32.5 t.
53193. DMCL. Lot No. 30256 Metro-Cammell 1957. 32.5 t.
53200–53208. DMBS. Lot No. 30259 Metro-Cammell 1957. 32.5 t.
53231. DMBS. Lot No. 30261 Metro-Cammell 1957. 32.5 t.
53291. DMBS. Lot No. 30270 Metro-Cammell 1957. 32.5 t.
53308. DMBS. Lot No. 30275 Metro-Cammell 1959. 32.5 t.
53321–53331. DMCL. Lot No. 30276 Metro-Cammell 1958. 32.5 t.
54342. DTCL. Lot No. 30468 Metro-Cammell 1959. 25.5 t.

960 991	**N**	RT	*SA*	LO	977895	(53308)	977896	(53331)
960 992		RT	*SA*	LO	977897	(53203)	977898	(53193)
960 993		RT	*SA*	LO	977899	(51427)	977900	(53321)
960 994		RT	*SA*	LO	977901	(53200)	977902	(53231)
960 995		RT	*SA*	LO	977903	(53208)	977904	(53291)

CLASS 101   TRACTOR UNIT

Converted 1986 from Class 101. DMBS–DMBS. For details of Class 101 see page 182.

51433. DMBS. Lot No. 30500 Metro-Cammell 1959. 32.5 t.
53167. DMBS. Lot No. 30254 Metro-Cammell 1957. 32.5 t.

–	**SO**	SO	*TE*	ZA	977391	(51433)	977392	(53167)

CLASS 101   LABORATORY UNIT

Converted 1990 from Class 101. DMBS–DMCL. For details of Class 101 see page 182. Laboratory Coach 19 Iris 2.

53222. DMBS. Lot No. 30261 Metro-Cammell 1957. 32.5 t.
53338. DMCL. Lot No. 30276 Metro-Cammell 1958. 32.5 t.

Non-standard livery: Grey.

–	**0**	SO	*RS*	ZA	977693	(53222)	977694	(53338)

CLASS 101   STORES VAN

Converted 1990 from Class 101. DTCL. For details of Class 101 see page 182.

DTCL. Lot No. 30648 Metro-Cammell 1959. 25.5 t.

		ML	*ST*	NL	042222	(54342)

CLASS 114   ROUTE LEARNING UNIT

Converted 1992 from Class 114. DMPMV–DTPMV.

Engines: Two Leyland TL11 of 152 kW (205 hp) per power car.
Transmission: Mechanical. Cardan shaft and freewheel to a four-speed epicyclic gearbox with a further cardan shaft to the final drive, each engine driving the inner axle of one bogie.
Gangways: Midland scissors type. Within unit only.
Bogies: DD9 (motor) and DT9c (trailer).
Dimensions: 20.45 x 2.82 m.

55929. DMPMV. Lot No. 30209 Derby 1958. 29.0 t.
54904. DTPMV. Lot No. 30210 Derby 1958. 29.2 t.

Non-standard livery: Grey, red and yellow.

–	**0**	E	*EW*	CF	977775	(55929)	977776	(54904)

CLASS 121/122   SANDITE UNITS

Converted 1970†, 1991‡, 1993 from Class 121, 122†. DMBS. For details of Class 121/122 see page 185.

55019. DMBS. Lot No. 30419 Gloucester 1958. 36.5 t.
55020–55028. DMBS. Lot No. 30518 Pressed Steel 1960. 38.0 t.

960 002	‡	**N**	RT	*SA*	RG	977722	(55020)
960 010		**N**	RT	*SA*	RG	977858	(55024)
960 011		**N**	RT	*SA*	LO	977859	(55025)
960 012		**N**	RT	*SA*	RG	977873	(55022)
960 013		**N**	RT	*SA*	RG	977866	(55030)
960 014		**N**	RT	*SA*	RG	977860	(55028)
960 015	l	**RT**	RT	*SA*	BY	975042	(55019)
960 021	‡	**N**	RT	*SA*	BY	977723	(55021)

CLASS 150   TRACK RECORDING UNIT

Built new York 1987. Class 150 derivative.

Non-standard livery: Blue and white with a red stripe below windows.

–	**0**	SO	*TE*	ZA	999600	999601

CLASS 205   TRACTOR UNIT

Converted 1994 from Class 205. DMBSO–DMBSO. For details of Class 205 see page 228.

DMBSO. Lot No. 30671 1960–62. 56 t.

930 301		RT	*SA*	SU	977939	(60145)	977940	(60149)

CLASS 205   SANDITE COACH

Converted 1994 from Class 205. TSO. For details of Class 205 see page 228. Works with 930 301.

TSO. Lot No. 30542 1958–59. 30 t.

–		RT	*SA*	SU	977870	(60660)

3.5. DMUs AWAITING DISPOSAL

The following withdrawn DMUs are awaiting disposal with the last known storage location shown.

Ex-CAPITAL STOCK

51359	BY		55402	ZA
51361	Kineton		55403	ZA
51368	Kineton		55709	NH
54350	Crewe Brook Sidings		59228	Crewe Brook Sidings
55202	ZA		59518	OM
55203	ZA		60200	ZG
55302	ZA		60201	ZG
55303	ZA			

Ex-SERVICE STOCK

975023	(55001)	LO
975025	(60755)	SL
977191	(56106)	Crewe Brook Sidings
977554	(54182)	Buxton LIP
977696	(60522)	EH
977697	(60523)	ZG
977698	(60152)	ZG
977699	(60153)	ZG

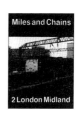

ELECTRIC MULTIPLE UNITS - INTRODUCTION

Electric Multiple Unit operation on Britain's main line railways has increased enormously since the end of the steam era, with most electrification schemes being carried out at 25 kV a.c. using overhead conductor wires. The notable exceptions to this are the lines of the former Southern Railway, where the existing 660-750 V d.c. third rail system has been extended, with the voltage increased to 850 V in certain areas. The other exception is the Merseyrail network, which has also been extended using the third rail system.

NUMBERING

BR design electric multiple unit vehicles are numbered in the series 61000-78999. Isle of Wight vehicles are numbered in a separate series. In this book, stock is generally listed in order of the unit or set number. The unit or set number is stated first, followed by any notes applicable to the particular set. These are followed by codes for livery, owner, operator and depot respectively. Finally the numbers of the individual cars in the set are given, in order. Please note that reformations can and do occur. For off-loan vehicles, the last storage location is given when known.

DESIGN CONSIDERATIONS

Unless stated otherwise, all multiple unit vehicles are of BR design, or designed by contractors for BR and have buckeye couplings and tread brakes. Seating is 3+2 in standard class open vehicles, 2+2 in first class open vehicles, 8 to a corridor standard class compartment and 6 to a corridor first class compartment. In express stock, open standards have 2+2 seating and open firsts have 2+1 seating.

VEHICLE CODES

The codes used by the BR Operating Department to describe the various different types of electric multiple unit vehicles and quoted in the class headings are as follows:

M	Motor
DM	Driving Motor
BDM	Battery Driving Motor
T	Trailer
DT	Driving Trailer
BDT	Battery Driving Trailer
B	Brake, i.e. vehicle with luggage space and guards compartment
F	First

S	Standard
C	Composite
RB	Buffet Car
RSM	Buffet Standard (Modular)
PMV	Parcels and Mails Van
H	handbrake fitted
LV	Luggage Van
K	Side corridor with lavatory
L	Open or semi-open Vehicle with lavatory
O	Open vehicle
so	Semi-open vehicle

The letters (A) and (B) may be added to the above codes to differentiate between two cars of the same operating type which have differences between them. The letter (T) denotes space for a catering trolley. Note that a consistent system is used, rather than the official operator codes which are sometimes inconsistent.

Notes:

(1) Compartment Stock (non-corridor) has no suffix.

(2) Semi-open composites generally have the first class accommodation in compartments and the standard class in open saloons.

(3) Unless stated otherwise, it is assumed that motor vehicles are fitted with pantographs. If the pantograph is on a trailer, then the trailer has the prefix 'P', e.g. PTSO - Pantograph trailer open standard.

A composite is a vehicle containing both First and Standard class accommodation.

A brake vehicle is a vehicle containing seperate specific accommodation for the guard (as opposed to the use of spare driving cabs on more recently-built units).

DIAGRAMS AND DESIGN CODES

For each type of vehicle, the official design code consists of a seven character code of two letters, four numbers and another letter, e.g. EC2040B. The first five characters of this are the diagram code and are given in the class heading or sub heading. These are explained as follows:

1st Letter

This is always 'E' for an electric multiple unit vehicle.

2nd Letter

as follows for various vehicle types:

A	Driving motor passenger vehicles.
B	Driving motor passenger vehicles with a brake compartment.
C	Non-Driving motor passenger vehicles.
D	Non-Driving trailer passenger vehicles with a brake compartment.

E	Driving Trailer passenger vehicles.
F	Battery Driving Trailer passenger vehicles.
G	Driving Trailer passenger vehicles with a brake compartment.
H	Trailer passenger vehicles.
I	Battery Driving Motor passenger vehicles.
J	Trailer passenger vehicles with a brake compartment.
N	Trailer passenger vehicles with a buffet compartment.
O	Battery Driving Trailer passenger vehicles with a brake compartment.
P	Trailer passenger vehicles with a handbrake.
X	Driving Motor Luggage Vans.

1st Figure

1	First class accommodation.
2	Standard class accommodation (including declassified seats).
3	Composite accommodation.
5	No passenger accommodation.

ACCOMMODATION

This information is given in class headings and sub headings in the form F/S nT, where F & S denote the number of first class and standard class seats followed by n which denotes the number of toilets. (e.g. 12/54 1T denotes 12 first class seats, 54 standard class seats and one toilet). In declassified vehicles, the capacity is still shown in terms of first and standard class seats to differentiate between the two physically different seat types available, although all seats are officially standard class in such instances. TD denotes a toilet suitable for a disabled person.

BUILD DETAILS

LOT NUMBERS

Each batch of vehicles is allocated a Lot (or batch) number when ordered and these are quoted in class headings and sub headings.

BUILDERS

These are shown in class headings. A list of builders is given on page 361.

LAYOUT

The layout in this section is as follows:

(1) Unit number.	(5) Operation code.
(2) Notes (if any).	(6) Depot code.
(3) Livery code.	(7) Individual car numbers.
(4) Owner code.	

Thus an example of the layout is as follows:

No.	Notes	Livery	Owner	Operation	Depot	Car 1	Car 2	Car 3	Car 4
317 398		**N**	A	WN	HE	77027	62687	71604	77075

For off-loan vehicles, the last storage location is given when known.

4.1. 25 kV a.c. OVERHEAD EMUs.

Note: All units are 25 kV overhead only except where stated otherwise.

CLASS 302

BDTCOL (declassified)–MBSO–TSOL–DTSO. All remaining units refurbished with new seats & fluorescent lighting.
Gangways: Within unit.
Traction Motors: Four EE536A 143.5 kW.
Dimensions: 19.50 x 2.82 m (outer cars), 19.36 x 2.82 m (inner cars).
Maximum Speed: 75 m.p.h.

75085–75205. BDTCOL. Lot No. 30436. York/Doncaster 1958–59. Dia. EF303. 24/52 1T. 39.5 t. B5 bogies.
75311–75325. BDTCOL. Lot No. 30440. York/Doncaster 1959. Dia. EF303. 24/52 1T. 39.5 t. B5 bogies.
61060–61091. MBSO. Lot No. 30434. York 1958–59. Dia. ED216. –/76. 55.3 t. Gresley Bogies.
61122–61193. MBSO. Lot No. 30438. York 1960. Dia. ED216. –/76. 55.3 t. Gresley Bogies.
70060–70091. TSOL. Lot No. 30437. York/Doncaster 1958–59. Dia. EH223. –/86 1T. 34.4 t. B4 bogies.
70122–70193. TSOL. Lot No. 30441. York 1959–61. Dia. EH223. –/86 1T. 34.4 t. B4 bogies.
75033–75079. DTSO. Lot No. 30435. York 1958–59. Dia. EE219. –/88. 33.4 t. B4 or B5 bogies.
75236–75250. DTSO. Lot No. 30439. York 1959–60. Dia. EE219. –/88. 33.4 t. B4 or B5 bogies.

302 201	N	F	LS	EM	75085	61060	70060	75033
302 203	N	F		Pig's Bay	75311	61122	70122	75236
302 204	N	F		EM	75088	61063	70063	75036
302 211	N	F		Pig's Bay	75095	61070	70070	75043
302 212	N	F		Pig's Bay	75096	61071	70071	75044
302 213	N	F		Pig's Bay	75097	61072	70072	75060
302 216	N	F	LS	EM	75100	61075	70075	75063
302 218	N	F		Pig's Bay	75191	61077	70077	75065
302 220	N	F		Pig's Bay	75193	61079	70079	75067
302 221	N	F		Pig's Bay	75194	61080	70080	75068
302 222	N	F		Pig's Bay	75195	61081	70081	75069
302 224	N	F		Pig's Bay	75197	61083	70083	75071
302 225	N	F	LS	EM	75198	61084	70084	75072
302 226	N	F		EM	75199	61085	70085	75073
302 227	N	F		EM	75325	61193	70193	75250
302 228	N	F	LS	EM	75201	61087	70087	75075
302 229	N	F		Pig's Bay	75202	61088	70088	75076
302 230	N	F	LS	EM	75205	61089	70091	75079

CLASS 303

DTSO–MBSO–BDTSO. Sliding doors.
Bogies: Gresley.
Gangways: Gangwayed within units only (non-gangwayed*).
Traction Motors: Four MV 155 kW.
Dimensions: 19.50 x 2.82 m (outer cars), 19.36 x 2.82 m (inner cars).
Maximum Speed: 75 m.p.h.

Class 303/0. Unrefurbished set*.

DTSO. Dia. EE206. –/83. 34.4 t.
MBSO. Dia. ED201. –/70. 56.4 t.
BDTSO. Dia. EF202. –/83. 38.4 t.

Non-standard Livery: Original Glasgow 'blue train' livery.
Note: 75752 carries "75758" and 75808 carries "75814"

Class 303/1. Refurbished with 2+2 seating and hopper-type window vents.

DTSO. Dia. EE241. –/56. 34.4 t.
MBSO. Dia. ED220. –/48. 56.4 t.
BDTSO. Dia. EF217. –/56. 38.4 t.

75566–75599. DTSO. Lot No. 30579 Pressed Steel 1959–60.
75747–75801. DTSO. Lot No. 30629 Pressed Steel 1960–61.
61481–61514. MBSO. Lot No. 30580 Pressed Steel 1959–60.
61813–61867. MBSO. Lot No. 30630 Pressed Steel 1960–61.
75601–75635. BDTSO. Lot No. 30581 Pressed Steel 1959–60.
75803–75857. BDTSO. Lot No. 30631 Pressed Steel 1960–61.

303 001	S	A	SR	GW	75566	61481	75601
303 003	S	A	SR	GW	75568	61483	75603
303 004	S	A	SR	GW	75569	61484	75604
303 006	S	A	SR	GW	75571	61486	75606
303 008	S	A	SR	GW	75573	61488	75608
303 009	S	A	SR	GW	75574	61489	75609
303 010	S	A	SR	GW	75575	61490	75610
303 011	S	A	SR	GW	75576	61491	75611
303 012	S	A	SR	GW	75577	61492	75612
303 013	S	A	SR	GW	75578	61493	75613
303 014	S	A	SR	GW	75579	61494	75614
303 016	S	A	SR	GW	75750	61496	75616
303 019	CC	A	SR	GW	75584	61499	75619
303 020	S	A	SR	GW	75585	61500	75620
303 021	CC	A	SR	GW	75586	61501	75621
303 023	CC	A	SR	GW	75588	61503	75623
303 024	S	A	SR	GW	75589	61504	75624
303 025	S	A	SR	GW	75590	61505	75625
303 027	S	A	SR	GW	75592	61507	75627
303 028	S	A	SR	GW	75600	61508	75635
303 032	S	A	SR	GW	75597	61512	75632
303 033	S	A	SR	GW	75595	61860	75817

303 034	S	A	SR	GW	75599	61514	75634
303 037	CC	A	SR	GW	75781	61813	75803
303 040	S	A	SR	GW	75581	61816	75806
303 043	S	A	SR	GW	75572	61819	75809
303 045	S	A	SR	GW	75755	61821	75811
303 047	S	A	SR	GW	75757	61823	75813
303 048	* 0	A	SR	GW (S)	75752	61824	75808
303 054	S	A	SR	GW	75764	61830	75820
303 055	S	A	SR	GW	75765	61831	75821
303 056	S	A	SR	GW	75766	61832	75822
303 058	S	A	SR	GW	75768	61834	75824
303 061	S	A	SR	GW	75771	61837	75827
303 065	S	A	SR	GW	75775	61841	75831
303 070	S	A	SR	GW	75780	61846	75836
303 077	S	A	SR	GW	75787	61853	75843
303 079	S	A	SR	GW	75789	61855	75845
303 080	S	A	SR	GW	75790	61856	75846
303 083	S	A	SR	GW	75793	61859	75849
303 085	S	A	SR	GW	75795	61861	75851
303 087	CC	A	SR	GW	75797	61863	75853
303 088	S	A	SR	GW	75798	61864	75854
303 089	S	A	SR	GW	75799	61865	75855
303 090	S	A	SR	GW	75800	61866	75856
303 091	S	A	SR	GW	75801	61867	75857

CLASS 305/2

BDTCOL (declassified)–MBSO–TSOL–DTSO or BDTCOL (declassified)–MBSO–
DTSO. All remaining units refurbished with fluorescent lighting and new seats.
Bogies: Gresley.
Gangways: Originally non-gangwayed, but now gangwayed within unit.
Traction Motors: Four GEC WT380 of 153 kW.
Dimensions: 19.53 x 2.82 m (outer cars), 19.36 x 2.82 m (inner cars).
Maximum Speed: 75 m.p.h.

BDTCOL. Dia. EF304. Lot No. 30566 York/Doncaster 1960. 24/52 1T. 36.5 t.
MBSO. Dia. ED216. Lot No. 30567 York/Doncaster 1960. –/76. 56.5 t.
TSOL. Dia. EH223. Lot No. 30568 York/Doncaster 1960. –/86 1T. 31.5 t.
DTSO. Dia. EE220. Lot No. 30569 York/Doncaster 1960. –/88. 32.7 t.

305 501	RR	A	SR	GW	75424	61410	70356	75443
305 502	RR	A	SR	GW	75425	61411	70357	75444
305 503	GM	A	NW	LG (U)	75426	61412		75445
305 506	GM	A	NW	LG (U)	75429	61415		75448
305 507	RR	A	NW	LG (U)	75430	61416		75449
305 508	RR	A	SR	GW	75431	61417	70363	75450
305 510	GM	A	NW	LG	75433	61419		75452
305 511	GM	A	NW	LG	75434	61420		75453
305 515	GM	A	NW	LG (U)	75438	61424		75457
305 516	GM	A	NW	LG (U)	75439	61425		75458
305 517	RR	A	SR	GW	75440	61426	70372	75459

305 518	**RR**	A		LG	75441	61427		75460
305 519	**RR**	A	*SR*	GW	75442	61428	70374	75461
Spare	**RR**	A		LG		61418		

CLASS 308

BDTCOL (declassified)–MBSO–TSOL–DTSO. Refurbished with new seats and fluorescent lighting. Originally 4-car units, but all TSOL now withdrawn.
Bogies: Gresley.
Gangways: Originally non-gangwayed, but now gangwayed within unit.
Traction Motors: Four English Electric 536A of 143.5 kW.
Dimensions: 19.36 x 2.82 m (outer cars), 19.35 x 2.82 m (inner cars).
Maximum Speed: 75 m.p.h.

75879–75886. BDTCOL. Dia. EF304. Lot No. 30652 York 1961. 24/52 1T. 36.3 t.
75897–75919. BDTCOL. Dia. EF304. Lot No. 30656 York 1961. 24/52 1T. 36.3 t.
61884–61891. MBSO. Dia. ED216. Lot No. 30653 York 1961. –/76. 55.0 t.
61893–61915. MBSO. Dia. ED216. Lot No. 30657 York 1961. –/76. 55.0 t.
75888–75895. DTSO. Dia. EE220. Lot No. 30655 York 1961. –/88. 33 t.
75930–75952. DTSO. Dia. EE220. Lot No. 30659 York 1961. –/88. 33 t.

308 134	**Y**	A	*NE*	NL	75879	61884	75888
308 136	**Y**	A	*NE*	NL	75881	61886	75890
308 137	**Y**	A	*NE*	NL	75882	61887	75891
308 138	**Y**	A	*NE*	NL	75883	61888	75892
308 141	**Y**	A	*NE*	NL	75886	61891	75895
308 143	**Y**	A	*NE*	NL	75897	61893	75930
308 144	**Y**	A	*NE*	NL	75880	61894	75931
308 145	**Y**	A	*NE*	NL	75899	61895	75932
308 147	**Y**	A	*NE*	NL	75901	61897	75934
308 152	**Y**	A	*NE*	NL	75913	61902	75939
308 153	**Y**	A	*NE*	NL	75907	61903	75940
308 154	**Y**	A	*NE*	NL	75908	61904	75941
308 155	**Y**	A	*NE*	NL	75909	61905	75942
308 157	**Y**	A	*NE*	NL	75915	61907	75944
308 158	**Y**	A	*NE*	NL	75912	61908	75945
308 159	**Y**	A	*NE*	NL	75906	61909	75946
308 161	**Y**	A	*NE*	NL	75911	61911	75948
308 162	**Y**	A	*NE*	NL	75916	61912	75949
308 163	**Y**	A	*NE*	NL	75917	61913	75950
308 164	**Y**	A	*NE*	NL	75918	61914	75951
308 165	**Y**	A	*NE*	NL	75919	61915	75952

CLASS 309/1   CLACTON EXPRESS STOCK

DMBSO(T)–TSOL–TCsoL–BDTSOL. Built 1962–3 as 2 car units. Made up to four cars by the conversion of loco-hauled stock in 1973. All now refurbished with fluorescent lighting, hopper ventilators and new seating.
Bogies: Commonwealth.
Gangways: Throughout.

Traction Motors: Four GEC of 210 kW.
Dimensions: 19.76 x 2.82 m (outer cars), 19.67 x 2.82 m (inner cars).
Maximum Speed: 100 m.p.h.

DMBSO(T). Dia. EB206. Lot No. 30684 York 1962–63. –/44. 60 t.
TSOL. Dia. EH227. Lot No. 30871 Wolverton 1973–74. –/64 2T. 35 t.
TCsoL. Dia. EH309. Lot No. 30872 Wolverton 1973–74. 24/28 1T. 36 t.
BDTSOL. Dia. EF213. Lot No. 30683 York 1960–62. –/60 1T. 40 t.

| 309 605 | N | A | | LM | 61944 | 71108 | 71113 | 75988 |
| 309 606 | N | A | | LM | 61945 | 71109 | 71112 | 75989 |

Former numbers of converted hauled stock:

71108 (26189) |71109 (26196) |71112 (16249) | 71113 (16244)

CLASS 309/2   CLACTON EXPRESS STOCK

BDTCsoL–MBSOL(T)–TSO–DTSOL. Built 1962–3. Units 309 613–309 617 formerly contained griddle cars, but these were withdrawn and their place was taken by the conversion of loco-hauled TSOs on refurbishment. All refurbished with fluorescent lighting, hopper ventilators and new seating.
Bogies: Commonwealth.
Gangways: Throughout.
Traction Motors: Four GEC of 210 kW.
Dimensions: 19.76 x 2.82 m (outer cars), 19.67 x 2.82 m (inner cars).
Maximum Speed: 100 m.p.h.

75639–44. BDTCsoL. Dia. EF301. Lot No. 30679 York 1962. 18/32 2T. 40 t.
75965. BDTCsoL. Dia. EF213. Lot No. 30675 York 1962. 18/32 2T. 40 t.
61927–31. MBSOL(T). Dia. ED209. Lot No. 30676 York 1962. –/44 2T. 58 t.
61934–38. MBSOL(T). Dia. ED209. Lot No. 30680 York 1962. –/44 2T. 58 t.
70256–59. TSO. Dia. EH229. Lot No. 30677 York 1962. –/68 35 t.
71756–60. TSO. Dia. EH228. Lot No. 31001 Wolverton 1984–87. –/68. 35 t.
75972–75. DTSOL. Dia. EF213. Lot No. 30678 York 1962. –/56 2T 37 t.
75978–82. DTSOL. Dia. EF213. Lot No. 30682 York 1962–1963. –/56 2T. 37 t.

Non-standard Livery: 309 624 is in Manchester Airport Air Express Livery (blue & white).

309 613	p	**RN**	A	NW	LG	75639	61934	71756	75978
309 616	p	**RN**	A	NW	LG	75642	61937	71759	75981
309 617	p	**RN**	A	NW	LG	75643	61938	71760	75982
309 623		**RN**	A	NW	LG	75641	61927	71758	75980
309 624		**0**	A	NW	LG	75965	61928	70256	75972
309 627		**RN**	A	NW	LG	75644	61931	70259	75975

Former numbers of converted hauled stock:

71756 (5068) | 71758 (5058) | 71759 (5062) | 71760 (5056)

CLASS 310

Disc brakes. All facelifted with new panels.
Bogies: B4.
Gangways: Within unit.
Traction Motors: Four EE546 of 201.5 kW.
Dimensions: 19.86 x 2.82 m (outer cars), 19.93 x 2.82 m (inner cars).
Maximum Speed: 75 m.p.h.

BDTSOL. Dia. EF211. Lot No. 30745 Derby 1965–67. –/80 2T. 37.3 t.
76228. BDTSOL. Formerly a DTCOL to Lot 30748. Dia. EF210. Seats –/68 2T.
76998. BDTSOL. Rebuilt from TSO 70756 to Lot 30747. Dia. EF214. Seats –/75 2T.
MBSO. Dia. ED219. Lot No. 30746 Derby 1965–67. –/68. 57.2 t.
TSO. Dia. EH232. Lot No. 30747 Derby 1965–67. –/98. 31.7 t.
DTCOL (310/0). Dia. EE306. Lot No. 30748 Derby 1965–67. 25/43 2T. 34.4 t.
DTSOL (310/1). Dia. EE237. Lot No. 30748 Derby 1965–67. –/75 2T. 34.4 t.

Class 310/0. BDTSOL–MBSO–TSO–DTCOL (declassified).

310 046	N	F	*LS*	EM	76130	62071	70731	76180
310 047	N	F	*LS*	EM	76131	62072	70732	76181
310 049	N	F	*LS*	EM	76133	62074	70734	76183
310 050	N	F	*LS*	EM	76134	62075	70735	76184
310 051	N	F	*LS*	EM	76135	62076	70736	76185
310 052	N	F	*LS*	EM	76136	62077	70737	76186
310 057	N	F	*LS*	EM	76141	62082	70742	76191
310 058	N	F	*LS*	EM	76142	62083	70743	76192
310 059	N	F	*LS*	EM	76143	62084	70744	76205
310 060	N	F	*LS*	EM	76144	62085	70745	76194
310 064	N	F	*LS*	EM	76148	62089	70749	76198
310 066	N	F	*LS*	EM	76228	62091	70751	76200
310 067	N	F	*LS*	EM	76151	62092	70752	76201
310 068	N	F	*LS*	EM	76152	62093	70753	76202
310 069	N	F	*LS*	EM	76153	62094	70754	76203
310 070	N	F	*LS*	EM	76154	62095	70755	76204
310 074	N	F	*LS*	EM	76145	62099	70759	76208
310 075	N	F	*LS*	EM	76159	62100	70760	76209
310 077	N	F	*LS*	EM	76161	62102	70762	76211
310 079	N	F	*LS*	EM	76163	62104	70764	76222
310 080	N	F	*LS*	EM	76164	62105	70765	76214
310 081	N	F	*LS*	EM	76165	62106	70766	76215
310 082	N	F	*LS*	EM	76166	62107	70767	76216
310 083	N	F	*LS*	EM	76167	62108	70768	76217
310 084	N	F	*LS*	EM	76168	62109	70769	76218
310 085	N	F	*LS*	EM	76169	62110	70770	76219
310 086	N	F	*LS*	EM	76170	62111	70771	76220
310 087	N	F	*LS*	EM	76171	62112	70772	76221
310 088	N	F	*LS*	EM	76172	62113	70773	76213
310 089	N	F	*LS*	EM	76173	62114	70774	76223
310 091	N	F	*LS*	EM	76175	62116	70776	76225

310 092		**N**	F	*LS*	EM	76176	62117	70777	76226
310 093		**N**	F	*LS*	EM	76177	62118	70778	76190
310 094		**N**	F	*LS*	EM	76998	62119	70780	76193
310 095		**N**	F	*LS*	EM	76179	62120	70779	76228

Name: Set 310 058 is named "Chafford Hundred".

Class 310/1. BDTSOL–MBSO–DTSOL (DTCOL (declassified*).

310 101		**RR**	F	*CT*	BY	76157	62098		76207
310 102		**RR**	F	*CT*	BY	76139	62080		76189
310 103		**RR**	F	*CT*	BY	76160	62101		76210
310 104		**RR**	F	*CT*	BY	76162	62103		76212
310 105		**RR**	F	*CT*	BY	76174	62115		76224
310 106		**RR**	F	*CT*	BY	76156	62097		76206
310 107		**RR**	F	*CT*	BY	76146	62087		76196
310 108		**RR**	F	*CT*	BY	76132	62073		76182
310 109		**RR**	F	*CT*	BY	76137	62078		76187
310 110		**RR**	F	*CT*	BY	76138	62079		76188
310 111		**RR**	F	*CT*	BY	76147	62088		76197
310 112	*	**RR**	F	*CT*	BY	76140	62086		76227
310 113	*	**RR**	F	*CT*	BY	76158	62090		76195

Spare TSO

70733	**RR**	F	KI		70757	**RR**	F	KI		70763	**RR**	F	KI
70747	**RR**	F	KI		70761	**RR**	F	ZN		70775	**RR**	F	ZN
70748	**RR**	F	KI										

CLASS 312

BDTSOL–MBSO–TSO–DTCOL (declassified*). Disc brakes.
Bogies: B4.
Gangways: Within unit.
Traction Motors: Four EE546 of 201.5 kW.
Dimensions: 19.86 x 2.82 m (outer cars), 19.93 x 2.82 m (inner cars).
Maximum Speed: 90 m.p.h.

Class 312/0. Standard design.

76994–97 BDTSOL. Dia. EF213. Lot No. 30891 York 1976. –/84 1T. 34.9 t.
62657–60 MBSO. Dia. ED214. Lot No. 30892 York 1976. –/68. 56 t.
71277–80 TSO. Dia. EH209. Lot No. 30893 York 1976. –/98. 30.5 t.
78045–48 DTCOL. Dia. EE305. Lot No. 30894 York 1976. 25/47 2T.
76949–74 BDTSOL. Dia. EF213. Lot No. 30863 York 1977–78. –/84 1T. 34.9 t.
62484–509 MBSO. Dia. ED212. Lot No. 30864 York 1977–78. –/68. 56 t.
71168–93 TSO. Dia. EH209. Lot No. 30865 York 1977–78. –/98. 30.5 t.
78000–25 DTCOL. Dia. EE305. Lot No. 30866 York 1977–78. 25/47 2T.

312 701	**N**	A	*GE*	IL	76949	62484	71168	78000
312 702	**N**	A	*GE*	IL	76950	62485	71169	78001
312 703	**N**	A	*GE*	IL	76951	62486	71170	78002
312 704	**GE**	A	*GE*	IL	76952	62487	71171	78003
312 705	**GE**	A	*GE*	IL	76953	62488	71172	78004

312 706	**GE**	A	*GE*	IL	76954	62489	71173	78005
312 707	**N**	A	*GE*	IL	76955	62490	71174	78006
312 708	**N**	A	*GE*	IL	76956	62491	71175	78007
312 709	**N**	A	*GE*	IL	76957	62492	71176	78008
312 710	**N**	A	*GE*	IL	76958	62493	71177	78009
312 711	**N**	A	*GE*	IL	76959	62494	71178	78010
312 712	**N**	A	*GE*	IL	76960	62495	71179	78011
312 713	**N**	A	*GE*	IL	76961	62496	71180	78012
312 714	**N**	A	*GE*	IL	76962	62497	71181	78013
312 715	**N**	A	*GE*	IL	76963	62498	71182	78014
312 716	**N**	A	*GE*	IL	76964	62499	71183	78015
312 717	**N**	A	*GE*	IL	76965	62500	71184	78016
312 718	**N**	A	*GE*	IL	76966	62501	71185	78017
312 719	**N**	A	*GE*	IL	76967	62502	71186	78018
312 720	**N**	A	*GE*	IL	76968	62503	71187	78019
312 721	**N**	A	*GE*	IL	76969	62504	71188	78020
312 722	**N**	A	*GE*	IL	76970	62505	71189	78021
312 723	**N**	A	*GE*	IL	76971	62506	71190	78022
312 724	**N**	A	*GE*	IL	76972	62507	71191	78023
312 725	*	**RR**	A	*CT*	TS (S)	76973	62509	71193	78025
312 726	*	**RR**	A	*CT*	TS (S)	76974	62508	71192	78024
312 727	*	**RR**	A	*CT*	TS (S)	76994	62657	71277	78045
312 728	*	**RR**	A	*CT*	TS (S)	76995	62658	71278	78046
312 729	*	**N**	A	*LS*	EM	76996	62659	71279	78047
312 730	*	**N**	A	*LS*	EM	76997	62660	71280	78048

Class 312/1. Can also operate on 6.25 kV a.c. overhead.

BDTSOL. Dia. EF213. Lot No. 30867 York 1975–76. –/84 2T. 34.9 t.
MBSO. Dia. ED213. Lot No. 30868 York 1975–76. –/68. 56 t.
TSO. Dia. EH209. Lot No. 30869 York 1975–76. –/98. 30.5 t.
DTCOL. Dia. EE305. Lot No. 30870 York 1975–76. 25/47 2T.

312 781	*	**N**	A	*LS*	EM	76975	62510	71194	78026
312 782	*	**N**	A	*LS*	EM	76976	62511	71195	78027
312 783	*	**N**	A	*LS*	EM	76977	62512	71196	78028
312 784	*	**N**	A	*LS*	EM	76978	62513	71197	78029
312 785	*	**N**	A	*LS*	EM	76979	62514	71198	78030
312 786	*	**N**	A	*LS*	EM	76980	62515	71199	78031
312 787	*	**N**	A	*LS*	EM	76981	62516	71200	78032
312 788	*	**N**	A	*LS*	EM	76982	62517	71201	78033
312 789	*	**N**	A	*LS*	EM	76983	62518	71202	78034
312 790	*	**N**	A	*LS*	EM	76984	62519	71203	78035
312 791	*	**N**	A	*LS*	EM	76985	62520	71204	78036
312 792	*	**N**	A	*LS*	EM	76986	62521	71205	78037
312 793	*	**N**	A	*LS*	EM	76987	62522	71206	78038
312 794	*	**N**	A	*LS*	EM	76988	62523	71207	78039
312 795	*	**N**	A	*LS*	EM	76989	62524	71208	78040
312 796	*	**N**	A	*LS*	EM	76990	62525	71209	78041
312 797	*	**N**	A	*LS*	EM	76991	62526	71210	78042
312 798	*	**N**	A	*LS*	EM	76992	62527	71211	78043
312 799	*	**N**	A	*LS*	EM	76993	62528	71212	78044

CLASS 313

DMSO–PTSO–BDMSO. Tightlock couplers. Sliding doors. Disc and rheostatic brakes.
System: 25 kV a.c. overhead/750 V d.c. third rail.
Bogies: BX1.
Gangways: Within unit. End doors.
Traction Motors: Four GEC G310AZ of 82.125 kW.
Dimensions: 19.80 x 2.82 m (outer cars), 19.92 x 2.82 m (inner cars).
Maximum Speed: 75 m.p.h.

DMSO. Dia. EA204. Lot No. 30879 York 1976–77. –/74. 36.4 t.
PTSO. Dia. EH210. Lot No. 30880 York 1976–77. –/84. 30.5 t.
BDMSO. Dia. EI201. Lot No. 30885 York 1976–77. –/74. 37.6 t.

Class 313/0. Standard Design.

313 018	N	F	WN	HE	62546	71230	62610
313 024	N	F	WN	HE	62552	71236	62616
313 025	N	F	WN	HE	62553	71237	62617
313 026	N	F	WN	HE	62554	71238	62618
313 027	N	F	WN	HE	62555	71239	62619
313 028	N	F	WN	HE	62556	71240	62620
313 029	N	F	WN	HE	62557	71241	62621
313 030	N	F	WN	HE	62558	71242	62622
313 031	N	F	WN	HE	62559	71243	62623
313 032	N	F	WN	HE	62560	71244	62624
313 033	N	F	WN	HE	62561	71245	62625
313 035	N	F	WN	HE	62563	71247	62627
313 036	N	F	WN	HE	62564	71248	62628
313 037	N	F	WN	HE	62565	71249	62629
313 038	N	F	WN	HE	62566	71250	62630
313 039	N	F	WN	HE	62567	71251	62631
313 040	N	F	WN	HE	62568	71252	62632
313 041	N	F	WN	HE	62569	71253	62633
313 042	N	F	WN	HE	62570	71254	62634
313 043	N	F	WN	HE	62571	71255	62635
313 044	N	F	WN	HE	62572	71256	62636
313 045	N	F	WN	HE	62573	71257	62637
313 046	N	F	WN	HE	62574	71258	62638
313 047	N	F	WN	HE	62575	71259	62639
313 048	N	F	WN	HE	62576	71260	62640
313 049	N	F	WN	HE	62577	71261	62641
313 050	N	F	WN	HE	62578	71262	62649
313 051	N	F	WN	HE	62579	71263	62643
313 052	N	F	WN	HE	62580	71264	62644
313 053	N	F	WN	HE	62581	71265	62645
313 054	N	F	WN	HE	62582	71266	62646
313 055	N	F	WN	HE	62583	/1267	62647
313 056	N	F	WN	HE	62584	71268	62648
313 057	N	F	WN	HE	62585	71269	62642

313 058	N	F	WN	HE	62586	71270	62650
313 059	N	F	WN	HE	62587	71271	62651
313 060	N	F	WN	HE	62588	71272	62652
313 061	N	F	WN	HE	62589	71273	62653
313 062	N	F	WN	HE	62590	71274	62654
313 063	N	F	WN	HE	62591	71275	62655
313 064	N	F	WN	HE	62592	71276	62656

Class 313/1. Extra shoegear for Silverlink services.
Note: These sets are being renumbered as they are modernised.

	(313 001)	N	F	NL	BY	62529	71213	62593
	(313 002)	N	F	NL	BY	62530	71214	62594
	(313 003)	N	F	NL	BY	62531	71215	62595
	(313 004)	N	F	NL	BY	62532	71216	62596
	(313 005)	N	F	NL	BY	62533	71217	62597
	(313 006)	N	F	NL	BY	62534	71218	62598
	(313 007)	N	F	NL	BY	62535	71219	62599
	(313 008)	N	F	NL	BY	62536	71220	62600
	(313 009)	N	F	NL	BY	62537	71221	62601
	(313 010)	N	F	NL	BY	62538	71222	62602
	(313 011)	N	F	NL	BY	62539	71223	62603
	(313 012)	N	F	NL	BY	62540	71224	62604
	(313 013)	N	F	NL	BY	62541	71225	62605
	(313 014)	N	F	NL	BY	62542	71226	62606
	(313 015)	N	F	NL	BY	62543	71227	62607
	(313 016)	N	F	NL	BY	62544	71228	62608
	(313 017)	N	F	NL	BY	62545	71229	62609
	(313 019)	N	F	NL	BY	62547	71231	62611
	(313 020)	N	F	NL	BY	62548	71232	62612
313 121	(313 021)	SL	F	NL	BY	62549	71233	62613
313 132	(313 022)	SL	F	NL	BY	62550	71234	62614
313 123	(313 023)	SL	F	NL	BY	62551	71235	62615
313 134	(313 034)	SL	F	NL	BY	62562	71246	62626

Name: PTSO 71232 of set 313 020 is named 'PARLIAMENT HILL'.

CLASS 314

DMSO–PTSO–DMSO. Thyristor control. Tightlock couplers. Sliding doors. Disc and rheostatic brakes.
Bogies: BX1.
Gangways: Within unit. End doors.
Traction Motors: Four GEC G310AZ (Brush TM61-53*) of 82.125 kW.
Dimensions: 19.80 x 2.82 m (outer cars), 19.92 x 2.82 m (inner cars).
Maximum Speed: 75 m.p.h.

64583–64614. DMSO. Dia. EA206. Lot No. 30912 York 1979. –/68. 34.5 t.
64588". DMSO. Dia. EA207. Lot No. 30908 York 1978–80. –/74. 35.63 t. Converted from Class 507 No. 64426. The original 64588 has been scrapped. This vehicle has an experimental seating layout.
PTSO. Dia. EH211. Lot No. 30913 York 1979. –/76. 33.0 t.

314 201	*	S	A	SR	GW	64583	71450	64584
314 202	*	S	A	SR	GW	64585	71451	64586
314 203	*	S	A	SR	GW	64587	71452	64588[ii]
314 204	*	S	A	SR	GW	64589	71453	64590
314 205	*	S	A	SR	GW	64591	71454	64592
314 206	*	S	A	SR	GW	64593	71455	64594
314 207		CC	A	SR	GW	64595	71456	64596
314 208		S	A	SR	GW	64597	71457	64598
314 209		S	A	SR	GW	64599	71458	64600
314 210		S	A	SR	GW	64601	71459	64602
314 211		S	A	SR	GW	64603	71460	64604
314 212		S	A	SR	GW	64605	71461	64606
314 213		S	A	SR	GW	64607	71462	64608
314 214		S	A	SR	GW	64609	71463	64610
314 215		CC	A	SR	GW	64611	71464	64612
314 216		S	A	SR	GW	64613	71465	64614

Name: PTSO 71452 of set 314 203 is named 'European Union'.

CLASS 315

DMSO–TSO–PTSO–DMSO. Thyristor control. Tightlock couplers. Sliding doors. Disc and rheostatic brakes.
Bogies: BX1.
Gangways: Within unit. End doors.
Traction Motors: Four Brush TM61-53 (GEC G310AZ*) of 82.125 kW.
Dimensions: 19.80 x 2.82 m (outer cars), 19.92 x 2.82 m (inner cars).
Maximum Speed: 75 m.p.h.

64461–64582. DMSO. Dia. EA207. Lot No. 30902 York 1980–81. –/74. 35 t.
71281–71341. TSO. Dia. EH216. Lot No. 30904 York 1980–81. –/86. 25.5 t.
71389–71449. PTSO. Dia. EH217. Lot No. 30903 York 1980–81. –/84. 32 t.

315 801	N	F	GE	IL	64461	71281	71389	64462
315 802	GE	F	GE	IL	64463	71282	71390	64464
315 803	GE	F	GE	IL	64465	71283	71391	64466
315 804	GE	F	GE	IL	64467	71284	71392	64468
315 805	N	F	GE	IL	64469	71285	71393	64470
315 806	GE	F	GE	IL	64471	71286	71394	64472
315 807	GE	F	GE	IL	64473	71287	71395	64474
315 808	N	F	GE	IL	64475	71288	71396	64476
315 809	GE	F	GE	IL	64477	71289	71397	64478
315 810	GE	F	GE	IL	64479	71290	71398	64480
315 811	N	F	GE	IL	64481	71291	71399	64482
315 812	GE	F	GE	IL	64483	71292	71400	64484
315 813	GE	F	GE	IL	64485	71293	71401	64486
315 814	N	F	GE	IL	64487	71294	71402	64488
315 815	GE	F	GE	IL	64489	71295	71403	64490
315 816	GE	F	GE	IL	64491	71296	71404	64492
315 817	N	F	GF	IL	64493	71297	71405	64494
315 818	N	F	GE	IL	64495	71298	71406	64496
315 819	N	F	GE	IL	64497	71299	71407	64498

315 820		**N**	F	*GE*	IL	64499	71300	71408	64500
315 821		**N**	F	*GE*	IL	64501	71301	71409	64502
315 822		**N**	F	*GE*	IL	64503	71302	71410	64504
315 823		**N**	F	*GE*	IL	64505	71303	71411	64506
315 824		**N**	F	*GE*	IL	64507	71304	71412	64508
315 825		**N**	F	*GE*	IL	64509	71305	71413	64510
315 826		**N**	F	*GE*	IL	64511	71306	71414	64512
315 827		**N**	F	*GE*	IL	64513	71307	71415	64514
315 828		**N**	F	*GE*	IL	64515	71308	71416	64516
315 829		**N**	F	*GE*	IL	64517	71309	71417	64518
315 830		**N**	F	*GE*	IL	64519	71310	71418	64520
315 831		**N**	F	*GE*	IL	64521	71311	71419	64522
315 832		**N**	F	*GE*	IL	64523	71312	71420	64524
315 833		**N**	F	*GE*	IL	64525	71313	71421	64526
315 834		**N**	F	*GE*	IL	64527	71314	71422	64528
315 835		**N**	F	*GE*	IL	64529	71315	71423	64530
315 836		**N**	F	*GE*	IL	64531	71316	71424	64532
315 837		**N**	F	*GE*	IL	64533	71317	71425	64534
315 838		**N**	F	*GE*	IL	64535	71318	71426	64536
315 839		**N**	F	*GE*	IL	64537	71319	71427	64538
315 840		**N**	F	*GE*	IL	64539	71320	71428	64540
315 841		**N**	F	*GE*	IL	64541	71321	71429	64542
315 842	*	**N**	F	*GE*	IL	64543	71322	71430	64544
315 843	*	**N**	F	*GE*	IL	64545	71323	71431	64546
315 844	*	**N**	F	*GE*	IL	64547	71324	71432	64548
315 845	*	**N**	F	*GE*	IL	64549	71325	71433	64550
315 846	*	**N**	F	*WN*	HE	64551	71326	71434	64552
315 847	*	**N**	F	*WN*	HE	64553	71327	71435	64554
315 848	*	**N**	F	*WN*	HE	64555	71328	71436	64556
315 849	*	**N**	F	*WN*	HE	64557	71329	71437	64558
315 850	*	**N**	F	*WN*	HE	64559	71330	71438	64560
315 851	*	**N**	F	*WN*	HE	64561	71331	71439	64562
315 852	*	**N**	F	*WN*	HE	64563	71332	71440	64564
315 853	*	**N**	F	*WN*	HE	64565	71333	71441	64566
315 854	*	**N**	F	*WN*	HE	64567	71334	71442	64568
315 855	*	**N**	F	*WN*	HE	64569	71335	71443	64570
315 856	*	**N**	F	*WN*	HE	64571	71336	71444	64572
315 857	*	**N**	F	*WN*	HE	64573	71337	71445	64574
315 858	*	**N**	F	*WN*	HE	64575	71338	71446	64576
315 859	*	**N**	F	*WN*	HE	64577	71339	71447	64578
315 860	*	**N**	F	*WN*	HE	64579	71340	71448	64580
315 861	*	**N**	F	*WN*	HE	64581	71341	71449	64582

CLASS 317

DTSO(A)–MSO–TCOL (declassified*)–DTSO(B). Thyristor control. Tightlock couplers. Sliding doors. Disc brakes.
Bogies: BP20 (MSO), BT13 (others).
Gangways: Throughout.
Traction Motors: Four GEC G315BZ of 247.5 kW.

Dimensions: 19.83 x 2.82 m (outer cars), 19.92 x 2.82 m (inner cars).
Maximum Speed: 100 m.p.h.

Class 317/1. Pressure ventilated.

DTSO(A) Dia. EE216. Lot No. 30955 York 1981–82. –/74. 29.44 t.
MSO. Dia. EC202. Lot No. 30958 York 1981–82. –/79. 49.76 t.
TCOL. Dia. EH307. Lot No. 30957 Derby 1981–82. 22/46 2T. 28.80 t. Controlled emission toilets (but decommisioned).
DTSO(B) Dia. EE235 (EE232†). Lot No. 30956 York 1981–82. –/70. (–/71†). 29.28 t.

317 301	*	**LS**	A	*LS*	EM	77024	62661	71577	77048
317 302	*	**LS**	A	*LS*	EM	77001	62662	71578	77049
317 303	*	**N**	A	*LS*	EM	77002	62663	71579	77050
317 304	*	**N**	A	*LS*	EM	77003	62664	71580	77051
317 305	*	**N**	A	*LS*	EM	77004	62665	71581	77052
317 306	*	**N**	A	*LS*	EM	77005	62666	71582	77053
317 307	*	**LS**	A	*LS*	EM	77006	62667	71583	77054
317 308	*	**LS**	A	*LS*	EM	77007	62668	71584	77055
317 309		**N**	A	*WN*	HE	77008	62669	71585	77056
317 310	*	**LS**	A	*LS*	EM	77009	62670	71586	77057
317 311	*	**LS**	A	*LS*	EM	77010	62671	71587	77058
317 312	*	**LS**	A	*LS*	EM	77011	62672	71588	77059
317 313	*	**LS**	A	*LS*	EM	77012	62673	71589	77060
317 314	*	**LS**	A	*LS*	EM	77013	62674	71590	77061
317 315		**N**	A	*WN*	HE	77014	62675	71591	77062
317 316		**N**	A	*WN*	HE	77015	62676	71592	77063
317 317	*	**LS**	A	*LS*	EM	77016	62677	71593	77064
317 318		**N**	A	*WN*	HE	77017	62678	71594	77065
317 319	*	**LS**	A	*LS*	EM	77018	62679	71595	77066
317 320		**N**	A	*WN*	HE	77019	62680	71596	77067
317 321		**N**	A	*WN*	HE	77020	62681	71597	77068
317 322	*	**N**	A	*LS*	EM	77021	62682	71598	77069
317 323		**N**	A	*WN*	HE	77022	62683	71599	77070
317 324		**N**	A	*WN*	HE	77023	62684	71600	77071
317 325		**N**	A	*WN*	HE	77000	62685	71601	77072
317 326		**N**	A	*WN*	HE	77025	62686	71602	77073
317 327		**N**	A	*WN*	HE	77026	62687	71603	77074
317 328		**N**	A	*WN*	HE	77027	62688	71604	77075
317 329	*	**LS**	A	*LS*	EM	77028	62689	71605	77076
317 330		**N**	A	*WN*	HE	77029	62690	71606	77077
317 331		**N**	A	*WN*	HE	77030	62691	71607	77078
317 332	*	**LS**	A	*LS*	EM	77031	62692	71608	77079
317 333		**N**	A	*WN*	HE	77032	62693	71609	77080
317 334		**N**	A	*WN*	HE	77033	62694	71610	77081
317 335		**N**	A	*WN*	HE	77034	62695	71611	77082
317 336		**N**	A	*WN*	HE	77035	62696	71612	77083
317 337	†	**N**	A	*WN*	HE	77036	62697	71613	77084
317 338	†	**N**	A	*WN*	HE	77037	62698	71614	77085
317 339	†	**N**	A	*WN*	HE	77038	62699	71615	77086
317 340	†	**N**	A	*WN*	HE	77039	62700	71616	77087

317 341	†	**N**	A	*WN*	HE	77040	62701	71617	77088
317 342	†	**N**	A	*WN*	HE	77041	62702	71618	77089
317 343	†	**N**	A	*WN*	HE	77042	62703	71619	77090
317 344	†	**N**	A	*WN*	HE	77043	62704	71620	77091
317 345	†	**N**	A	*WN*	HE	77044	62705	71621	77092
317 346	†	**N**	A	*WN*	HE	77045	62706	71622	77093
317 347	†	**N**	A	*WN*	HE	77046	62707	71623	77094
317 348	†	**N**	A	*WN*	HE	77047	62708	71624	77095

Class 317/2. Convection heating.

77200–19. DTSO(A). Dia. EE224. Lot No. 30994 York 1985–86. –/74. 29.31 t.
77280–83. DTSO(A). Dia. EE224. Lot No. 31007 York 1987. –/74. 29.31 t.
62846–65. MSO. Dia. EC205. Lot No. 30996 York 1985–86. –/79. 50.08 t.
62886–89. MSO. Dia. EC205. Lot No. 31009 York 1987. –/79. 50.08 t.
71734–53. TCOL. Dia. EH308. Lot No. 30997 Yk 1985–86. 22/46 2T. 28.28 t.
71762–65. TCOL. Dia. EH308. Lot No. 31010 York 1987. 22/46 2T. 28.28 t.
77220–39. DTSO(B). Dia. EE225. Lot No. 30995 York 1985–86. 29.28 t. –/71.
77284–87. DTSO(B). Dia. EE225. Lot No. 31008 York 1987. 29.28 t. –/71.

317 349	**N**	A	*WN*	HE	77200	62846	71734	77220
317 350	**N**	A	*WN*	HE	77201	62847	71735	77221
317 351	**N**	A	*WN*	HE	77202	62848	71736	77222
317 352	**N**	A	*WN*	HE	77203	62849	71739	77223
317 353	**N**	A	*WN*	HE	77204	62850	71738	77224
317 354	**N**	A	*WN*	HE	77205	62851	71737	77225
317 355	**N**	A	*WN*	HE	77206	62852	71740	77226
317 356	**N**	A	*WN*	HE	77207	62853	71742	77227
317 357	**N**	A	*WN*	HE	77208	62854	71741	77228
317 358	**N**	A	*WN*	HE	77209	62855	71743	77229
317 359	**N**	A	*WN*	HE	77210	62856	71744	77230
317 360	**N**	A	*WN*	HE	77211	62857	71745	77231
317 361	**N**	A	*WN*	HE	77212	62858	71746	77232
317 362	**N**	A	*WN*	HE	77213	62859	71747	77233
317 363	**N**	A	*WN*	HE	77214	62860	71748	77234
317 364	**N**	A	*WN*	HE	77215	62861	71749	77235
317 365	**N**	A	*WN*	HE	77216	62862	71750	77236
317 366	**N**	A	*WN*	HE	77217	62863	71752	77237
317 367	**N**	A	*WN*	HE	77218	62864	71751	77238
317 368	**N**	A	*WN*	HE	77219	62865	71753	77239
317 369	**N**	A	*WN*	HE	77280	62886	71762	77284
317 370	**N**	A	*WN*	HE	77281	62887	71763	77285
317 371	**N**	A	*WN*	HE	77282	62888	71764	77286
317 372	**N**	A	*WN*	HE	77283	62889	71765	77287

Names:

TCOL No. 71735 of set 317 350 is named 'HARLOW 50 years 1947–1997'.
TCOL No. 71746 of set 317 361 is named 'Kings Lynn Festival'.
TCOL No. 71752 of set 317 366 is named 'Letchworth Garden City'.
TCOL No. 71764 of set 317 371 is named 'Stevenage new town 50 years 1946–1996'.
TCOL No. 71765 of set 317 372 is named 'Welwyn Garden City'.

CLASS 318

DTSOL–MSO–DTSO. Thyristor control. Tightlock couplers. Sliding doors. Disc brakes.
Bogies: BP20 (MSO), BT13 (others).
Gangways: Throughout.
Traction Motors: Four Brush TM 2141 of 268 kW.
Dimensions: 19.83 x 2.82 m (outer cars), 19.92 x 2.82 m (inner cars).
Maximum Speed: 90 m.p.h.

77240–59. DTSOL. Dia. EE227. Lot No. 30999 York 1985–86. –/66 1T. 30.01 t.
77288. DTSOL. Dia. EE227. Lot No. 31020 York 1986–87. –/66 1T. 30.01 t.
62866–85. MSO. Dia. EC207. Lot No. 30998 York 1985–86. –/79. 50.90 t.
62890. MSO. Dia. EC207. Lot No. 31019 York 1987. –/79. 50.90 t.
77260–79. DTSO. Dia. EE228. Lot No. 31000 York 1985–86. –/71. 26.60 t.
77289. DTSO. Dia. EE228. Lot No. 31021 York 1987. –/71. 26.60 t.

318 250	S	F	*SR*	GW	77260	62866	77240
318 251	S	F	*SR*	GW	77261	62867	77241
318 252	S	F	*SR*	GW	77262	62868	77242
318 253	S	F	*SR*	GW	77263	62869	77243
318 254	S	F	*SR*	GW	77264	62870	77244
318 255	S	F	*SR*	GW	77265	62871	77245
318 256	S	F	*SR*	GW	77266	62872	77246
318 257	S	F	*SR*	GW	77267	62873	77247
318 258	S	F	*SR*	GW	77268	62874	77248
318 259	S	F	*SR*	GW	77269	62875	77249
318 260	S	F	*SR*	GW	77270	62876	77250
318 261	S	F	*SR*	GW	77271	62877	77251
318 262	S	F	*SR*	GW	77272	62878	77252
318 263	S	F	*SR*	GW	77273	62879	77253
318 264	S	F	*SR*	GW	77274	62880	77254
318 265	S	F	*SR*	GW	77275	62881	77255
318 266	S	F	*SR*	GW	77276	62882	77256
318 267	S	F	*SR*	GW	77277	62883	77257
318 268	S	F	*SR*	GW	77278	62884	77258
318 269	S	F	*SR*	GW	77279	62885	77259
318 270	S	F	*SR*	GW	77289	62890	77288

Names:

DTSOL No. 77240 of set 318 250 is named 'GEOFF SHAW'.
DTSOL No. 77256 of set 318 266 is named 'STRATHCLYDER'.

CLASS 319

GTO Thyristor control. Tightlock couplers. Sliding doors. Disc brakes.
System: 25 kV a.c. overhead/750 V d.c. third rail.
Bogies: P7-4 (MSO), T3-7 (others).
Gangways: Within unit. End doors.
Traction Motors: Four GEC G315BZ of 268 kW.

Dimensions: 19.83 x 2.82 m (outer cars), 19.92 x 2.82 m (inner cars).
Maximum Speed: 100 m.p.h.

Class 319/0. DTSO (A)–MSO–TSOL–DTSO (B).

DTSO (A). Dia. EE233. Lot No. 31022 (odd nos.) York 1987–8. –/82. 30.12 t.
MSO. Dia. EC209. Lot No. 31023 York 1987–8. –/82. 51 t.
TSOL. Dia. EH234. Lot No. 31024 York 1987–8. –/77 2T. 51 t.
DTSO (B). Dia. EE234. Lot No. 31025 (even nos.) York 1987–8. –/78. 30 t.

319 001	**CX**	P	*SC*	SU	77291	62891	71772	77290
319 002	**CX**	P	*SC*	SU	77293	62892	71773	77292
319 003	**CX**	P	*SC*	SU	77295	62893	71774	77294
319 004	**CX**	P	*SC*	SU	77297	62894	71775	77296
319 005	**CX**	P	*SC*	SU	77299	62895	71776	77298
319 006	**CX**	P	*SC*	SU	77301	62896	71777	77300
319 007	**CX**	P	*SC*	SU	77303	62897	71778	77302
319 008	**CX**	P	*SC*	SU	77305	62898	71779	77304
319 009	**CX**	P	*SC*	SU	77307	62899	71780	77306
319 010	**CX**	P	*SC*	SU	77309	62900	71781	77308
319 011	**CX**	P	*SC*	SU	77311	62901	71782	77310
319 012	**CX**	P	*SC*	SU	77313	62902	71783	77312
319 013	**CX**	P	*SC*	SU	77315	62903	71784	77314

Names:

TSOL 71776 of set 319 005 is named 'Partnership For Progress'.
TSOL 71779 of set 319 008 is named 'Cheriton'.
TSOL 71780 of set 319 009 is named 'Coquelles'.

Class 319/2. Units converted from Class 319/0 for Connex express services from London to Brighton. DTSO–MSO–TSOL–DTCO.

DTSO. Dia. EE244. Lot No. 31022 (odd nos.) York 1987–8. –/64. 30.2 t.
MSO. Dia. EN262. Lot No. 31023 York 1987–8. –/60 (including 12 seats in a 'snug' under the pantograph area. External sliding doors sealed adjacent to this area. 51 t.
TSOL. Dia. EH212. Lot No. 31024 York 1987–8. –/52 1T 1TD. 34 t.
DTCO. Dia. EE374. Lot No. 31025 (even nos.) York 1987–8. 18/36. 30 t.

319 214	(319 014)	**CX**	P	*SC*	SU	77317	62904	71785	77316
319 215	(319 015)	**CX**	P	*SC*	SU	77319	62905	71786	77318
319 216	(319 016)	**CX**	P	*SC*	SU	77321	62906	71787	77320
319 217	(319 017)	**CX**	P	*SC*	SU	77323	62907	71788	77322
319 218	(319 018)	**CX**	P	*SC*	SU	77325	62908	71789	77324
319 219	(319 019)	**CX**	P	*SC*	SU	77327	62909	71790	77326
319 220	(319 020)	**CX**	P	*SC*	SU	77329	62910	71791	77328

Names:

TSOL 71786 of set 319 215 is named 'London'.
TSOL 71788 of set 319 217 is named 'Brighton'.
TSOL 71789 of set 319 218 is named 'Croydon'.

Class 319/1. DTCO–MSO–TSOL–DTSO.
Class 319/3. Units converted from Class 319/1 by replacing first class sets with standard class seats. DTSO (A)–MSO–TSOL–DTSO (B).

DTCO. Dia. EE310. Lot No. 31063 York 1990. 16/54. 29 t.
DTSO(A). Dia. EE2??. Lot No. 31063 York 1990. –/77. 29 t.
MSO. Dia. EC214. Lot No. 31064 York 1990. –/79. 50.6 t.
TSOL. Dia. EH238. Lot No. 31065 York 1990. –/74 2T. 31 t.
DTSO(B). Dia. EE240. Lot No. 31066 York 1990. –/78. 29.7 t.

	(319 161)	**NW**	P	*TL*	SU	77459	63043	71929	77458
	(319 162)	**NW**	P	*TL*	SU	77461	63044	71930	77460
319 363	(319 163)	**TN**	P	*TL*	SU	77463	63045	71931	77462
	(319 164)	**NW**	P	*TL*	SU	77465	63046	71932	77464
	(319 165)	**NW**	P	*TL*	SU	77467	63047	71933	77466
	(319 166)	**NW**	P	*TL*	SU	77469	63048	71934	77468
	(319 167)	**NW**	P	*TL*	SU	77471	63049	71935	77470
319 368	(319 168)	**TN**	P	*TL*	SU	77473	63050	71936	77472
	(319 169)	**NW**	P	*TL*	SU	77475	63051	71937	77474
	(319 170)	**NW**	P	*TL*	SU	77477	63052	71938	77476
	(319 171)	**NW**	P	*TL*	SU	77479	63053	71939	77478
	(319 172)	**NW**	P	*TL*	SU	77481	63054	71940	77480
	(319 173)	**NW**	P	*TL*	SU	77483	63055	71941	77482
	(319 174)	**NW**	P	*TL*	SU	77485	63056	71942	77484
	(319 175)	**NW**	P	*TL*	SU	77487	63057	71943	77486
	(319 176)	**NW**	P	*TL*	SU	77489	63058	71944	77488
	(319 177)	**NW**	P	*TL*	SU	77491	63059	71945	77490
	(319 178)	**NW**	P	*TL*	SU	77493	63060	71946	77492
	(319 179)	**NW**	P	*TL*	SU	77495	63061	71947	77494
	(319 180)	**NW**	P	*TL*	SU	77497	63062	71948	77496
	(319 181)	**NW**	P	*TL*	SU	77973	63093	71979	77974
	(319 182)	**NW**	P	*TL*	SU	77975	63094	71970	77976
319 383	(319 183)	**TN**	P	*TL*	SU	77977	63095	71981	77978
	(319 184)	**NW**	P	*TL*	SU	77979	63096	71982	77980
319 385	(319 185)	**TN**	P	*TL*	SU	77981	63097	71983	77982
	(319 186)	**NW**	P	*TL*	SU	77983	63098	71984	77984

Class 319/4. Units converted from Class 319/0. Refurbished with carpets. DTSO(A) converted to composite. DTCO–MSO–TSOL–DTSO.
77331–381. DTCO. Dia. EE233. Lot No. 31022 (odd nos.) York 1987–8. 12/54. 30.12 t.
77431–457. DTCO. Dia. EE233. Lot No. 31038 (odd nos.) York 1988. 12/54. 30.12 t.
62911–936. MSO. Dia. EC209. Lot No. 31023 York 1987–8. –/77. 51 t.
62961–974. MSO. Dia. EC209. Lot No. 31039 York 1988. –/77. 51 t.
71792–817. TSOL. Dia. EH234. Lot No. 31024 York 1987–8. –/72 2T. 51 t.
71866–879. TSOL. Dia. EH234. Lot No. 31040 York 1988. –/72 2T. 51 t.
77330–380. DTSO. Dia. EE234. Lot No. 31025 (even nos.) York 1987–8. –/74. 30 t.
77430–456. DTSO. Dia. EE234. Lot No. 31041 (even nos.) York 1988. –/74. 30 t.

319 421	(319 021)	**TN**	P	*TL*	SU	77331	62911	71792	77330
	(319 022)	**TL**	P	*TL*	SU	77333	62912	71793	77332

	(319 023)	**N**	P	*TL*	SU	77335	62913	71794	77334
	(319 024)	**N**	P	*TL*	SU	77337	62914	71795	77336
	(319 025)	**N**	P	*TL*	SU	77339	62915	71796	77338
	(319 026)	**N**	P	*TL*	SU	77341	62916	71797	77340
	(319 027)	**N**	P	*TL*	SU	77343	62917	71798	77342
	(319 028)	**N**	P	*TL*	SU	77345	62918	71799	77344
319 429	(319 029)	**TN**	P	*TL*	SU	77347	62919	71800	77346
	(319 030)	**TL**	P	*TL*	SU	77349	62920	71801	77348
	(319 031)	**TL**	P	*TL*	SU	77351	62921	71802	77350
	(319 032)	**TL**	P	*TL*	SU	77353	62922	71803	77352
	(319 033)	**TL**	P	*TL*	SU	77355	62923	71804	77354
	(319 034)	**TL**	P	*TL*	SU	77357	62924	71805	77356
	(319 035)	**TL**	P	*TL*	SU	77359	62925	71806	77358
	(319 036)	**TL**	P	*TL*	SU	77361	62926	71807	77360
	(319 037)	**TL**	P	*TL*	SU	77363	62927	71808	77362
	(319 038)	**TL**	P	*TL*	SU	77365	62928	71809	77364
319 439	(319 039)	**TN**	P	*TL*	SU	77367	62929	71810	77366
	(319 040)	**TL**	P	*TL*	SU	77369	62930	71811	77368
	(319 041)	**TL**	P	*TL*	SU	77371	62931	71812	77370
	(319 042)	**TL**	P	*TL*	SU	77373	62932	71813	77372
	(319 043)	**TL**	P	*TL*	SU	77375	62933	71814	77374
	(319 044)	**TL**	P	*TL*	SU	77377	62934	71815	77376
	(319 045)	**TL**	P	*TL*	SU	77379	62935	71816	77378
	(319 046)	**TL**	P	*TL*	SU	77381	62936	71817	77380
	(319 047)	**TL**	P	*TL*	SU	77431	62961	71866	77430
	(319 048)	**TL**	P	*TL*	SU	77433	62962	71867	77432
	(319 049)	**TL**	P	*TL*	SU	77435	62963	71868	77434
	(319 050)	**TL**	P	*TL*	SU	77437	62964	71869	77436
	(319 051)	**TL**	P	*TL*	SU	77439	62965	71870	77438
	(319 052)	**TL**	P	*TL*	SU	77441	62966	71871	77440
	(319 053)	**TL**	P	*TL*	SU	77443	62967	71872	77442
	(319 054)	**TL**	P	*TL*	SU	77445	62968	71873	77444
	(319 055)	**TL**	P	*TL*	SU	77447	62969	71874	77446
	(319 056)	**TL**	P	*TL*	SU	77449	62970	71875	77448
	(319 057)	**TL**	P	*TL*	SU	77451	62971	71876	77450
	(319 058)	**TL**	P	*TL*	SU	77453	62972	71877	77452
	(319 059)	**TL**	P	*TL*	SU	77455	62973	71878	77454
319 460	(319 060)	**TN**	P	*TL*	SU	77457	62974	71879	77456

Names:

TSOL 71801 of set 319 030 is named 'Harlington Festival'.
TSOL 71874 of set 319 055 is named 'Brixton Challenge'.

CLASS 320

DTSO (A)–MSO–DTSO (B). Thyristor control. Tightlock couplers. Sliding doors. Disc brakes.
Bogies: P7-4 (MSO), T3-7 (others).
Gangways: Within unit.
Traction Motors: Brush TM2141B of 268 kW.
Dimensions: 19.83 x 2.82 m (outer cars), 19.92 x 2.82 m (inner car).

Maximum Speed: 75 m.p.h.

DTSO (A). Dia. EE238. Lot No. 31060 York 1990. –/77. 30.7 t.
MSO. Dia. EC212. Lot No. 31062 York 1990. –/77. 52.1 t.
DTSO (B). Dia. EE239. Lot No. 31061 York 1990. –/76 31.7 t.

320 301	S	F	*SR*	GW	77899	63021	77921
320 302	S	F	*SR*	GW	77900	63022	77922
320 303	S	F	*SR*	GW	77901	63023	77923
320 304	S	F	*SR*	GW	77902	63024	77924
320 305	S	F	*SR*	GW	77903	63025	77925
320 306	CC	F	*SR*	GW	77904	63026	77926
320 307	CC	F	*SR*	GW	77905	63027	77927
320 308	CC	F	*SR*	GW	77906	63028	77928
320 309	CC	F	*SR*	GW	77907	63029	77929
320 310	CC	F	*SR*	GW	77908	63030	77930
320 311	CC	F	*SR*	GW	77909	63031	77931
320 312	CC	F	*SR*	GW	77910	63032	77932
320 313	CC	F	*SR*	GW	77911	63033	77933
320 314	CC	F	*SR*	GW	77912	63034	77934
320 315	CC	F	*SR*	GW	77913	63035	77935
320 316	CC	F	*SR*	GW	77914	63036	77936
320 317	CC	F	*SR*	GW	77915	63037	77937
320 318	S	F	*SR*	GW	77916	63038	77938
320 319	S	F	*SR*	GW	77917	63039	77939
320 320	CC	F	*SR*	GW	77918	63040	77940
320 321	CC	F	*SR*	GW	77919	63041	77941
320 322	S	F	*SR*	GW	77920	63042	77942

Names:

MSO 63025 of set 320 305 is named 'GLASGOW SCHOOL OF ART'.
MSO 63026 of set 320 306 is named 'MODEL RAIL SCOTLAND'.
MSO 63041 of set 320 321 is named 'The Rt. Hon. John Smith, QC, MP'.
MSO 63042 of set 320 322 is named 'FESTIVE GLASGOW ORCHID'.

CLASS 321

DTCO (DTSO on Class 321/9)–MSO–TSOL–DTSO. Thyristor control. Tightlock couplers. Sliding doors. Disc brakes.
Bogies: P7-4 (MSO), T3-7 (others).
Gangways: Within unit.
Traction Motors: Brush TM2141C (268 kW).
Dimensions: 19.83 x 2.82 m (outer cars), 19.92 x 2.82 m (inner cars).
Maximum Speed: 100 m.p.h.
Non-standard livery: NS (Netherlands Railways Inter-City livery (Yellow and deep blue).

Note: Lot numbers and diagrams were officially changed on 09/02/90.

Class 321/3. Units built for Liverpool Street workings.

DTCO. Dia. EE308. Lot No. 31053 York 1988–90. 16/56. 29.3 t.
MSO. Dia. EC210. Lot No. 31054 York 1988–90. –/79. 51.5 t.

TSOL. Dia. EH235. Lot No. 31055 York 1988–90. –/74 2T. 28 t.
DTSO. Dia. EE236. Lot No. 31056 York 1988–90. –/78. 29.1 t.

321 301	**NW**	F	*GE*	IL	78049	62975	71880	77853
321 302	**NW**	F	*GE*	IL	78050	62976	71881	77854
321 303	**NW**	F	*GE*	IL	78051	62977	71882	77855
321 304	**NW**	F	*GE*	IL	78052	62978	71883	77856
321 305	**NW**	F	*GE*	IL	78053	62979	71884	77857
321 306	**NW**	F	*GE*	IL	78054	62980	71885	77858
321 307	**NW**	F	*GE*	IL	78055	62981	71886	77859
321 308	**NW**	F	*GE*	IL	78056	62982	71887	77860
321 309	**NW**	F	*GE*	IL	78057	62983	71888	77861
321 310	**NW**	F	*GE*	IL	78058	62984	71889	77862
321 311	**NW**	F	*GE*	IL	78059	62985	71890	77863
321 312	**NW**	F	*GE*	IL	78060	62986	71891	77864
321 313	**NW**	F	*GE*	IL	78061	62987	71892	77865
321 314	**NW**	F	*GE*	IL	78062	62988	71893	77866
321 315	**NW**	F	*GE*	IL	78063	62989	71894	77867
321 316	**NW**	F	*GE*	IL	78064	62990	71895	77868
321 317	**NW**	F	*GE*	IL	78065	62991	71896	77869
321 318	**NW**	F	*GE*	IL	78066	62992	71897	77870
321 319	**NW**	F	*GE*	IL	78067	62993	71898	77871
321 320	**NW**	F	*GE*	IL	78068	62994	71899	77872
321 321	**GE**	F	*GE*	IL	78069	62995	71900	77873
321 322	**NW**	F	*GE*	IL	78070	62996	71901	77874
321 323	**NW**	F	*GE*	IL	78071	62997	71902	77875
321 324	**NW**	F	*GE*	IL	78072	62998	71903	77876
321 325	**NW**	F	*GE*	IL	78073	62999	71904	77877
321 326	**NW**	F	*GE*	IL	78074	63000	71905	77878
321 327	**NW**	F	*GE*	IL	78075	63001	71906	77879
321 328	**GE**	F	*GE*	IL	78076	63002	71907	77880
321 329	**NW**	F	*GE*	IL	78077	63003	71908	77881
321 330	**GE**	F	*GE*	IL	78078	63004	71909	77882
321 331	**NW**	F	*GE*	IL	78079	63005	71910	77883
321 332	**NW**	F	*GE*	IL	78080	63006	71911	77884
321 333	**NW**	F	*GE*	IL	78081	63007	71912	77885
321 334	**0**	F	*GE*	IL	78082	63008	71913	77886
321 335	**GE**	F	*GE*	IL	78083	63009	71914	77887
321 336	**NW**	F	*GE*	IL	78084	63010	71915	77888
321 337	**NW**	F	*GE*	IL	78085	63011	71916	77889
321 338	**NW**	F	*GE*	IL	78086	63012	71917	77890
321 339	**NW**	F	*GE*	IL	78087	63013	71918	77891
321 340	**GE**	F	*GE*	IL	78088	63014	71919	77892
321 341	**GE**	F	*GE*	IL	78089	63015	71920	77893
321 342	**NW**	F	*GE*	IL	78090	63016	71921	77894
321 343	**NW**	F	*GE*	IL	78091	63017	71922	77895
321 344	**NW**	F	*GE*	IL	78092	63018	71923	77896
321 345	**NW**	F	*GE*	IL	78093	63019	71924	77897
321 346	**NW**	F	*GE*	IL	78094	63020	71925	77898
321 347	**GE**	F	*GE*	IL	78131	63105	71991	78280
321 348	**NW**	F	*GE*	IL	78132	63106	71992	78281

321 349	**NW**	F	*GE*	IL	78133	63107	71993	78282
321 350	**GE**	F	*GE*	IL	78134	63108	71994	78283
321 351	**NW**	F	*GE*	IL	78135	63109	71995	78284
321 352	**NW**	F	*GE*	IL	78136	63110	71996	78285
321 353	**NW**	F	*GE*	IL	78137	63111	71997	78286
321 354	**NW**	F	*GE*	IL	78138	63112	71998	78287
321 355	**NW**	F	*GE*	IL	78139	63113	71999	78288
321 356	**NW**	F	*GE*	IL	78140	63114	72000	78289
321 357	**GE**	F	*GE*	IL	78141	63115	72001	78290
321 358	**NW**	F	*GE*	IL	78142	63116	72002	78291
321 359	**NW**	F	*GE*	IL	78143	63117	72003	78292
321 360	**NW**	F	*GE*	IL	78144	63118	72004	78293
321 361	**NW**	F	*GE*	IL	78145	63119	72005	78294
321 362	**NW**	F	*GE*	IL	78146	63120	72006	78295
321 363	**GE**	F	*GE*	IL	78147	63121	72007	78296
321 364	**GE**	F	*GE*	IL	78148	63122	72008	78297
321 365	**NW**	F	*GE*	IL	78149	63123	72009	78298
321 366	**NW**	F	*GE*	IL	78150	63124	72010	78299

Names:
TSOL No. 71891 of set 321 312 is named 'Southend-on-Sea'.
TSOL No. 71913 of set 321 334 is named 'Amsterdam'.
TSOL No. 71915 of set 321 336 is named 'Geoffrey Freeman Allen'.
TSOL No. 71995 of set 321 351 is named 'GURKHA'.

Class 321/4. Units built for WCML workings.

DTCO. Dia. EE309. Lot No. 31067 York 1989–90. 28/40. 29.3 t.
MSO. Dia. EC210. Lot No. 31068 York 1989–90. –/79. 51.5 t.
TSOL. Dia. EH235. Lot No. 31069 York 1989–90. –/74 2T. 28 t.
DTSO. Dia. EE236. Lot No. 31070 York 1989–90. –/78. 29.1 t.

Note: The DTCO's of sets allocated to IL have 12 first class seats declassified.

321 401	**NW**	F	*NL*	BY	78095	63064	71949	77943
321 402	**NW**	F	*NL*	BY	78096	63064	71950	77944
321 403	**NW**	F	*NL*	BY	78097	63065	71951	77945
321 404	**NW**	F	*NL*	BY	78098	63066	71952	77946
321 405	**NW**	F	*NL*	BY	78099	63067	71953	77947
321 406	**NW**	F	*NL*	BY	78100	63068	71954	77948
321 407	**NW**	F	*NL*	BY	78101	63069	71955	77949
321 408	**NW**	F	*NL*	BY	78102	63070	71956	77950
321 409	**NW**	F	*NL*	BY	78103	63071	71957	77951
321 410	**NW**	F	*NL*	BY	78104	63072	71958	77952
321 411	**NW**	F	*NL*	BY	78105	63073	71959	77953
321 412	**NW**	F	*NL*	BY	78106	63074	71960	77954
321 413	**NW**	F	*NL*	BY	78107	63075	71961	77955
321 414	**NW**	F	*NL*	BY	78108	63076	71962	77956
321 415	**NW**	F	*NL*	BY	78109	63077	71963	77957
321 416	**NW**	F	*NL*	BY	78110	63078	71964	77958
321 417	**NW**	F	*NL*	BY	78111	63079	71965	77959
321 418	**NW**	F	*NL*	BY	78112	63080	71968	77962
321 419	**NW**	F	*NL*	BY	78113	63081	71967	77961

321 420	**NW**	F		BY	78114	63082	71966	77960
321 421	**NW**	F	*NL*	BY	78115	63083	71969	77963
321 422	**NW**	F	*NL*	BY	78116	63084	71970	77964
321 423	**NW**	F	*NL*	BY	78117	63085	71971	77965
321 424	**NW**	F	*NL*	BY	78118	63086	71972	77966
321 425	**NW**	F	*NL*	BY	78119	63087	71973	77967
321 426	**NW**	F	*NL*	BY	78120	63088	71974	77968
321 427	**NW**	F	*NL*	BY	78121	63089	71975	77969
321 428	**SL**	F	*NL*	BY	78122	63090	71976	77970
321 429	**SL**	F	*NL*	BY	78123	63091	71977	77971
321 430	**NW**	F	*NL*	BY	78124	63092	71978	77972
321 431	**SL**	F	*NL*	BY	78151	63125	72011	78300
321 432	**SL**	F	*NL*	BY	78152	63126	72012	78301
321 433	**SL**	F	*NL*	BY	78153	63127	72013	78302
321 434	**NW**	F	*NL*	BY	78154	63128	72014	78303
321 435	**NW**	F	*NL*	BY	78155	63129	72015	78304
321 436	**NW**	F	*NL*	BY	78156	63130	72016	78305
321 437	**NW**	F	*NL*	BY	78157	63131	72017	78306
321 438	**NW**	F	*GE*	IL	78158	63132	72018	78307
321 439	**GE**	F	*GE*	IL	78159	63133	72019	78308
321 440	**NW**	F	*GE*	IL	78160	63134	72020	78309
321 441	**GE**	F	*GE*	IL	78161	63135	72021	78310
321 442	**NW**	F	*GE*	IL	78162	63136	72022	78311
321 443	**GE**	F	*GE*	IL	78125	63099	71985	78274
321 444	**GE**	F	*GE*	IL	78126	63100	71986	78275
321 445	**NW**	F	*GE*	IL	78127	63101	71987	78276
321 446	**NW**	F	*GE*	IL	78128	63102	71988	78277
321 447	**NW**	F	*GE*	IL	78129	63103	71989	78278
321 448	**NW**	F	*GE*	IL	78130	63104	71990	78279

Name: TSOL 71955 of set 321 407 is named 'HERTFORDSHIRE WRVS'.

Class 321/9. Units owned by West Yorkshire PTE although managed by Porterbrook Leasing Company. DTSO (A)–MSO–TSOL–DTSO (B).

DTSO (A). Dia. EE277. Lot No. 31108 York 1991. –/77. 29.3 t.
MSO. Dia. EC216. Lot No. 31109 York 1991. –/79. 51.5 t.
TSOL. Dia. EH240. Lot No. 31110 York 1991. –/74 2T. 28 t.
DTSO (B). Dia. EE277. Lot No. 31111 York 1991. –/77. 29.1 t.

321 901	**Y**	P	*NE*	NL	77990	63153	72128	77993
321 902	**Y**	P	*NE*	NL	77991	63154	72129	77994
321 903	**Y**	P	*NE*	NL	77992	63155	72130	77995

CLASS 322

DTCO–MSO–TSOL–DTSO. Units built for use on Stansted Airport services. Thyristor control. Tightlock couplers. Sliding doors. Disc brakes.
Bogies: P7-4 (MSO), T3-7 (others).
Gangways: Within unit.
Traction Motors: Brush TM2141C (268 kW).
Dimensions: 19.83 x 2.82 m (outer cars), 19.92 x 2.82 m (inner cars).

Maximum Speed: 100 m.p.h.
Non-Standard Livery: Stansted Skytrain livery (grey with a yellow stripe).

DTCO. Dia. EE313. Lot No. 31094 York 1990. 35/22. 30.43 t.
MSO. Dia. EC215. Lot No. 31092 York 1990. –/70. 52.27 t.
TSOL. Dia. EH239. Lot No. 31093 York 1990. –/60 2T. 29.51 t.
DTSO. Dia. EE242. Lot No. 31091 York 1990. –/65. 29.77 t.

322 481	**0**	F	*WN*	HE	78163	63137	72023	77985
322 482	**0**	F	*WN*	HE	78164	63138	72024	77986
322 483	**0**	F	*WN*	HE	78165	63139	72025	77987
322 484	**0**	F	*WN*	HE	78166	63140	72026	77988
322 485	**0**	F	*WN*	HE	78167	63141	72027	77989

CLASS 323

DMSO (A)–TSOL–DMSO (B). Aluminium alloy bodies. Thyristor control.
Tightlock couplers. Sliding doors. Disc brakes.
Bogies: RFS BP62 (motor cars) and BT52 (trailer car).
Gangways: Within unit.
Traction Motors: Four Holec DMKT 52/24 of 146 kW per car.
Dimensions: 23.37 x 2.80 m (motor cars), 23.44 x 2.80 m (trailer cars).
Maximum Speed: 90 m.p.h.

DMSO (A). Dia. EA272. Lot No. 31112 Hunslet 1992–3. –/98 (–/82*). 41.0 t.
TSOL. Dia. EH296. Lot No. 31113 Hunslet 1992–3. –/88 1T. (–/80 1T*). 23.37 t.
DMSO (B). Dia. EA272. Lot No. 31114 Hunslet 1992–3. –/98 (–/82*). 39.4t.

323 201		**CE**	P	*CT*	BY	64001 72201 65001
323 202		**CE**	P	*CT*	BY	64002 72202 65002
323 203		**CE**	P	*CT*	BY	64003 72203 65005
323 204		**CE**	P	*CT*	BY	64004 72204 65004
323 205		**CE**	P	*CT*	BY	64005 72205 65003
323 206		**CE**	P	*CT*	BY	64006 72206 65006
323 207		**CE**	P	*CT*	BY	64007 72207 65007
323 208		**CE**	P	*CT*	BY	64008 72208 65008
323 209		**CE**	P	*CT*	BY	64009 72209 65009
323 210		**CE**	P	*CT*	BY	64010 72210 65010
323 211		**CE**	P	*CT*	BY	64011 72211 65011
323 212		**CE**	P	*CT*	BY	64012 72212 65012
323 213		**CE**	P	*CT*	BY	64013 72213 65013
323 214		**CE**	P	*CT*	BY	64014 72214 65014
323 215		**CE**	P	*CT*	BY	64015 72215 65015
323 216		**CE**	P	*CT*	BY	64016 72216 65016
323 217		**CE**	P	*CT*	BY	64017 72217 65017
323 218		**CE**	P	*CT*	BY	64018 72218 65018
323 219		**CE**	P	*CT*	BY	64019 72219 65019
323 220		**CE**	P	*CT*	BY	64020 72220 65020
323 221		**CE**	P	*CT*	BY	64021 72221 65021
323 222		**CE**	P	*CT*	BY	64022 72222 65022
323 223	*	**GM**	P	*NW*	LG	64023 72223 65023
323 224	*	**GM**	P	*NW*	LG	64024 72224 65024

323 225	*	**GM**	P	*NW*	LG	64025	72225	65025
323 226		**GM**	P	*NW*	LG	64026	72226	65026
323 227		**GM**	P	*NW*	LG	64027	72227	65027
323 228		**GM**	P	*NW*	LG	64028	72228	65028
323 229		**GM**	P	*NW*	LG	64029	72229	65029
323 230		**GM**	P	*NW*	LG	64030	72230	65030
323 231		**GM**	P	*NW*	LG	64031	72231	65031
323 232		**GM**	P	*NW*	LG	64032	72232	65032
323 233		**NT**	P	*NW*	LG	64033	72233	65033
323 234		**GM**	P	*NW*	LG	64034	72234	65034
323 235		**GM**	P	*NW*	LG	64035	72235	65035
323 236		**GM**	P	*NW*	LG	64036	72236	65036
323 237		**GM**	P	*NW*	LG	64037	72237	65037
323 238		**GM**	P	*NW*	LG	64038	72238	65038
323 239		**GM**	P	*NW*	LG	64039	72239	65039
323 240		**CE**	P	*CT*	BY	64040	72340	65040
323 241		**CE**	P	*CT*	BY	64041	72341	65041
323 242		**CE**	P	*CT*	BY	64042	72342	65042
323 243		**CE**	P	*CT*	BY	64043	72343	65043

CLASS 325

DTPMV (A)–MPMV–TPMV–DTPMV (B). New postal units based on Class 319. GTO thyristor control. Roller shutter doors and compatibility with diesel locomotive haulage. Disc brakes.
System: 25 kV a.c. overhead/750 V d.c. third rail.
Bogies: P7-4 (MPMV), T3-7 (others).
Gangways: None.
Traction Motors: Four GEC G315BY of 247.5 kW.
Dimensions: 19.83 x 2.82 m (outer cars), 19.92 x 2.82 m (inner cars).
Maximum Speed: 100 m.p.h.

68300–68330 (Even Nos.). DTPMV (A). Dia. EE501. Lot No. 31144 ABB Derby 1995.   . t.
MPMV. Dia. EC501. Lot No. 31145 ABB Derby 1995.   . t.
TPMV. Dia. EH501. Lot No. 31146 ABB Derby 1995.   . t.
68301–68331 (Odd Nos.). DTPMV (B). Dia. EE501. Lot No. 31144 ABB Derby 1995.   . t.

325 001	**RM**	R	*EW*	CE	68300	68340	68360	68301
325 002	**RM**	R	*EW*	CE	68302	68341	68361	68303
325 003	**RM**	R	*EW*	CE	68304	68342	68362	68305
325 004	**RM**	R	*EW*	CE	68306	68343	68363	68307
325 005	**RM**	R	*EW*	CE	68308	68344	68364	68309
325 006	**RM**	R	*EW*	CE	68310	68345	68365	68311
325 007	**RM**	R	*EW*	CE	68312	68346	68366	68313
325 008	**RM**	R	*EW*	CE	68314	68347	68367	68315
325 009	**RM**	R	*EW*	CE	68316	68348	68368	68317
325 010	**RM**	R	*EW*	CE	68318	68349	68369	68319
325 011	**RM**	R	*EW*	CE	68320	68350	68370	68321
325 012	**RM**	R	*EW*	CE	68322	68351	68371	68323

325 013	**RM**	R	*EW*	CE	68324	68352	68372	68325
325 014	**RM**	R	*EW*	CE	68326	68353	68373	68327
325 015	**RM**	R	*EW*	CE	68328	68354	68374	68329
325 016	**RM**	R	*EW*	CE	68330	68355	68375	68331

Name: 325 008 is named 'Peter Howarth C.B.E.'

CLASS 332   HEATHROW EXPRESS

DMCFO–PTSOL–TSO–DMSO* or DMSO–PTSOL–TSO–DMFO. Initially running as 3-car sets DMSO–PTSOL–DMSO. New units under construction for Heathrow Express service. Air conditioned. IGBT control. Tightlock couplers. Sliding doors. Disc brakes. Actual formations may be changed.

Bogies: CAF.
Gangways: Within unit.
Traction Motors: 4 Siemens per motor car.
Dimensions: 23.00 m x   . m.
Maximum Speed: 100 m.p.h.

DMCFO. CAF 1997–8. 14/– + wheelchair space.   . t.
DMSO. CAF 1997–8. –/48 + 4 tip-up.   . t.
PTSOL. CAF 1997–8. –/44 + 11 tip-up.   . t.
TSO. CAF 1997–8. –/56 + 11 tip-up.   . t.
DMFO. CAF 1997–8. 26/–.   . t.

Note: Owing to an error by the procurement consultants, the motor cars of these units have numbers in the series normally used for driving trailers and the PTSOLs have numbers in the series normally used for motor cars.

332 001	**HE**	B	*HE*		78401	63400	72400	78400
332 002	**HE**	B	*HE*		78403	63401	72401	78402
332 003	**HE**	B	*HE*		78405	63402	72402	78404
332 004	**HE**	B	*HE*		78407	63403	72403	78406
332 005	**HE**	B	*HE*	OH	78409	63404	72404	78408
332 006	**HE**	B	*HE*	OH	78411	63405	72405	78410
332 007	**HE**	B	*HE*	OH	78413	63406	72406	78412
332 008	**HE**	B	*HE*	OH	78415	63407	72407	78414
332 009	**HE**	B	*HE*		78417	63408	72408	78416
332 010	**HE**	B	*HE*		78419	63409	72409	78418
332 011	**HE**	B	*HE*		78421	63410	72410	78420
332 012	**HE**	B	*HE*		78423	63411	72411	78422
332 013	**HE**	B	*HE*		78425	63412	72412	78424
332 014	**HE**	B	*HE*		78427	63413	72413	78426

334   GEC-ALSTHOM JUNIPER   Scotrail Jc

CLASS 357   ADTRANZ ELECTROSTAR

DMSO (A)–PTSOL–MSO–DMSO (B). New units under construction for use on
LTS Rail. Air conditioned. Aluminium bodies. IGBT control. Tightlock couplers.
Sliding doors. Disc brakes. Regenerative braking. Provision for 750 V d.c.
supply if required.
Bogies: ADtranz P3-25 (motor cars), T3-25 (trailer car).
Gangways: Within unit.
Traction Motors: Two of 250 kW each per motor car.
Dimensions: 20.75 m x 2.80 m (motor cars), 20.10 x 2.80 m (inner cars).
Maximum Speed: 100 m.p.h.

DMSO (A). Dia. EA   . ADtranz Derby 1998–. –/71.   . t.
PTSOL. Dia. EH   ADtranz Derby 1998–. –/62.   . t.
MSO. Dia. EH   ADtranz Derby 1998–. –/78.   . t.
DMSO (B). Dia. EA   ADtranz Derby 1998–. –/71   . t.

357 001	P	LS	67651	74051	74151	67751
357 002	P	LS	67652	74052	74152	67752
357 003	P	LS	67653	74053	74153	67753
357 004	P	LS	67654	74054	74154	67754
357 005	P	LS	67655	74055	74155	67755
357 006	P	LS	67656	74056	74156	67756
357 007	P	LS	67657	74057	74157	67757
357 008	P	LS	67658	74058	74158	67758
357 009	P	LS	67659	74059	74159	67759
357 010	P	LS	67660	74060	74160	67760
357 011	P	LS	67661	74061	74161	67761
357 012	P	LS	67662	74062	74162	67762
357 013	P	LS	67663	74063	74163	67763
357 014	P	LS	67664	74064	74164	67764
357 015	P	LS	67665	74065	74165	67765
357 016	P	LS	67666	74066	74166	67766
357 017	P	LS	67667	74067	74167	67767
357 018	P	LS	67668	74068	74168	67768
357 019	P	LS	67669	74069	74169	67769
357 020	P	LS	67670	74070	74170	67770
357 021	P	LS	67671	74071	74171	67771
357 022	P	LS	67672	74072	74172	67772
357 023	P	LS	67673	74073	74173	67773
357 024	P	LS	67674	74074	74174	67774
357 025	P	LS	67675	74075	74175	67775
357 026	P	LS	67676	74076	74176	67776
357 027	P	LS	67677	74077	74177	67777
357 028	P	LS	67678	74078	74178	67778
357 029	P	LS	67679	74079	74179	67779
357 030	P	LS	67680	74080	74180	67780
357 031	P	LS	67681	74081	74181	67781
357 032	P	LS	67682	74082	74182	67782
357 033	P	LS	67683	74083	74183	67783
357 034	P	LS	67684	74084	74184	67784

357 035	P	*LS*		67685	74085	74185	67785
357 036	P	*LS*		67686	74086	74186	67786
357 037	P	*LS*		67687	74087	74187	67787
357 038	P	*LS*		67688	74088	74188	67788
357 039	P	*LS*		67689	74089	74189	67789
357 040	P	*LS*		67690	74090	74190	67790
357 041	P	*LS*		67691	74091	74191	67791
357 042	P	*LS*		67692	74092	74192	67792
357 043	P	*LS*		67693	74093	74193	67793
357 044	P	*LS*		67694	74094	74194	67794

CLASS 365   NETWORKER EXPRESS

DMCO–TSOL–PTSOL–DMSO. New units with aluminium bodies. GTO thyristor control. Tightlock couplers. Sliding doors. Disc rheostatic and regenerative braking.
System: 25 kV a.c. overhead/750 V d.c. third rail.
Bogies: P7 (power cars), T3 (trailers).
Gangways: Within unit.
Traction Motors: Four GEC-Alsthom G354CX three-phase induction motors.
Dimensions: 20.89 x 2.81 m (outer cars), 20.06 x 2.81 m (inner cars).
Maximum Speed: 100 m.p.h.

DMCO. Dia. EA301. Lot No. 31133 ABB York 1994–5. 12/56. 46.7 t.
TSOL. Dia. EH298. Lot No. 31134 ABB York 1994–5. –/59 + 5 tip-up 1TD. 32.9 t.
PTSOL. Dia. EH298. Lot No. 31135 ABB York 1994–5. –/68 1T. 34.6 t.
DMSO. Dia. EA208. Lot No. 31136 ABB York 1994–5. –/72. 46 t.

365 501	**CN**	F	*SE*	RE	65894	72241	72240	65935
365 502	**CN**	F	*SE*	RE	65895	72243	72242	65936
365 503	**CN**	F	*SE*	RE	65896	72245	72244	65937
365 504	**CN**	F	*SE*	RE	65897	72247	72246	65938
365 505	**CN**	F	*SE*	RE	65898	72249	72248	65939
365 506	**CN**	F	*SE*	RE	65899	72251	72250	65940
365 507	**CN**	F	*SE*	RE	65900	72253	72252	65941
365 508	**CN**	F	*SE*	RE	65901	72255	72254	65942
365 509	**CN**	F	*SE*	RE	65902	72257	72256	65943
365 510	**CN**	F	*SE*	RE	65903	72259	72258	65944
365 511	**CN**	F	*SE*	RE	65904	72261	72260	65945
365 512	**CN**	F	*SE*	RE	65905	72263	72262	65946
365 513	**CN**	F	*SE*	RE	65906	72265	72264	65947
365 514	**CN**	F	*SE*	RE	65907	72267	72266	65948
365 515	**CN**	F	*SE*	RE	65908	72269	72268	65949
365 516	**CN**	F	*SE*	RE	65909	72271	72270	65950
365 517	**NW**	F	*WN*	HE	65910	72273	72272	65951
365 518	**NW**	F	*WN*	HE	65911	72275	72274	65952
365 519	**NW**	F	*WN*	HE	65912	72277	72276	65953
365 520	**NW**	F	*WN*	HE	65913	72279	72278	65054
365 521	**NW**	F	*WN*	HE	65914	72281	72280	65955
365 522	**NW**	F	*WN*	HE	65915	72283	72282	65956
365 523	**NW**	F	*WN*	HE	65916	72285	72284	65957

365 524	**NW**	F	*WN*	HE	65917	72287	72286	65958
365 525	**NW**	F	*WN*	HE	65918	72289	72288	65959
365 526	**NW**	F	*WN*	HE	65919	72291	72290	65960
365 527	**NW**	F	*WN*	HE	65920	72293	72292	65961
365 528	**NW**	F	*WN*	HE	65921	72295	72294	65962
365 529	**NW**	F	*WN*	HE	65922	72297	72296	65963
365 530	**NW**	F	*WN*	HE	65923	72299	72298	65964
365 531	**NW**	F	*WN*	HE	65924	72301	72300	65965
365 532	**NW**	F	*WN*	HE	65925	72303	72302	65966
365 533	**NW**	F	*WN*	HE	65926	72305	72304	65967
365 534	**NW**	F	*WN*	HE	65927	72307	72306	65968
365 535	**NW**	F	*WN*	HE	65928	72309	72308	65969
365 536	**NW**	F	*WN*	HE	65929	72311	72310	65970
365 537	**NW**	F	*WN*	HE	65930	72313	72312	65971
365 538	**NW**	F	*WN*	HE	65931	72315	72314	65972
365 539	**NW**	F	*WN*	HE	65932	72317	72316	65973
365 540	**NW**	F	*WN*	HE	65933	72319	72318	65974
365 541	**NW**	F	*WN*	HE	65934	72321	72320	65975

CLASS 375   ADTRANZ ELECTROSTAR

DMSO(A)–PTSOL–MSO–DMSO(B). New units under construction for use on
Connex South Eastern. Air conditioned. Aluminium bodies. IGBT control.
Tightlock couplers. Swing plug doors. Disc brakes. Regenerative braking. Full
details not yet available.
System: 25 kV a.c. overhead/750 V d.c. third rail.
Bogies: .
Gangways: Throughout.
Traction Motors:
Dimensions:
Maximum Speed: 100 m.p.h.

DMSO(A). Dia. EA   . ADtranz Derby 1998. .   . t.
PTSOL. Dia. EH   ADtranz Derby 1998. .   . t.
MSO. Dia. EH   ADtranz Derby 1998. .   . t.
DMSO(B). Dia. EA   ADtranz Derby 1998.   . t.

375 001	CX	*SE*
375 002	CX	*SE*
375 003	CX	*SE*
375 004	CX	*SE*
375 005	CX	*SE*
375 006	CX	*SE*
375 007	CX	*SE*
375 008	CX	*SE*
375 009	CX	*SE*
375 010	CX	*SE*
375 011	CX	*SE*
375 012	CX	*SE*
375 013	CX	*SE*
375 014	CX	*SE*

375 015	CX	*SE*
375 016	CX	*SE*
375 017	CX	*SE*
375 018	CX	*SE*
375 019	CX	*SE*
375 020	CX	*SE*
375 021	CX	*SE*
375 022	CX	*SE*
375 023	CX	*SE*
375 024	CX	*SE*
375 025	CX	*SE*
375 026	CX	*SE*
375 027	CX	*SE*
375 028	CX	*SE*
375 029	CX	*SE*
375 030	CX	*SE*

4.2. 750 V d.c. THIRD RAIL EMUs

These classes operate on the third rail system at 750–850 V d.c. Except where stated otherwise, all multiple units can run in multiple with one another. Buffet cars have electric cooking. In addition to the class number, the old SR designations e.g. 4 Cig are quoted where appropriate. Outer couplings are buckeyes on units built before 1982 with bar couplings within the units. Newer units have tightlock outer couplings.

CLASS 438   4 TC

DTSO–TFK–TBSK–DTSO. Converted from loco-hauled stock. Unpowered units which worked push & pull with class 431/2 tractor units and class 33/1 and 73 locos. Express stock.

Electrical Equipment: 1966-type.
Bogies: B5 (SR) bogies.
Gangways: Throughout.
Dimensions: 19.66 x 2.82 m.
Maximum Speed: 90 m.p.h.

DTSO. Dia. EE266. Lot No. 30764 York 1966–67. –/64. 32 t.
TFK. Dia. EH160. Lot No. 30766 York 1966–67. 42/– 2T. 33.5 t.
TBSK. Dia. EJ260. Lot No. 30765 York 1966–67. –/32 1T. 35.5 t.

410	B	P	ZG	76288	70859	70812	76287
417	B	RO	TM	76302	70860	70826	76301
Spare	N	P	ZG	76327			

Former numbers of converted hauled stock:

70812 (34987)	70860 (13019)	76288 (4391)	76302 (4382)
70826 (34980)	76287 (4379)	76301 (4375)	76327 (4018)
70859 (13040)			

CLASS 421/5   'GREYHOUND' 4 Cig (PHASE 2)

DTCsoL (A)–MBSO–TSO–DTCsoL (B). Express stock. All facelifted with new trim and fluorescent lighting in saloons.

Note: The following details apply to all Class 421 (phase 2) sets.

Diagram Numbers: EE369, ED264, EH287, EE369.
Electrical Equipment: 1963-type.
Bogies: Two Mk. 6 motor bogies (MBSO). B5 (SR) bogies (trailer cars).
Gangways: Throughout.
Traction Motors: Four EE507 of 185 kW.
Dimensions: 19.75 x 2.82 m.
Maximum Speed: 90 m.p.h.

76561–76567. DTCsoL(A). Lot No. 30802 York 1970. 18/36 2T. 35.5 t.
76581–76610. DTCsoL(A). Lot No. 30806 York 1970. 18/36 2T. 35.5 t.
76717–76787. DTCsoL(A). Lot No. 30814 York 1970–72. 18/36 2T. 35.5 t.
76859. DTCsoL(A). Lot No. 30827 York 1972. 18/36 2T. 35.5 t.
62277–62283. MBSO. Lot No. 30804 York 1970. –/56. 49 t.
62287–62316. MBSO. Lot No. 30808 York 1970. –/56. 49 t.
62355–62425. MBSO. Lot No. 30816 York 1970. –/56. 49 t.
62430. MBSO. Lot No. 30829 York 1972. –/56. 49 t.
70967–70996. TSO. Lot No. 30809 York 1970–71. –/72. 31.5t.
71035–71105. TSO. Lot No. 30817 York 1970. –/72. 31.5t.
71106. TSO. Lot No. 30830 York 1972. –/72. 31.5t.
71926–71928. TSO. Lot No. 30805 York 1970. –/72. 31.5t.
76571–76577. DTCsoL(B). Lot No. 30802 York 1970. 24/28 2T. 35 t.
76611–76640. DTCsoL(B). Lot No. 30807 York 1970. 24/28 2T. 35 t.
76788–76858. DTCsoL(B). Lot No. 30815 York 1970–72. 24/28 2T. 35 t.
76859. DTCsoL(B). Lot No. 30828 York 1972. 18/36 2T. 35 t.

These sets are known as 'Greyhound' units and are fitted with an additional stage of field weakening to improve the maximum attainable speed. This term is traditional on the lines of the former London & South Western railway, as it was formerly applied to their Class T9 4–4–0 express steam locomotives.

1301	N	F	SW	FR	76595	62301	70981	76625
1302	N	F	SW	FR	76584	62290	70970	76614
1303	N	F	SW	FR	76581	62287	70967	76611
1304	N	F	SW	FR	76583	62289	70969	76613
1305	N	F	SW	FR	76717	62355	71035	76788
1306	ST	F	SW	FR	76723	62361	71041	76794
1307	N	F	SW	FR	76586	62292	70972	76616
1308	N	F	SW	FR	76627	62298	70978	76622
1309	N	F	SW	FR	76594	62300	70980	76624
1310	N	F	SW	FR	76567	62283	71926	76577
1311	N	F	SW	FR	76561	62277	71927	76571
1312	ST	F	SW	FR	76562	62278	71928	76572
1313	ST	F	SW	FR	76596	62302	70982	76626
1314	ST	F	SW	FR	76588	62294	70974	76618
1315	ST	F	SW	FR	76608	62314	70994	76638
1316	ST	F	SW	FR	76585	62291	70971	76615
1317	ST	F	SW	FR	76597	62303	70983	76592
1318	N	F	SW	FR	76590	62296	70976	76620
1319	N	F	SW	FR	76591	62297	70977	76621
1320	ST	F	SW	FR	76593	62299	70979	76623
1321	ST	F	SW	FR	76589	62295	70975	76619
1322	ST	F	SW	FR	76587	62293	70973	76617

Former numbers of converted buffet cars:

71926 (69315) | 71927 (69330) | 71928 (69331)

Note: No new Lot Nos were issued for the above conversions.

CLASS 424/3   3 Cop (PHASE 2)

DTSOL(A)–MBSO–DTSOL(B). Express stock.

Diagram Numbers: EE3??, ED264, EE3??.
Electrical Equipment: 1963-type.
Bogies: Two Mk. 6 motor bogies (MBSO). B5 (SR) bogies (trailer cars).
Gangways: Throughout.
Traction Motors: Four EE507 of 185 kW.
Dimensions: 19.75 x 2.82 m.
Maximum Speed: 90 m.p.h.

76563–76570. DTSOL(A). Lot No. 30802 York 1970. –/60 1T. 35.5 t.
76602. DTSOL(A). Lot No. 30806 York 1972. –/60 1T. 35.5 t.
76728–76750. DTSOL(A). Lot No. 30814 York 1970–72. –60 1T. 35 t.
62279–62286. MBSO. Lot No. 30804 York 1970. –/56 + 1 tip-up. 49 t.
62308. MBSO. Lot No. 30808 York 1970. –/56 + 1 tip-up. 49 t.
62366–62388. MBSO. Lot No. 30816 York 1970. –/56 + 1 tip-up. 49 t.
76573–76580. DTSOL(B). Lot No. 30802 York 1970. –/60 1T. 35 t.
76632. DTSOL(B). Lot No. 30807 York 1970. –/60 1T. 35 t.
76799–76821. DTSOL(B). Lot No. 30815 York 1970–72. –/60 1T. 35 t.

1401	(2208)	CX	P	SC	BI	76568	62284	76578
1402	(2204)	CX	P	SC	BI	76564	62280	76574
1403	(2203)	CX	P	SC	BI	76563	62279	76573
1404	(2206)	CX	P	SC	BI	76602	62308	76632
1405	(2205)	CX	P	SC	BI	76565	62281	76575
1406	(2252)	CX	P	SC	BI	76728	62366	76799
1407	(2257)	CX	P	SC	BI	76729	62367	76800
1408	(2261)	CX	P	SC	BI	76750	62388	76821
1409	(2209)	CX	P	SC	BI	76569	62285	76579
1410	(2253)	CX	P	SC	BI	76734	62372	76805
1411	(2210)	CX	P	SC	BI	76570	62286	76580

CLASS 411/5   REFURBISHED 4 Cep

DMSO (A)–TBCK–TSOL–DMSO (B). Kent Coast Express Stock. Refurbished and renumbered from the 71/72xx series. Fitted with hopper ventilators, Inter-City 70 seats and fluorescent lighting.

Electrical Equipment: 1957-type.
Bogies: One Mk. 4 (Mk 3B§) motor bogie (DMSO). Commonwealth trailer bogies.
Gangways: Throughout.
Traction Motors: Two EE507 of 185 kW.
Dimensions: 19.75 x 2.82 m.
Maximum Speed: 90 m.p.h.

†–70345 is a TBFK with one compartment declassified. It is from the original refurbished unit (1500), has a different interior colour scheme and does not have hopper ventilators.

DMSO (A). Dia. EA263. –/64. 44.15 t.
TBCK. Dia. EJ361. 24/6 2T. 36.17 t.
TSOL. Dia. EH282. –/64 2T. 33.78 t.
DMSO (B). Dia. EA264. –/64. 43.54 t.

Lot numbers are as follows, all cars being built at Ashford/Eastleigh:

61229–61240. 30449 1958.	**70241.** 30640 1961.
61306–61409. 30454 1958–59.	**70261–70302.** 30455 1958–59.
61696–61811. 30619 1960–61.	**70304–70355.** 30456 1958–59.
61868–61869. 30638 1960–61.	**70504–70551.** 30620 1960–61.
61950–61959. 30708 1963.	**70553–70610.** 30621 1960–61.
70043–70044. 30639 1961.	**70654–70657.** 30709 1963.
70229–70234. 30450 1958.	**70660–70664.** 30710 1963.
70235–70239. 30451 1958.	

1507	**N**	P	*SW*	FR	61363	70332	70289	61362
1509	**N**	P	*SE*	RE	61335	70318	70275	61334
1510	**N**	P	*SE*	RE	61365	70333	70290	61364
1511	**N**	P	*SE*	RE	61367	70334	70291	61366
1512	**ST**	P	*SW*	FR	61321	70311	70268	61320
1517	**N**	P	*SE*	RE	61317	70309	70266	61316
1519	**ST**	P	*SW*	FR	61403	70352	70516	61402
1520	**N**	P	*SE*	RE	61343	70327	70284	61380
1527	**N**	P	*SE*	RE	61237	70239	70233	61238
1530	**N**	P	*SE*	RE	61331	70316	70273	61330
1531	**ST**	P	*SW*	FR	61233	70237	70231	61234
1532	**ST**	P	*SW*	FR	61391	70346	71626	61390
1533	**ST**	P	*SW*	BM	61393	70347	71627	61385
1534	**ST**	P	*SW*	FR	61405	70353	71628	61404
1535	**N**	P	*SW*	FR	61397	70349	71629	61396
1536	**N**	P	*SE*	RE	61399	70350	71631	61398
1537	**N**	P	*SW*	FR	61229	70235	70229	61230
1538	**ST**	P	*SW*	FR	61307	70304	70261	61306
1539	**ST**	P	*SW*	FR	61401	70351	71632	61400
1541	**N**	P	*SE*	RE	61409	70355	71633	61408
1543	**N**	P	*SE*	RE	61323	70312	70297	61322
1544	**ST**	P	*SW*	FR	61315	70302	70265	61349
1547	**ST**	P	*SW*	FR	61329	70578	70272	61328
1548	**ST**	P	*SW*	FR	61375	70338	70295	61374
1549	**N**	P	*SE*	RE	61339	70320	70277	61338
1550	**N**	P	*SE*	RE	61313	70307	70264	61312
1551	**N**	P	*SE*	RE	61325	70313	70270	61324
1553	**N**	P	*SE*	RE	61728	70306	70263	61350
1554	**N**	P	*SE*	RE	61369	70335	70292	61368
1555	**N**	P	*SE*	RE	61311	70326	70283	61310
1556	**N**	P	*SE*	RE	61371	70336	70293	61370
1557	**N**	P	*SE*	RE	61337	70331	70288	61360
1559	**N**	P	*SE*	RE	61377	70339	70296	61376
1560	**N**	P	*SE*	RE	61307	70344	70301	61386
1561	**N**	P	*SE*	RE	61231	70604	70230	61232
1562	**N**	P	*SE*	RE	61407	70236	70241	61406

▲ Strathclyde PTE liveried Class 303 No. 303 065 departs from Glasgow Central with a service for Neilson on 29th August 1996. **Colin J. Marsden**

▼ A Greater Manchester PTE liveried Class 305 No. 305 506 approaches Dinting Crossing with the 15.08 Hadfield–Manchester Piccadilly. The date is 24th September 1997, shortly before the Hatfield/Glossop branches went over to Class 323 operation. **Nic Joynson**

▲ Class 308 No. 308 153, in West Yorkshire PTE livery, leaves Skipton as the 17.43 to Leeds on 17th August 1996. **Dave McAlone**

▼ North West Regional Railways liveried Class 309 'Clacton' unit No. 309 617 passes Slindon, Staffordshire on 25th July 1997 as the 19.06 Birmingham New Street–Manchester Piccadilly. **Hugh Ballantyne**

▲ Class 310 No. 310 087, in Network SouthEast livery, passes Shadwell on 2nd September 1997 with the 11.30 London Fenchurch Street–Shoeburyness. The line on the left is used by Docklands Light Railway cars. **Kevin Conkey**

▼ Displaying the new Great Eastern livery to good effect, Class 312 No. 312 706 crosses the River Stour at Manningtree with a Harwich Town–Ipswich service on 12th November 1997. **John A. Day**

▲ Silverlink liveried Class 313 No. 313 134 at Watford Junction on 18th October 1997. **Kevin Conkey**

▼ The first unit to be painted into Great Eastern livery, Class 315 No. 315 809 at Shenfield during the livery launch on 15th April 1997. **Colin J. Marsden**

LTS Rail has created its own livery by painting the red stripe on some of its Network SouthEast liveried units light green. One of the units, Class 317 No. 317 301 is pictured here approaching Shadwell on 2nd September 1997 with the 10.05 London Fenchurch Street–Grays.

Kevin Conkey

▲ A southbound 'Thameslink' service approaches Coulsdon on 26th July 1996, formed by Class 319 No. 319 049. The unit carries the original Thameslink livery which has now been abandoned in favour of a new design.　**Dave McAlone**

▼ Class 320 No 320 306, in carmine and cream livery, at Glasgow Central on 18th February 1997. This livery has now been adopted by Strathclyde PTE.
Colin J. Marsden

▲ Class 321 No. 321 444 at Ipswich with the 11.30 to London Liverpool Street on 7th July 1997. Slight differences can be seen between the Great Eastern livery carried on this class and that carried on Classes 312 and 315.   **John A. Day**

▼ Stansted Skytrain liveried Class 322 No. 322 485 approaches Bethnal Green on 16th August 1997 with the 11.00 London Liverpool Street–Stansted service.
Kevin Conkey

▲ Class 323 No. 323 233, in North Western Trains livery, arrives ECS at Manchester Piccadilly on 27th October 1997 prior to working the 11.43 to Glossop.
Peter Fox

▼ Heathrow Express Class 332 Nos. 332 003 and 332 004 at Old Oak Common on 23rd June 1997. **Colin J. Marsden**

Royal Mail liveried Class 325 No. 325 010 passes Low Gill on 11th July 1997 with a Glasgow to Crewe empty stock working.

Kevin Conkey

▲ Connex liveried Class 365s Nos. 365 505 and 365 504 enter Chatham with the 11.05 London Victoria–Ramsgate fast service. The unit has had yellow vinyls applied to the lower bodyside to hide the Network SouthEast stripes but retains the blue around the windows. **David Brown**

▼ Class 411/5 (4 Cep) units. Nos. 1699 and 1698 depart from Redhill with the 07.45 Three Bridges–London Bridge on 28th July 1997. **David Brown**

▲ Class 421/3 (4 Cig) No. 1726 leads 2259 and 1719 passing Earlswood with the 15.17 Victoria–Bournemouth/Littlehampton/Eastbourne on 21st July 1997.
David Brown

▼ Stagecoach liveried Class 442 No. 2402 leaves Clapham Junction on 8th May 1996 with the 11.50 London Waterloo–Poole. **Kevin Conkey**

▲ Class 423/0 No. 3433 runs through Surbiton on 10th July 1996 with an ECS for Hampton Court. **Colin J. Marsden**

▼ Gatwick Express liveried Class 488/3 TSOL No. 72715 of set 8316 approaches Clapham Junction on 8th May 1996 as part of the 10.50 Gatwick Airport–London Victoria. **Kevin Conkey**

South West Trains Stagecoach liveried Class 455/7 No. 5726 and Network SouthEast liveried No. 5717 depart from New Malden on 6th September 1997 with the 08.33 London Waterloo–Richmond–Waterloo service. **David Brown**

▲ On the last day of Wimbledon to West Croydon services, 31st May 1997, Class 456 No. 456 017 stands at West Croydon with the 09.15 to Wimbledon.

Chris Wilson

▼ Class 466 No. 466 040 and Class 465 No. 465 224 leave London Bridge on 22nd March 1997 with the 12.20 London Charing Cross–Orpington.

Kevin Conkey

▲ Merseytravel liveried Class 507 No. 507 005 at Ainsdale with a Southport to Liverpool Central service on 10th August 1997. **Martyn Hilbert**

▼ Island Line Class 483 No. 006 at Sandown with the 15.45 Ryde Pier Head–Shanklin on 21st July 1997. **Martyn Hilbert**

Eurostar Class 373 set No. 3019 leads the 08.53 London Waterloo–Paris Nord at Westenhanger on 4th April 1997.

Michael J. Collins

1563	§	**N**	P	*SE*	RE	61740	70575	70526	61741
1564	§	**N**	P	*SE*	RE	61788	70599	70550	61789
1565	§	**N**	P	*SE*	RE	61762	70586	71711	61763
1566	§	**N**	P	*SE*	RE	61722	70566	70517	61723
1568	§	**ST**	P	*SW*	FR	61766	70588	70539	61767
1570	§	**N**	P	*SE*	RE	61738	70574	70525	61739
1571	§	**N**	P	*SE*	RE	61806	70608	71636	61807
1572	§	**N**	P	*SE*	RE	61734	70572	70523	61735
1573	§	**N**	P	*SW*	FR	61726	70568	70519	61727
1574	§	**N**	P	*SE*	RE	61792	70601	71635	61793
1575	§	**N**	P	*SE*	RE	61768	70583	70540	61769
1576	§	**N**	P	*SE*	RE	61770	70590	70541	61771
1577	§	**N**	P	*SE*	RE	61718	70564	70515	61719
1578	§	**ST**	P	*SW*	FR	61700	70555	70506	61701
1580	§	**N**	P	*SE*	RE	61756	70589	70534	61757
1581	§	**ST**	P	*SW*	FR	61784	70597	70548	61785
1582	§	**N**	P	*SE*	RE	61748	70603	71630	61797
1584	§	**N**	P	*SE*	RE	61752	70581	70532	61753
1585	§	**N**	P	*SE*	RE	61710	70560	70511	61711
1586	§	**N**	P	*SE*	RE	61714	70562	70513	61715
1587	§	**N**	P	*SE*	RE	61764	70587	71625	61765
1588	§	**N**	P	*SE*	RE	61720	70044	70520	61721
1589	§	**ST**	P	*SW*	FR	61742	70576	70527	61743
1590	§	**N**	P	*SE*	RE	61696	70553	70504	61697
1591	§	**N**	P	*SE*	RE	61790	70600	70551	61791
1592	§	**N**	P	*SE*	RE	61778	70594	70545	61779
1593	§	**N**	P	*SE*	RE	61730	70570	70521	61731
1594	§	**N**	P	*SE*	RE	61754	70582	70533	61755
1595	§	**N**	P	*SE*	RE	61704	70557	70508	61705
1597	§	**N**	P	*SE*	RE	61708	70559	70510	61709
1599	§	**N**	P	*SE*	RE	61706	70558	70509	61707
1602	§	**N**	P	*SE*	RE	61958	70565	70279	61959
1607	§	**N**	P	*SE*	RE	61698	70554	70505	61699
1609	§	**N**	P	*SE*	RE	61744	70577	70528	61745
1610	§	**N**	P	*SE*	RE	61750	70580	70531	61751
1611	§	**N**	P	*SE*	RE	61758	70584	70537	61759
1612	§	**N**	P	*SW*	FR	61794	70602	70535	61795
1613	§	**N**	P	*SE*	RE	61760	70585	70536	61761
1614	§	**N**	P	*SE*	RE	61702	70556	70507	61703
1615	§	**N**	P	*SE*	RE	61956	70657	70664	61957
1616	§	**N**	P	*SE*	RE	61950	70654	70543	61951
1617	§	**N**	P	*SE*	RE	61800	70605	70661	61801
1618	§	**N**	P	*SE*	RE	61868	70043	70663	61869
1619	§	**N**	P	*SE*	RE	61952	70655	70662	61953
Spare		**N**	P		ZD	61383			
Spare		**J**	P		BM		70302		
Spare		**N**	P		ZG			70345	
Spare		**N**	P		ZA			70660	

Units fitted with B5(SR) bogies.

1697	**N**	P	*SC*	BI	61373	70337	70294	61372
1698	**N**	P	*SC*	BI	61355	70343	70300	61384
1699	§ **N**	P	*SC*	BI	61712	70561	70512	61713

Former numbers of converted hauled stock:

71625 (4381)	71628 (3844)	71631 (4436)	71635 (3990)
71626 (3916)	71629 (3992)	71632 (4063)	71636 (4065)
71627 (3921)	71630 (3988)	71633 (4072)	71711 (3994)

Note: No new Lot Nos. were issued for the above conversions.

CLASS 421/3   4 Cig (PHASE 1)

DTCsoL(A)–MBSO–TSO–DTCsoL(B). Express stock. Fitted with electric parking brake. Facelifted with new trim and fluorescent lighting in saloons.

Electrical Equipment: 1963-type.
Bogies: Two Mk. 4 motor bogies (MBSO). B5 (SR) bogies (trailer cars).
Gangways: Throughout.
Traction Motors: Four EE507 of 185 kW.
Dimensions: 19.75 x 2.82 m.
Maximum Speed: 90 m.p.h.

DTCsoL(A). Dia. EE364. Lot No. 30741 York 1964–65. 18/36 2T. 35.5 t.
MBSO. Dia. ED260. Lot No. 30742 York 1964–65. –/56. 49 t.
70695–70730. TSO. Dia. EH275. Lot No. 30730 York 1964–65. –/72. 31.5 t.
71044–71097. TSO. Dia. EH275. Lot No. 30817 York 1970. –/72. 31.5 t.
71766–71770. TSO. Dia. EH275. Lot No. 30784 York 1964–65. –/72. 31.5 t.
DTCsoL(B). Dia. EE363. Lot No. 30740 York 1964–65. 24/28 2T.

* Units reformed from Class 422 to enable all Class 422 power cars to have Mk. 6 motor bogies. Phase 1 units with phase 2 TSOs.

1701	**N**	A		ZG	76087	62028	70706	76033
1702	**N**	A	*SC*	BI	76101	62042	70720	76047
1703	**N**	A	*SC*	BI	76097	62038	70716	76043
1704	**CX**	A	*SC*	BI	76092	62033	70711	76038
1705	**N**	A	*SC*	BI	76076	62017	70695	76022
1706	**N**	A	*SC*	BI	76094	62035	70713	76040
1707	**N**	A	*SC*	BI	76084	62025	70703	76030
1708	**N**	A	*SC*	BI	76110	62051	70729	76056
1709	**N**	A	*SC*	BI	76103	62044	70722	76049
1710	**CX**	A	*SC*	BI	76078	62019	70697	76024
1711	**N**	A	*SC*	BI	76114	62055	71766	76060
1712	**N**	A	*SC*	BI	76079	62020	70698	76025
1713	**N**	A	*SC*	BI	76128	62069	71767	76074
1714	**N**	A	*SC*	BI	76077	62018	70696	76023
1717	**N**	A	*SC*	BI	76083	62024	70702	76029
1719	**CW**	A	*SC*	BI	76116	62057	70719	76062
1720	**N**	A	*SC*	BI	/6098	62039	70704	76044
1721	**CX**	A	*SC*	BI	76090	62031	70709	76036
1722	**CX**	A	*SC*	BI	76106	62047	70725	76052

1724	**CX**	A	*SC*	BI	76120	62061	71770	76066
1725	**CX**	A	*SC*	BI	76088	62029	70707	76034
1726	**CX**	A	*SC*	BI	76109	62050	70728	76055
1727	**CX**	A	*SC*	BI	76111	62052	70730	76057
1731	**N**	A	*SC*	BI	76095	62036	70714	76041
1733	* **CX**	A	*SC*	BI	76122	62063	71047	76068
1734	* **N**	A	*SC*	BI	76063	62054	71044	76059
1735	* **N**	A	*SC*	BI	76117	62058	71050	76051
1736	* **N**	A		ZG	76124	62065	71052	76070
1737	* **N**	A		ZG	76121	62062	71058	76067
1738	* **N**	A	*SC*	BI	76129	62064	71046	76069
1739	* **N**	A	*SC*	BI	76123	62070	71066	76075
1740	* **CX**	A	*SC*	BI	76126	62067	71097	76072
1741	**CX**	A	*SC*	BI	76089	62030	70708	76035
1742	**N**	A		ZG	76086	62027	70705	76032
1743	* **CX**	A	*SC*	BI	76118	62059	71065	76064
1744	* **CX**	A	*SC*	BI	76127	62068	71066	76073
1745	**N**	A	*SC*	BI	76085	62026	70704	76031
1746	**N**	A	*SC*	BI	76091	62032	70710	76037
1747	**N**	A	*SC*	BI	76026	62034	70712	76093
1748	* **N**	A		ZG	76115	62056	71067	76061
1750	**N**	A	*SC*	BI	76080	62021	70699	76039
1751	**N**	A	*SC*	BI	76125	62066	71051	76071
1752	**N**	A	*SC*	BI	76119	62060	70717	76065
1753	**N**	A	*SC*	BI	76102	62043	70721	76048
1799	* **N**	A		ZF		62053	71068	76058

Former numbers of converted buffet cars:

71766 (69303) | 71768 (69317) | 71769 (69305) | 71770 (69308)
71767 (69314)

Note: No new lot numbers were issued for the above conversions

CLASS 421/4   4 Cig (PHASE 2)

DTCsoL(A)–MBSO–TSO–DTCsoL(B). Express stock. Facelifted with new trim and fluorescent lighting in saloons. For details see Class 421/5.

1801	**N**	P	*SC*	BI	76777	62415	71095	76848
1802	**N**	P	*SC*	BI	76754	62392	71072	76825
1803	**N**	A	*SC*	BI	76780	62418	71098	76851
1804	**CX**	A	*SC*	BI	76778	62416	71096	76849
1805	**N**	A	*SC*	BI	76782	62420	71100	76853
1806	**N**	F	*SE*	RE	76783	62421	71101	76854
1807	**N**	F	*SE*	RE	76784	62422	71102	76855
1808	**N**	F	*SE*	RE	76785	62423	71103	76856
1809	**N**	F	*SE*	RE	76786	62424	71104	76857
1810	**N**	F	*SE*	RE	76787	62425	71105	76858
1811	**N**	F	*SE*	RE	76781	62419	71099	76852
1812	**N**	F	*SE*	RE	76757	62395	71075	76828
1813	**N**	F	*SE*	RE	76859	62430	71106	76860

1831	**N**	A	*SC*	BI	76598	62304	70984	76628
1832	**CX**	A	*SC*	BI	76719	62357	71037	76790
1833	**CX**	A	*SC*	BI	76582	62288	70968	76612
1834	**CX**	A	*SC*	BI	76566	62282	70988	76576
1835	**CX**	A	*SC*	BI	76601	62307	70987	76631
1837	**CX**	A	*SC*	BI	76722	62360	71040	76793
1839	**N**	F	*SE*	RE	76607	62313	70993	76637
1840	**N**	F	*SE*	RE	76724	62362	71042	76795
1841	**N**	F	*SE*	RE	76603	62309	70989	76633
1842	**N**	F	*SE*	RE	76725	62363	71043	76796
1843	**N**	F	*SE*	RE	76731	62369	71049	76802
1845	**CX**	A	*SC*	BI	76599	62305	70985	76629
1846	**CX**	A	*SC*	BI	76737	62375	71055	76808
1847	**N**	A	*SC*	BI	76600	62306	70986	76630
1848	**N**	A	*SC*	BI	76605	62311	70991	76635
1850	**CX**	A	*SC*	BI	76718	62356	71036	76789
1851	**N**	A	*SC*	BI	76721	62355	71039	76792
1853	**N**	A	*SC*	BI	76606	62312	70992	76636
1854	**N**	A	*SC*	BI	76738	62376	71056	76809
1855	**N**	A	*SC*	BI	76720	62358	71038	76791
1856	**N**	A	*SC*	BI	76739	62377	71057	76810
1857	**N**	A	*SC*	BI	76610	62316	70996	76640
1858	**N**	A	*SC*	BI	76604	62310	70990	76634
1859	**N**	A	*SC*	BI	76727	62365	71045	76798
1860	**N**	A	*SC*	BI	76752	62390	71070	76823
1861	**N**	A	*SC*	BI	76735	62373	71053	76806
1862	**CX**	A	*SC*	BI	76736	62374	71054	76807
1863	**N**	A	*SC*	BI	76742	62380	71060	76813
1864	**N**	A	*SC*	BI	76741	62379	71059	76812
1865	**N**	A	*SC*	BI	76745	62383	71063	76639
1866	**N**	A	*SC*	BI	76743	62381	71061	76814
1867	**N**	A	*SC*	BI	76744	62382	71062	76815
1868	**N**	A	*SC*	BI	76751	62389	71069	76822
1869	**N**	A	*SC*	BI	76753	62391	71071	76804
1870	**N**	F	*SE*	RE	76108	62305	71089	76842
1871	**N**	F	*SE*	RE	76756	62394	71074	76827
1872	**N**	F	*SE*	RE	76771	62396	71076	76829
1873	**N**	F	*SE*	RE	76759	62397	71077	76830
1874	**N**	A	*SC*	BI	76755	62393	71073	76826
1876	**N**	F	*SE*	RE	76761	62399	71079	76832
1877	**N**	F	*SE*	RE	76763	62401	71081	76834
1878	**N**	F	*SE*	RE	76768	62406	71086	76839
1879	**N**	F	*SE*	RE	76760	62398	71078	76831
1880	**N**	F	*SW*	FR	76770	62400	71088	76841
1881	**N**	F	*SW*	FR	76762	62400	71080	76833
1882	**N**	F	*SW*	FR	76765	62403	71083	76836
1883	**N**	F	*SW*	FR	76764	62402	71082	76835
1884	**N**	F	*SW*	FR	76767	62405	71085	76838
1805	**N**	F	*SW*	FR	76769	62407	71087	76840
1886	**N**	F	*SW*	FR	76772	62410	71090	76843
1887	**N**	F	*SW*	FR	76766	62404	71084	76837

1888	**N**	F	*SW*	FR	76773	62411	71091	76844
1889	**N**	F	*SW*	FR	76774	62412	71092	76845
1890	**N**	F	*SW*	FR	76775	62413	71093	76846
1891	**N**	F	*SW*	FR	76776	62414	71094	76847
Spare		A		ZG			70995	
Spare		F		ZG				76824

CLASS 421/9   4 Cig (PHASE 1)

DTCsoL(A)–MBSO–TSO–DTCsoL(B). Express stock. Fitted with electric parking brake.
Facelifted with new trim and fluorescent lighting in saloons. For details see Class 421/3. These units are fitted with ex-Class 432 Mark 6 motor bogies.

1901	**N**	P	*SC*	BI	76082	62023	70701	76028
1902	**N**	P	*SC*	BI	76100	62041	71768	76046
1903	**CX**	A	*SC*	BI	76081	62022	70700	76027
1904	**CX**	A	*SC*	BI	76107	62048	70726	76053
1905	**CX**	A	*SC*	BI	76099	62040	70718	76045
1906	**CX**	A	*SC*	BI	76105	62046	70724	76113
1907	**CX**	A	*SC*	BI	76104	62045	70723	76050
1908	**N**	A	*SC*	BI	76096	62037	70715	76042

CLASS 422/3   Facelifted 4 Big (PHASE 2/1)

DTCsoL (A)–MBSO–TSRB–DTCsoL (B). Express stock. Units reformed from Class 421 to ensure that all Class 422 power cars have Mk. 6 motor bogies. Phase 2 units with phase 1 TSRBs (except for 69333 which is a phase 2 TSRB).

Diagram Numbers: EE369, ED264, EN260, EE369.
Electrical Equipment: 1963-type.
Bogies: Two Mk. 6 motor bogies (MBSO). B5 (SR) bogies (trailer cars).
Gangways: Throughout.
Traction Motors: Four EE507 of 185 kW.
Dimensions: 19.75 x 2.82 m.
Maximum Speed: 90 m.p.h.

69301–69318. TSRB. Lot No. 30744 York 1966. –/40. 35 t.
69332–69339. TSRB. Lot No. 30805 York 1970. –/40. 35 t.

2251	**N**	P	*SC*	BI	76726	62364	69302	76797
2254	**N**	P	*SC*	BI	76732	62370	69306	76803
2255	**N**	P	*SC*	BI	76740	62378	69310	76811
2256	**N**	P	*SC*	BI	76747	62385	69307	76818
2258	**N**	P	*SC*	BI	76746	62384	69316	76817
2259	**N**	P	*SC*	BI	76748	62386	69318	76819
2260	**N**	P	*SC*	BI	76749	62387	69304	76820
2262	**N**	P	*SC*	BI	76779	62417	69333	76850

Spare TSRB (Phase 1)

| 69301 | P | ZG | 69311 | P | ZG | 69312 | P | ZG |
| 69313 | P | ZG | | | | | | |

Spare TSRB (Phase 2)

69332	P	ZG	69336	P	ZG	69338	P	ZG
69334	P	ZG	69337	P	ZG	69339	P	ZG
69335	P	ZG						

CLASS 412   REFURBISHED 4 Bep

DMSO (A)–TBCK–TRSB–DMSO (B). Kent Coast Express Stock. Refurbished and renumbered from the 70xx series. Fitted with hopper ventilators, Inter-City 70 seats and fluorescent lighting.

Electrical Equipment: 1957-type.
Bogies: Mk 6 motor bogies and B5(SR) trailer bogies.
Gangways: Throughout.
Traction Motors: Four EE507 of 185 kW.
Dimensions: 19.75 x 2.82 m.
Maximum Speed: 90 m.p.h.

DMSO (A). Dia. EA263. –/64. 44.15 t.
TBCK. Dia. EJ361. 24/6 2T. 36.17 t.
TRSB. Dia. EN261. –/24 1T + 9 longitudinal buffet chairs. 35.5 t.
DMSO (B). Dia. EA264. –/64. 43.54 t.
Lot numbers are as follows, all cars being built at Ashford/Eastleigh:

61736–61809. 30619 1960–61.	**70354.** 30456 1959.
61954–61955. 30708 1963.	**70573–70609.** 30621 1960–61.
69341–69347. 30622 1961.	**70656.** 30709 1963.

2301	**ST**	P	*SW*	FR	61804	70607	69341	61805
2302	**ST**	P	*SW*	FR	61774	70592	69342	61809
2303	**ST**	P	*SW*	FR	61954	70656	69347	61955
2304	**ST**	P	*SW*	FR	61736	70573	69344	61737
2305	**ST**	P	*SW*	FR	61798	70354	69345	61799
2306	**ST**	P	*SW*	FR	61808	70609	69346	61775
2307	**ST**	P	*SW*	FR	61802	70606	69343	61803

Former numbers of converted buffet cars:

| 69341 (69014) | 69343 (69018) | 69345 (69013) | 69347 (69015) |
| 69342 (69019) | 69344 (69012) | 69346 (69016) | |

Note: No new Lot Nos. were issued for the above conversions.

CLASS 442   WESSEX EXPRESS STOCK

DTFsoL–TSOL(A)–MBRSM–TSOL(B)–DTSOL. Express stock built for Waterloo–Bournemouth–Weymouth service. Now also used on certain Portsmouth Harbour services. Air conditioned (heat pump system). Power-operated sliding plug doors. Can be hauled and heated by any BR ETH fitted locomotive.

Multiple working with class 33/1 and 73 locomotives.

Electrical Equipment: 1986-type.
Bogies: Mk 6 motor bogies (MBRSM). T4 trailer bogies.
Gangways: Throughout.
Traction Motors: Four EE546 of 300 kW recovered from class 432.
Dimensions: 23.00 x 2.74 m (inner cars), 23.15 x 2.74 m (outer cars).
Maximum Speed: 100 m.p.h.

DTFsoL. Dia. EE160. Lot No. 31030 Derby 1988–89. 50/– 1T. (36 in six compartments and 14 2+2 in one saloon). Public Telephone. 39.06 t.
TSOL (A). Dia. EH288. Lot No. 31032 Derby 1988–89. –/80 2T. 35.26 t.
MBRSM. Dia. ED265. Lot No. 31034 Derby 1988–89. –/14. 54.10 t.
TSOL (B). Dia. EH289. Lot No. 31033 Derby 1988–89. –/76 2T + wheelchair space. + 2 tip-up seats. 35.36 t.
DTSOL. Dia. EE273. Lot No. 31031 Derby 1988–89. –/78 1T. 39.06 t.

2401	**NW**	A	SW	BM	77382	71818	62937	71842	77406
2402	**ST**	A	SW	BM	77383	71819	62938	71843	77407
2403	**NW**	A	SW	BM	77384	71820	62941	71,844	77408
2404	**NW**	A	SW	BM	77385	71821	62939	71845	77409
2405	**NW**	A	SW	BM	77386	71822	62944	71846	77410
2406	**NW**	A	SW	BM	77389	71823	62942	71847	77411
2407	**NW**	A	SW	BM	77388	71824	62943	71848	77412
2408	**NW**	A	SW	BM	77387	71825	62945	71849	77413
2409	**NW**	A	SW	BM	77390	71826	62946	71850	77414
2410	**NW**	A	SW	BM	77391	71827	62948	71851	77415
2411	**NW**	A	SW	BM	77392	71828	62940	71858	77422
2412	**NW**	A	SW	BM	77393	71829	62947	71853	77417
2413	**NW**	A	SW	BM	77394	71830	62949	71854	77418
2414	**NW**	A	SW	BM	77395	71831	62950	71855	77419
2415	**NW**	A	SW	BM	77396	71832	62951	71856	77420
2416	**NW**	A	SW	BM	77397	71833	62952	71857	77421
2417	**NW**	A	SW	BM	77398	71834	62953	71852	77416
2418	**NW**	A	SW	BM	77399	71835	62954	71859	77423
2419	**NW**	A	SW	BM	77400	71836	62955	71860	77424
2420	**NW**	A	SW	BM	77401	71837	62956	71861	77425
2421	**NW**	A	SW	BM	77402	71838	62957	71862	77426
2422	**NW**	A	SW	BM	77403	71839	62958	71863	77427
2423	**NW**	A	SW	BM	77404	71840	62959	71864	77428
2424	**NW**	A	SW	BM	77405	71841	62960	71865	77429

Names of MBRSM:

62937	BEAULIEU	62947	SPECIAL OLYMPICS
62938	COUNTY OF HAMPSHIRE	62948	MERIDIAN TONIGHT
62939	BOROUGH OF WOKING	62951	MARY ROSE
62941	THE NEW FOREST	62952	Mum in a Million, Doreen Scanlon
62942	VICTORY	62954	WESSEX CANCER TRUST
62943	THOMAS HARDY	62955	BBC SOUTH TODAY
62944	CITY OF PORTSMOUTH	62956	CITY SOUTHAMPTON
62945	COUNTY OF DORSET	62958	OPERATION OVERLORD
62946	BOURNEMOUTH ORCHESTRAS	62959	COUNTY OF SURREY

CLASS 423   **4 Vep**

DTCsoL–MBSO–TSO–DTCsoL (DTSO*). Outer suburban stock. Facelifted with fluorescent lighting.

Electrical Equipment: 1963-type.
Bogies: Two Mk. 4 motor bogies (MBSO). B5 (SR) bogies (trailer cars).
Gangways: Throughout.
Traction Motors: Four EE507 of 185 kW.
Dimensions: 19.75 x 2.82 m.
Maximum Speed: 90 m.p.h.

62121–40. MBSO. Dia. ED266. Lot No. 30760 Derby 1967. –/76. 49 t.
62182–216. MBSO. Dia. ED266. Lot No. 30773 York 1967–68. –/76. 49 t.
62217–66. MBSO. Dia. ED266. Lot No. 30794 York 1968–69. –/76. 49 t.
62267–76. MBSO. Dia. ED266. Lot No. 30800 York 1970. –/76. 49 t.
62317–54. MBSO. Dia. ED266. Lot No. 30813 York 1970–73. –/76. 49 t.
62435–75. MBSO. Dia. ED266. Lot No. 30851 York 1973–74. –/76. 49 t.
70781–800. TSO. Dia. EH291. Lot No. 30761 York 1967. –/98. 31.5 t.
70872–906. TSO. Dia. EH291. Lot No. 30772 York 1967–68. –/98. 31.5 t.
70907–56. TSO. Dia. EH291. Lot No. 30793 York 1968–69. –/98. 31.5 t.
70957–66. TSO. Dia. EH291. Lot No. 30801 York 1970. –/98. 31.5 t.
70997–71034. TSO. Dia. EH291. Lot No. 30812 York 1970–73. –/98. 31.5 t.
71115–55. TSO. Dia. EH291. Lot No. 30852 York 1973–74. –/98. 31.5 t.
76230–69. DTCsoL. Dia. EE373. Lot No. 30758 York 1967. 18/46 1T. 35 t.
76275. DTSO (Class 438). Dia. EE266. Lot No. 30764 York 1966. –/64 1T. 32 t.
(Converted from hauled TSO 3929).
76333–402. DTCsoL. Dia. EE373. Lot No. 30771 Yk 1967–68. 18/46 1T. 35 t.
76441–540. DTCsoL. Dia. EE373. Lot No. 30792 Yk 1968–69. 18/46 1T. 35 t.
76541–60. DTCsoL. Dia. EE373. Lot No. 30799 York 1970. 18/46 1T. 35 t.
76641–716. DTCsoL. Dia. EE373. Lot No. 30811 Yk 1970–73. 18/46 1T. 35 t.
76861–942. DTCsoL. Dia. EE368. Lot No. 30853 Yk 1973–74. 18/46 1T. 35 t.

3401	ST	F	SW	WD	76230	62276	70781	76231
3402	N	F	SW	WD	76233	62123	70782	76232
3403	N	F	SW	WD	76234	62254	70783	76235
3404	N	F	SW	WD	76378	62261	70894	76236
3405	N	F	SW	WD	76239	62271	70785	76238
3406	ST	F	SW	WD	76241	62130	70786	76240
3407	ST	F	SW	WD	76243	62348	70787	76242
3408	N	F	SW	WD	76244	62435	70788	76245
3409	N	F	SW	WD	76246	62239	70789	76247
3410	ST	F	SW	WD	76369	62442	70790	76249
3411	ST	F	SW	WD	76251	62342	70791	76250
3412	N	A	SE	RE	76252	62340	70792	76253
3413	N	F	SW	WD	76255	62441	70793	76254
3414	N	F	SW	WD	76257	62446	70794	76248
3415	N	F	SW	WD	76258	62451	70795	76259
3416	N	A	SE	RE	76261	62451	70796	76260
3417	ST	F	SW	WD	76262	62236	70797	76263
3418	N	F	SW	WD	76265	62133	70875	76264

3419	**ST**	F	*SW*	WD	76267	62354	70799	76266
3420	**ST**	F	*SW*	WD	76269	62349	70800	76268
3421	**N**	A	*SE*	RE	76889	62449	71129	76890
3422	**N**	A	*SE*	RE	76372	62201	70891	76371
3423	**N**	A	*SE*	RE	76452	62222	70912	76451
3424	**N**	A	*SE*	RE	76354	62185	70882	76353
3425	**N**	F	*SW*	WD	76338	62192	70874	76358
3426	**N**	F	*SW*	WD	76386	62208	70898	76385
3427	**N**	F	*SW*	WD	76374	62184	70892	76373
3428	**N**	F	*SW*	WD	76454	62223	70913	76453
3429	**N**	F	*SW*	WD	76334	62202	70872	76333
3430	**N**	F	*SW*	WD	76348	62189	70879	76347
3431	**N**	F	*SW*	WD	76458	62182	70915	76457
3432	**N**	F	*SW*	WD	76400	62225	70905	76399
3433	**N**	F	*SW*	WD	76444	62215	70908	76443
3434	**N**	F	*SW*	WD	76462	62218	70917	76461
3435	**CX**	P	*SC*	BI	76342	62228	70876	76341
3436	**CX**	P	*SC*	BI	76350	62190	70880	76349
3437	**CX**	P	*SC*	BI	76346	62186	70878	76345
3438	**N**	P	*SC*	BI	76530	62262	70951	76529
3439	**N**	P	*SC*	BI	76402	62227	70906	76401
3442	**N**	P	*SC*	BI	76492	62216	70932	76491
3445	**N**	A	*SE*	RE	76450	62242	70911	76449
3446	**N**	A	*SE*	RE	76532	62243	70952	76531
3447	**N**	A	*SE*	RE	76380	62199	70895	76379
3448	**N**	A	*SE*	RE	76376	62221	70886	76375
3449	**N**	A	*SE*	RE	76336	62205	70873	76335
3450	**N**	A	*SE*	RE	76460	62203	70916	76459
3451	**N**	A	*SE*	RE	76488	62240	70930	76487
3452	**N**	A	*SE*	RE	76340	62183	71021	76690
3453	**N**	A	*SE*	RE	76382	62226	70896	76381
3454	**N**	A	*SE*	RE	76390	62200	70798	76389
3455	**N**	F	*SW*	WD	76388	62206	70899	76387
3456	**N**	F	*SW*	WD	76458	62210	70914	76455
3457	**N**	F	*SW*	WD	76392	62197	70901	76391
3458	**N**	F	*SW*	WD	76394	62209	70902	76393
3459	**ST**	F	*SW*	WD	76396	62224	70903	76395
3462	**N**	P	*SC*	BI	76536	62213	70954	76535
3463	**N**	P	*SC*	BI	76398	62266	70904	76397
3464	**N**	P	*SC*	BI	76442	62265	70907	76441
3466	**N**	F	*SW*	WD	76464	62214	70918	76463
3467	**N**	F	*SW*	WD	76446	62217	70909	76445
3468	**N**	F	*SW*	WD	76448	62267	70910	76447
3469	**N**	F	*SW*	WD	76546	62219	70959	76545
3470	**N**	F	*SW*	WD	76496	62220	70934	76495
3471	**N**	A	*SE*	RE	76498	62269	70935	76497
3472	**N**	A	*SE*	RE	76500	62244	70936	76499
3473	**N**	A	*SE*	RE	76502	62245	70937	76339
3474	**N**	A	*SE*	RE	76504	62246	70938	76503
3475	**N**	A	*SE*	RE	76552	62270	70962	76551
3476	**N**	P	*SC*	BI	76548	62247	70960	76547

3478	N	P	SC	BI	76653	62125	71003	76654
3479	N	F	SW	WD	76655	62272	71004	76656
3480	N	F	SW	WD	76474	62323	70923	76473
3481	N	F	SW	WD	76648	62324	70900	76647
3482	N	F	SW	WD	76657	62320	71005	76658
3483	N	F	SW	WD	76661	62233	71007	76662
3484	N	F	SW	WD	76476	62325	70924	76475
3485	N	F	SW	WD	76508	62327	70940	76507
3486	N	F	SW	WD	76478	62234	70925	76477
3487	N	A	SE	RE	76645	62250	70941	76509
3488	N	F	SW	WD	76663	62235	71008	76664
3489	N	F	SW	WD	76665	62251	71009	76666
3490	N	F	SW	WD	76695	62328	71024	76696
3491	N	A	SE	RE	76337	62436	70927	76481
3492	N	A	SE	RE	76667	62344	71010	76668
3493	N	A	SE	RE	76669	62237	71011	76670
3494	N	A	SE	RE	76675	62330	71014	76676
3495	N	A	SE	RE	76699	62331	71026	76700
3496	N	A	SE	RE	76673	62334	71013	76674
3497	N	A	SE	RE	76671	62346	71012	76672
3498	N	A	SE	RE	76701	62333	71027	76702
3499	N	A	SE	RE	76901	62347	71135	76902
3500	N	A	SE	RE	76470	62455	70921	76469
3501	CX	P	SC	BI	76512	62332	70942	76511
3503	CX	P	SC	BI	76681	62231	71017	76682
3504	CX	P	SC	BI	76711	62351	71032	76712
3505	CX	P	SC	BI	76472	62352	70922	76471
3506	N	P	SC	BI	76554	62317	70963	76553
3507	N	P	SC	BI	76558	62232	70965	76557
3508	ST	F	SW	WD	76643	62273	70998	76644
3509	ST	F	SW	WD	76560	62275	70966	76559
3510	ST	F	SW	WD	76641	62318	70997	76642
3511	N	A	SE	RE	76893	62135	70999	76646
3512	CX	P	SC	BI	76679	62337	71016	76680
3513	N	P	SC	BI	76691	62336	71022	76692
3514	CX	P	SC	BI	76683	62136	71018	76684
3515	CX	P	SC	BI	76544	62319	70958	76543
3516	N	F	SW	WD	76693	62268	71023	76694
3517	CX	P	SC	BI	76685	62338	71019	76686
3518	CX	P	SC	BI	76689	62343	70887	76363
3519	CX	F	SW	WD	76556	62274	70964	76555
3520	N	F	SW	WD	76697	62131	71025	76698
3521	N	A	SE	RE	76484	62345	70928	76483
3522	CX	P	SC	BI	76705	62341	71029	76706
3523	N	F	SW	WD	76651	62139	71002	76652
3524	N	F	SW	WD	76466	62322	70919	76370
3526	N	P	SC	BI	76524	62255	70948	76523
3527	N	P	SC	BI	76520	62326	70946	76519
3528	N	P	SC	BI	76518	62258	70945	76517
3529	N	F	SW	WD	76659	62257	71006	76660
3530	N	F	SW	WD	76468	62256	70920	76467

3531	**N**	F	*SW*	WD	76649	62230	71001	76650
3532	**N**	P	*SC*	BI	76528	62321	70950	76527
3533	**N**	P	*SC*	BI	76364	62260	70949	76525
3534	**N**	P	*SC*	BI	76506	62259	70939	76505
3535	**CX**	P	*SC*	BI	76677	62335	71015	76678
3536	**N**	F	*SW*	WD	76384	62207	70897	76383
3537	**N**	P	*SC*	BI	76514	62249	70943	76513
3539	**N**	F	*SW*	WD	76861	62122	71115	76862
3540	**N**	F	*SW*	WD	76863	62128	71116	76864
3541	**N**	P	*SC*	BI	76703	62238	71028	76704
3542	**ST**	F	*SW*	WD	76480	62127	70926	76479
3543	**N**	A	*SE*	RE	76899	62137	71134	76900
3544	**N**	A	*SE*	RE	76892	62454	71131	76894
3545	**N**	A	*SE*	RE	76875	62121	71122	76876
3546	**CX**	P	*SC*	BI	76687	62339	71020	76688
3547	**N**	A	*SE*	RE	76895	62126	71132	76896
3548	**CX**	A	*SE*	RE	76903	62452	71136	76904
3549	**CX**	P	*SC*	B1	76707	62132	71030	76708
3550	**CX**	P	*SC*	BI	76490	62350	70931	76489
3551	**CX**	P	*SC*	BI	76465	62456	71033	76714
3552	**ST**	F	*SW*	WD	76715	62353	71034	76716
3553	**N**	A	*SE*	RE	76913	62241	71141	76914
3554	**CX**	A	*SE*	RE	76905	62461	71137	76906
3555	**ST**	F	*SW*	WD	76865	62140	71117	76866
3556	**CX**	A	*SE*	RE	76885	62457	71127	76886
3557	**ST**	F	*SW*	WD	76869	62437	71119	76870
3558	**ST**	F	*SW*	WD	76352	62447	70881	76351
3559	**N**	F	*SW*	WD	76486	62439	70929	76485
3560	**N**	A	*SE*	RE	76897	62191	71133	76898
3561	**N**	F	*SW*	WD	76867	62453	71118	76868
3562	**N**	A	*SE*	RE	76907	62129	71138	76908
3563	**N**	F	*SW*	WD	76873	62438	71121	76874
3564	**N**	A	*SE*	RE	76883	62458	71126	76884
3565	**N**	A	*SE*	RE	76877	62134	71123	76878
3566	**N**	A	*SE*	RE	76915	62443	71142	76916
3567	**N**	F	*SW*	WD	76871	62138	71120	76872
3568	**N**	A	*SE*	RE	76887	62440	71128	76888
3569	**N**	F	*SW*	WD	76344	62448	70877	76343
3570	**N**	A	*SE*	RE	76909	62187	71139	76910
3571	**N**	A	*SE*	RE	76927	62463	71148	76928
3572	**N**	A	*SE*	RE	76879	62468	71124	76880
3573	**N**	A	*SE*	RE	76919	62444	71144	76920
3574	**N**	A	*SE*	RE	76929	62464	71149	76930
3575	**N**	A	*SE*	RE	76931	62469	71150	76932
3576	**N**	F	*SW*	WD	76362	62196	70890	76361
3577	**N**	A	*SE*	RE	76933	62459	71151	76934
3578	**N**	F	*SW*	WD	76356	62193	70883	76355
3579	**N**	A	*SE*	RE	76935	62471	71152	76936
3580	**N**	F	*SW*	WD	76360	62195	70885	76359
3581	**N**	F	*SW*	WD	76366	62198	70888	76365
3582	* **N**	A	*SE*	RE	76891	62472	71130	76275

3583	N	A	SE	RE	76937	62450	71153	76938
3584	N	A	SE	RE	76881	62473	71125	76882
3585	N	A	SE	RE	76939	62445	71154	76940
3586	N	A	SE	RE	76921	62474	71145	76922
3587	N	A	SE	RE	76925	62465	71147	76926
3588	N	A	SE	RE	76923	62467	71146	76924
3589	N	A	SE	RE	76911	62466	71140	76912
3590	N	A	SE	RE	76941	62460	71155	76942
3591	N	A	SE	RE	76917	62475	71143	76918
3801	N	P	SE	RE	76522	62229	70947	76521
3802	N	P	SE	RE	76534	62188	70953	76533
3803	N	P	SE	RE	76494	62263	70933	76493
3804	N	P	SE	RE	76368	62204	70889	76367
3805	N	P	SE	RE	76540	62211	70956	76539
3806	N	P	SE	RE	76538	62212	70955	76537
3807	N	P	SE	RE	76542	62264	70957	76541
3808	N	P	SE	RE	76550	62248	70961	76549
3809	N	P	SE	RE	76516	62253	70944	76515
3810	N	P	SE	RE	76709	62252	71031	76710
Spare	N	A		WD (U)		62470		
Spare		A		ZG	76510			

CLASS 455/7

DTSO–MSO–TSO–DTSO. Sliding doors. Disc brakes. Fluorescent lighting. Second series with TSOs originally in class 508. Pressure ventilation.

Bogies: BT13 (DTSO), BP27 (MSO), BX1 (TSO).
Gangways: Through gangwayed.
Traction Motors: Four EE507 of 185 kW.
Dimensions: 19.83 x 2.82 m. (outer cars), 19.92 x 2.82 m (inner cars).
Maximum Speed: 75 m.p.h.

DTSO. Dia. EE218. Lot No. 30976 York 1984–85. –/74. 29.5 t.
MSO. Dia. EC203. Lot No. 30975 York 1984–85. –/84. 45 t.
TSO. Dia. EH219. Lot No. 30944 York 1977–80. –/86. 25.48 t.

5701	N	P	SW	WD	77727	62783	71545	77728
5702	N	P	SW	WD	77729	62784	71547	77730
5703	N	P	SW	WD	77731	62785	71540	77732
5704	N	P	SW	WD	77733	62786	71548	77734
5705	N	P	SW	WD	77735	62787	71565	77736
5706	N	P	SW	WD	77737	62788	71534	77738
5707	N	P	SW	WD	77739	62789	71536	77740
5708	N	P	SW	WD	77741	62790	71560	77742
5709	N	P	SW	WD	77743	62791	71532	77744
5710	N	P	SW	WD	77745	62792	71566	77746
5711	N	P	SW	WD	77747	62793	71542	77748
5712	N	P	SW	WD	77749	62794	71546	77750
5713	ST	P	SW	WD	77751	62795	71567	77752
5714	ST	P	SW	WD	77753	62796	71539	77754

5715	**ST**	P	*SW*	WD	77755	62797	71535	77756
5716	**ST**	P	*SW*	WD	77757	62798	71564	77758
5717	**ST**	P	*SW*	WD	77759	62799	71528	77760
5718	**ST**	P	*SW*	WD	77761	62800	71557	77762
5719	**ST**	P	*SW*	WD	77763	62801	71558	77764
5720	**ST**	P	*SW*	WD	77765	62802	71568	77766
5721	**ST**	P	*SW*	WD	77767	62803	71553	77768
5722	**ST**	P	*SW*	WD	77769	62804	71533	77770
5723	**ST**	P	*SW*	WD	77771	62805	71526	77772
5724	**N**	P	*SW*	WD	77773	62806	71561	77774
5725	**ST**	P	*SW*	WD	77775	62807	71541	77776
5726	**ST**	P	*SW*	WD	77777	62808	71556	77778
5727	**ST**	P	*SW*	WD	77779	62809	71562	77780
5728	**ST**	P	*SW*	WD	77781	62810	71527	77782
5729	**ST**	P	*SW*	WD	77783	62811	71550	77784
5730	**ST**	P	*SW*	WD	77785	62812	71551	77786
5731	**ST**	P	*SW*	WD	77787	62813	71555	77788
5732	**ST**	P	*SW*	WD	77789	62814	71552	77790
5733	**ST**	P	*SW*	WD	77791	62815	71549	77792
5734	**ST**	P	*SW*	WD	77793	62816	71531	77794
5735	**ST**	P	*SW*	WD	77795	62817	71563	77796
5736	**ST**	P	*SW*	WD	77797	62818	71554	77798
5737	**ST**	P	*SW*	WD	77799	62819	71544	77800
5738	**ST**	P	*SW*	WD	77801	62820	71529	77802
5739	**ST**	P	*SW*	WD	77803	62821	71537	77804
5740	**ST**	P	*SW*	WD	77805	62822	71530	77806
5741	**ST**	P	*SW*	WD	77807	62823	71559	77808
5742	**ST**	P	*SW*	WD	77809	62824	71543	77810
5750	**ST**	P	*SW*	WD	77811	62825	71538	77812

Names:

5711	SPIRIT OF RUGBY
5735	The Royal Borough of Kingston
5750	Wimbledon Train Care

CLASS 455/8

DTSO–MSO–TSO–DTSO. Sliding doors. Disc brakes. Fluorescent lighting.
First series. Pressure ventilation.

Bogies: BP20 (MSO), BT13 (trailer cars).
Gangways: Through gangwayed.
Traction Motors: Four EE507 of 185 kW.
Dimensions: 19.83 x 2.82 m. (outer cars), 19.92 x 2.82 m (inner cars).
Maximum Speed: 75 m.p.h.

DTSO. Dia. EE218. Lot No. 30972 York 1982–84. –/74. 29.5 t.
MSO. Dia. EC203. Lot No. 30973 York 1982–84. –/84. 45.6 t.
TSO. Dia. EH221. Lot No. 30974 York 1982–84. –/84. 27.1 t.

5801	N	F	SC	SU	77579	62709	71637	77580
5802	N	F	SC	SU	77581	62710	71664	77582
5803	N	F	SC	SU	77583	62711	71639	77584
5804	CX	F	SC	SU	77585	62712	71640	77586
5805	CX	F	SC	SU	77587	62713	71641	77588
5806	N	F	SC	SU	77589	62714	71642	77590
5807	N	F	SC	SU	77591	62715	71643	77592
5808	N	F	SC	SU	77593	62716	71644	77594
5809	N	F	SC	SU	77595	62717	71645	77596
5810	N	F	SC	SU	77597	62718	71646	77598
5811	N	F	SC	SU	77599	62719	71647	77600
5812	N	F	SC	SU	77601	62720	71648	77602
5813	N	F	SC	SU	77603	62721	71649	77604
5814	N	F	SC	SU	77605	62722	71650	77606
5815	N	F	SC	SU	77607	62723	71651	77608
5816	N	F	SC	SU	77609	62724	71652	77633
5817	N	F	SC	SU	77611	62725	71653	77612
5818	N	F	SC	SU	77613	62726	71654	77614
5819	N	F	SC	SU	77615	62727	71655	77616
5820	N	F	SC	SU	77617	62728	71656	77618
5821	N	F	SC	SU	77619	62729	71657	77620
5822	N	F	SC	SU	77621	62730	71658	77622
5823	N	F	SC	SU	77623	62731	71659	77624
5824	N	F	SC	SU	77637	62732	71660	77626
5825	N	F	SC	SU	77627	62733	71661	77628
5826	N	F	SC	SU	77629	62734	71662	77630
5827	N	F	SC	SU	77610	62735	71663	77632
5828	N	F	SC	SU	77634	62736	71638	77631
5829	N	F	SC	SU	77635	62737	71665	77636
5830	N	F	SC	SU	77625	62743	71666	77638
5831	N	F	SC	SU	77639	62719	71667	77640
5832	N	F	SC	SU	77641	62740	71668	77642
5833	N	F	SC	SU	77643	62741	71669	77644
5834	N	F	SC	SU	77645	62742	71670	77646
5835	N	F	SC	SU	77647	62738	71671	77648
5836	N	F	SC	SU	77649	62744	71672	77650
5837	N	F	SC	SU	77651	62745	71673	77652
5838	N	F	SC	SU	77653	62746	71674	77654
5839	N	F	SC	SU	77655	62747	71675	77656
5840	N	F	SC	SU	77657	62748	71676	77658
5841	N	F	SC	SU	77659	62749	71677	77660
5842	N	F	SC	SU	77661	62750	71678	77662
5843	N	F	SC	SU	77663	62751	71679	77664
5844	N	F	SC	SU	77665	62752	71680	77666
5845	N	F	SC	SU	77667	62753	71681	77668
5846	N	F	SC	SU	77669	62754	71682	77670
5847	N	F	SC	SU	77671	62755	71683	77672
5848	ST	F	SC	SU	77673	62756	71684	77674
5849	N	F	SC	SU	77675	62757	71685	77676
5850	N	F	SC	SU	77677	62758	71686	77678
5851	N	F	SC	SU	77679	62759	71687	77680

5852	**N**	F	*SC*	SU	77681	62760	71688	77682
5853	**N**	F	*SC*	SU	77683	62761	71689	77684
5854	**N**	F	*SC*	SU	77685	62762	71690	77686
5855	**N**	F	*SC*	SU	77687	62763	71691	77688
5856	**N**	F	*SC*	SU	77689	62764	71692	77690
5857	**N**	F	*SC*	SU	77691	62765	71693	77692
5858	**ST**	F	*SC*	SU	77693	62766	71694	77694
5859	**N**	F	*SC*	SU	77695	62767	71695	77696
5860	**N**	F	*SC*	SU	77697	62768	71696	77698
5861	**N**	F	*SC*	SU	77699	62769	71697	77700
5862	**N**	F	*SC*	SU	77701	62770	71698	77702
5863	**N**	F	*SC*	SU	77703	62771	71699	77704
5864	**N**	F	*SC*	SU	77705	62772	71700	77706
5865	**N**	F	*SC*	SU	77707	62773	71701	77708
5866	**N**	F	*SC*	SU	77709	62774	71702	77710
5867	**N**	F	*SC*	SU	77711	62775	71703	77712
5868	**N**	F	*SC*	SU	77713	62776	71704	77714
5869	**N**	F	*SC*	SU	77715	62777	71705	77716
5870	**N**	F	*SC*	SU	77717	62778	71706	77718
5871	**N**	F	*SC*	SU	77719	62779	71707	77720
5872	**N**	F	*SC*	SU	77721	62780	71708	77722
5873	**N**	F	*SC*	SU	77723	62781	71709	77724
5874	**N**	F	*SC*	SU	77725	62782	71710	77726

CLASS 455/9

DTSO–MSO–TSO–DTSO. Sliding doors. Disc brakes. Fluorescent lighting. Third series. Convection heating.

Bogies: BP20 (MSO), BT13 (trailer cars).
Gangways: Through gangwayed.
Traction Motors: Four EE507 of 185 kW.
Dimensions: 19.83 x 2.82 m. (outer cars), 19.92 x 2.82 m (inner cars).
Maximum Speed: 75 m.p.h.

DTSO. Dia. EE226. Lot No. 30991 York 1985. –/74. 29.5 t.
MSO. Dia. EC206. Lot No. 30992 York 1985. –/84. 45.6 t.
TSO. Dia. EH224. Lot No. 30993 York 1985. –/84. 27.1 t.
TSO n. Dia. EH224. Lot No. 30932 Derby 1981. –/84. 27.1 t.

* Chopper control.
§ Tread brakes.
c "Crossrail" interiors.
n Prototype vehicle converted from a Class 210 DEMU.

5901		**N**	P	*SW*	WD	77813	62826	71714	77814
5902		**N**	P	*SW*	WD	77815	62827	71715	77816
5903		**N**	P	*SW*	WD	77817	62828	71716	77818
5904		**N**	P	*SW*	WD	77819	62829	71717	77820
5905	c	**N**	P	*SW*	WD	77821	62830	71731	77822
5906		**N**	P	*SW*	WD	77823	62831	71719	77824
5907		**N**	P	*SW*	WD	77825	62832	71720	77826

5908		**N**	P	*SW*	WD	77827	62833	71721	77828
5909		**N**	P	*SW*	WD	77829	62834	71722	77830
5910		**N**	P	*SW*	WD	77831	62835	71723	77832
5911		**N**	P	*SW*	WD	77833	62836	71724	77834
5912	*	**N**	P	*SW*	WD	77835	62837	71725	77836
5913	§	**N**	P	*SW*	WD	77837	62838	71726	77838
5914	§	**N**	P	*SW*	WD	77839	62839	71727	77840
5915	§	**N**	P	*SW*	WD	77841	62840	71728	77842
5916	*	**N**	P	*SW*	WD	77843	62841	71729	77844
5917	*	**N**	P	*SW*	WD	77845	62842	71730	77846
5918	*c	**N**	P	*SW*	WD	77847	62843	71732	77848
5919	*	**N**	P	*SW*	WD	77849	62844	71718	77850
5920	*	**N**	P	*SW*	WD	77851	62845	71733	77852
Spare	n	**N**	P	*SW*	WD			67400	

CLASS 488   VICTORIA–GATWICK TRAILER SETS

TFOLH–TSOL (Class 488/3 only)–TSOLH. Converted 1983–84 from loco-hauled Mk. 2F FOs and TSOs for Victoria–Gatwick service. Express stock. Air conditioned. Fluorescent lighting. Conversion consisted of a modified seating layout and the removal of one toilet to provide additional luggage space.

Bogies: B4.
Gangways: Throughout.
Dimensions: 20.12 x 2.82 m.
Maximum Speed: 90 m.p.h.

72500–72509. TFOLH. Dia. EP101. Lot No. 30859 Derby 1973–74. 41/– 1T. 35 t.
72602–14/6–8/20–44/6/7. TSOLH. Dia. EP201. Lot No. 30860 Derby 1973–74. –/48 1T. 35 t.
72615/19/45. TSOLH. Dia. EP201. Lot No. 30846 Derby 1973. –/48 1T. 35 t.
72701–72718. TSOL. Dia. EH285. Lot No. 30860 Derby 1973–74. –/48 1T. 35 t.

CLASS 488/2. TFOLH–TSOLH. Note: TFOLH fitted with public telephone.

8201	**GX**	P	*GX*	SL	72500 (3413)	72638 (6068)
8202	**GX**	P	*GX*	SL	72501 (3382)	72617 (6086)
8203	**GX**	P	*GX*	SL	72502 (3321)	72640 (6097)
8204	**GX**	P	*GX*	SL	72503 (3407)	72641 (6079)
8205	**GX**	P	*GX*	SL	72504 (3406)	72628 (6058)
8206	**GX**	P	*GX*	SL	72505 (3415)	72629 (6048)
8207	**GX**	P	*GX*	SL	72506 (3335)	72642 (6076)
8208	**GX**	P	*GX*	SL	72507 (3412)	72643 (6040)
8209	**GX**	P	*GX*	SL	72508 (3409)	72644 (6039)
8210	**GX**	P	*GX*	SL	72509 (3398)	72635 (6128)

CLASS 488/3. TSOLH–TSOL–TSOLH.

8302	**GX**	P	*GX*	SL	72602 (6130)	72701 (6088)	72604 (6087)
8303	**GX**	P	*GX*	SL	72603 (6093)	72702 (6099)	72608 (6077)
8304	**GX**	P	*GX*	SL	72606 (6084)	72703 (6075)	72611 (6083)
8305	**GX**	P	*GX*	SL	72605 (6082)	72704 (6132)	72609 (6080)

8306	**GX**	P	*GX*	SL	72607 (6020)	72705 (6032)	72610 (6074)
8307	**GX**	P	*GX*	SL	72612 (6156)	72706 (6143)	72613 (6126)
8308	**GX**	P	*GX*	SL	72614 (6090)	72707 (6127)	72615 (5938)
8309	**GX**	P	*GX*	SL	72616 (6007)	72708 (6095)	72639 (6070)
8310	**GX**	P	*GX*	SL	72618 (6044)	72709 (5982)	72619 (5909)
8311	**GX**	P	*GX*	SL	72620 (6140)	72710 (6003)	72621 (6108)
8312	**GX**	P	*GX*	SL	72622 (6004)	72711 (6109)	72623 (6118)
8313	**GX**	P	*GX*	SL	72624 (5972)	72712 (6091)	72625 (6085)
8314	**GX**	P	*GX*	SL	72626 (6017)	72713 (6023)	72627 (5974)
8315	**GX**	P	*GX*	SL	72636 (6071)	72714 (6092)	72645 (5942)
8316	**GX**	P	*GX*	SL	72630 (6094)	72715 (6019)	72631 (6096)
8317	**GX**	P	*GX*	SL	72632 (6072)	72716 (6114)	72633 (6129)
8318	**GX**	P	*GX*	SL	72634 (6089)	72717 (6069)	72637 (6098)
8319	**GX**	P	*GX*	SL	72646 (6078)	72718 (5979)	72647 (6081)

CLASS 489   VICTORIA–GATWICK GLV

Converted 1983–84 from class 414/3 (2 Hap) DMBSOs to work with class 488.

Bogies: Mk 4.
Gangways: Gangwayed at inner end only.
Traction Motors: Two EE507 of 185 kW.
Dimensions: 19.49 x 2.82 m.
Maximum Speed: 90 m.p.h.

DMLV. Dia. EX561. Lot No. 30452 Ashford/Eastleigh 1959. 40.5 t.

9101	**GX**	P	*GX*	SL	68500(61269)
9102	**GX**	P	*GX*	SL	68501(61281)
9103	**GX**	P	*GX*	SL	68502(61273)
9104	**GX**	P	*GX*	SL	68503(61277)
9105	**GX**	P	*GX*	SL	68504(61286)
9106	**GX**	P	*GX*	SL	68505(61299)
9107	**GX**	P	*GX*	SL	68506(61292)
9108	**GX**	P	*GX*	SL	68507(61267)
9109	**GX**	P	*GX*	SL	68508(61272)
9110	**GX**	P	*GX*	SL	68509(61280)

CLASS 456

DMSO–DTSO. Sliding doors. Disc brakes. Fluorescent lighting.

Bogies: P7 (motor) and T3 trailer.
Gangways: Within set.
Traction Motors: Two EE507 of 185 kW.
Dimensions: 19.83 x 2.82 m.
Maximum Speed: 75 m.p.h.

DMSO. Dia. EA267. Lot No. 31073 York 1990–1. –/79. 41.1 t.
DTSO. Dia. EE276. Lot No. 31074 York 1990–1. –/51. 31.4 t.

456 001	**CX**	P	*SC*	SU	64735	78250
456 002	**N**	P	*SC*	SU	64736	78251
456 003	**N**	P	*SC*	SU	64737	78252
456 004	**N**	P	*SC*	SU	64738	78253
456 005	**N**	P	*SC*	SU	64739	78254
456 006	**N**	P	*SC*	SU	64740	78255
456 007	**N**	P	*SC*	SU	64741	78256
456 008	**N**	P	*SC*	SU	64742	78257
456 009	**N**	P	*SC*	SU	64743	78258
456 010	**N**	P	*SC*	SU	64744	78259
456 011	**N**	P	*SC*	SU	64745	78260
456 012	**N**	P	*SC*	SU	64746	78261
456 013	**N**	P	*SC*	SU	64747	78262
456 014	**N**	P	*SC*	SU	64748	78263
456 015	**N**	P	*SC*	SU	64749	78264
456 016	**N**	P	*SC*	SU	64750	78265
456 017	**N**	P	*SC*	SU	64751	78266
456 018	**N**	P	*SC*	SU	64752	78267
456 019	**N**	P	*SC*	SU	64753	78268
456 020	**N**	P	*SC*	SU	64754	78269
456 021	**N**	P	*SC*	SU	64755	78270
456 022	**N**	P	*SC*	SU	64756	78271
456 023	**N**	P	*SC*	SU	64757	78272
456 024	**CX**	P	*SC*	SU	64758	78273

Note: DTSO 78273 of set 456024 is named 'Sir Cosmo Bonsor'

CLASS 458   GEC-ALSTHOM JUNIPER Jo3 4 Jop

DMSO(A)–PTSOL–TSO–DMSO(B). New units under construction for use on South West Trains. Steel bodies. IGBT control. Tightlock couplers. Sliding doors. Disc brakes.
Bogies: .
Gangways: Throughout.
Traction Motors: 2 per motor car.
Dimensions: 20.00 m x   . m.
Maximum Speed: 100 m.p.h.

DMCO(A). Dia. EA   . Metro-Cammell 1998. .   . t.
PTSOL. Dia. EH   Metro-Cammell 1998. .   . t.
TSO. Dia. EH   Metro-Cammell 1998. .   . t.
DMCO(B). Dia. EA   Metro-Cammell 1998.   . t.

458 001	**ST**	P	*SW*		67601	74001	74101	67701
458 002	**ST**	P	*SW*		67602	74002	74102	67702
458 003	**ST**	P	*SW*		67603	74003	74103	67703
458 004	**ST**	P	*SW*		67604	74004	74104	67704
458 005	**ST**	P	*SW*		67605	74005	74105	67705
458 006	**ST**	P	*SW*		67606	74006	74106	67706
458 007	**ST**	P	*SW*		67607	74007	74107	67707

458 008	**ST**	P	SW		67608	74008	74108	67708
458 009	**ST**	P	SW		67609	74009	74109	67709
458 010	**ST**	P	SW		67610	74010	74110	67710
458 011	**ST**	P	SW		67611	74011	74111	67711
458 012	**ST**	P	SW		67612	74012	74112	67712
458 013	**ST**	P	SW		67613	74013	74113	67713
458 014	**ST**	P	SW		67614	74014	74114	67714
458 015	**ST**	P	SW		67615	74015	74115	67715
458 016	**ST**	P	SW		67616	74016	74116	67716
458 017	**ST**	P	SW		67617	74017	74117	67717
458 018	**ST**	P	SW		67618	74018	74118	67718
458 019	**ST**	P	SW		67619	74019	74119	67719
458 020	**ST**	P	SW		67620	74020	74120	67720
458 021	**ST**	P	SW		67621	74021	74121	67721
458 022	**ST**	P	SW		67622	74022	74122	67722
458 023	**ST**	P	SW		67623	74023	74123	67723
458 024	**ST**	P	SW		67624	74024	74124	67724
458 025	**ST**	P	SW		67625	74025	74125	67725
458 026	**ST**	P	SW		67626	74026	74126	67726
458 027	**ST**	P	SW		67627	74027	74127	67727
458 028	**ST**	P	SW		67628	74028	74128	67728
458 029	**ST**	P	SW		67629	74029	74129	67729
458 030	**ST**	P	SW		67630	74030	74130	67730

CLASS 460   GEC-ALSTHOM JUNIPER   Jо2 8 Gat

DMLFO–TFOL–TCOL–2MSO–TSOL–MSO–DMSO. New units under construction for use on Gatwick Express. Steel bodies. IGBT control. Tightlock couplers for emergency use only. Sliding doors. Disc brakes.
Bogies: .
Gangways: Throughout.
Traction Motors: 2 per motor car.
Dimensions: 20.00 m x   . m.
Maximum Speed: 100 m.p.h.

DMLFO. Dia. EA   . Metro-Cammell 1999. .   . t.
TFOL. Dia. EH   Metro-Cammell 1999. .   t.
TCOL. Dia. EH   Metro-Cammell 1999. .   t.
MSO. Dia. EH   Metro-Cammell 1999. .   . t.
TSOL. Dia. EH   Metro-Cammell 1999. .   . t.
DMSO. Dia. EA   Metro-Cammell 1999. .   . t.

460 001	P	GX	67801	74201	74211	74221	74231	74241	74251	67811
460 002	P	GX	67802	74202	74212	74222	74232	74242	74252	67812
460 003	P	GX	67803	74203	74213	74223	74233	74243	74253	67813
460 004	P	GX	67804	74204	74214	74224	74234	74244	74254	67814
460 005	P	GX	67805	74205	74215	74225	74235	74245	74255	67815
460 006	P	GX	67806	74206	74216	74226	74236	74246	74256	67816
460 007	P	GX	67807	74207	74217	74227	74237	74247	74257	67817
460 008	P	GX	67808	74208	74218	74228	74238	74248	74258	67818

CLASS 465 NETWORKER

DMSO (A)–TSO–TSOL–DMSO (B). New units with Aluminium bodies. Sliding doors. Disc, rheostatic and regenerative brakes.

Electrical Equipment: Networker.
Bogies: P3 (motor cars) and T3 (trailers).
Gangways: Within set.
Traction Motors: Four Brush totally-enclosed squirrel-caged three-phase induction motors (Class 465/0 & 465/1) or four GEC Alsthom G352BY motors (Class 465/2) per car driven by four GTO inverters.
Dimensions: 20.89 x 2.82 m (outer cars), 20.06 x 2.81 m (inner cars).
Maximum Speed: 75 m.p.h.

64759–64808. DMSO(A). Dia. EA268. Lot No. 31100 BREL York 1991–3. –/86. 38.9 t.
64809–64858. DMSO(B). Dia. EA268. Lot No. 31100 BREL York 1993–2. –/86. 39 t.
65700–65749. DMSO(A). Dia. EA269. Lot No. 31103 Metro-Cammell 1991–3. –/86. 38.8 t.
65750–65799. DMSO(B). Dia. EA269. Lot No. 31103 Metro-Cammell 1991–3. –/86. 38.9 t.
65800–65846. DMSO(A). Dia. EA268. Lot No. 31130 ABB York 1993–4. –/86. 38.9 t.
65847–65893. DMSO(B). Dia. EA268. Lot No. 31130 ABB York 1993–4. –/86. 39 t.
72028–72126 (even Nos.). TSO. Dia. EH293. Lot No. 31102 BREL York 1991–3. –/86. 29.5 t.
72029–72127 (odd Nos.). TSOL. Dia. EH292. Lot No. 31101 BREL York 1991–3. –/86. 28.6 t.
72719–72817 (odd Nos.). TSOL. Dia. EH294. Lot No. 31104 Metro-Cammell 1991–2. –/86. 30.2 t.
72720–72818 (even Nos.). TSO. Dia. EH295. Lot No. 31105 Metro-Cammell 1991–2. –/86. 29.1 t.
72900–72992 (even Nos.). TSO. Dia. EH293. Lot No. 31102 ABB York 1993–4. –/86. 29.5 t.
72901–72993 (odd Nos.). TSOL. Dia. EH294. Lot No. 31101 ABB York 1993–4. –/86. 28.6 t.

Class 465/0. Built by ABB.

465 001	NW	F	SE	SG	64759	72028	72029	64809
465 002	NW	F	SE	SG	64760	72030	72031	64810
465 003	CX	F	SE	SG	64761	72032	72033	64811
465 004	NW	F	SE	SG	64762	72034	72035	64812
465 005	NW	F	SE	SG	64763	72036	72037	64813
465 006	NW	F	SE	SG	64764	72038	72039	64814
465 007	NW	F	SE	SG	64765	72040	72041	64815
465 008	NW	F	SE	SG	64766	72042	72043	64816
465 009	NW	F	SE	SG	64767	72044	72045	64817
465 010	NW	F	SE	SG	64768	72046	72047	64818
465 011	NW	F	SE	SG	64769	72048	72049	64819

465 012	**NW**	F	*SE*	SG	64770	72050	72051	64820
465 013	**NW**	F	*SE*	SG	64771	72052	72053	64821
465 014	**NW**	F	*SE*	SG	64772	72054	72055	64822
465 015	**NW**	F	*SE*	SG	64773	72056	72057	64823
465 016	**NW**	F	*SE*	SG	64774	72058	72059	64824
465 017	**NW**	F	*SE*	SG	64775	72060	72061	64825
465 018	**NW**	F	*SE*	SG	64776	72062	72063	64826
465 019	**NW**	F	*SE*	SG	64777	72064	72065	64827
465 020	**NW**	F	*SE*	SG	64778	72066	72067	64828
465 021	**NW**	F	*SE*	SG	64779	72068	72069	64829
465 022	**NW**	F	*SE*	SG	64780	72070	72071	64830
465 023	**NW**	F	*SE*	SG	64781	72072	72073	64831
465 024	**NW**	F	*SE*	SG	64782	72074	72075	64832
465 025	**NW**	F	*SE*	SG	64783	72076	72077	64833
465 026	**NW**	F	*SE*	SG	64784	72078	72079	64834
465 027	**NW**	F	*SE*	SG	64785	72080	72081	64835
465 028	**NW**	F	*SE*	SG	64786	72082	72083	64836
465 029	**NW**	F	*SE*	SG	64787	72084	72085	64837
465 030	**NW**	F	*SE*	SG	64788	72086	72087	64838
465 031	**NW**	F	*SE*	SG	64789	72088	72089	64839
465 032	**NW**	F	*SE*	SG	64790	72090	72091	64840
465 033	**NW**	F	*SE*	SG	64791	72092	72093	64841
465 034	**NW**	F	*SE*	SG	64792	72094	72095	64842
465 035	**NW**	F	*SE*	SG	64793	72096	72097	64843
465 036	**NW**	F	*SE*	SG	64794	72098	72099	64844
465 037	**NW**	F	*SE*	SG	64795	72100	72101	64845
465 038	**NW**	F	*SE*	SG	64796	72102	72103	64846
465 039	**NW**	F	*SE*	SG	64797	72104	72105	64847
465 040	**NW**	F	*SE*	SG	64798	72106	72107	64848
465 041	**NW**	F	*SE*	SG	64799	72108	72109	64849
465 042	**NW**	F	*SE*	SG	64800	72110	72111	64850
465 043	**NW**	F	*SE*	SG	64801	72112	72113	64851
465 044	**NW**	F	*SE*	SG	64802	72114	72115	64852
465 045	**NW**	F	*SE*	SG	64803	72116	72117	64853
465 046	**NW**	F	*SE*	SG	64804	72118	72119	64854
465 047	**NW**	F	*SE*	SG	64805	72120	72121	64855
465 048	**NW**	F	*SE*	SG	64806	72122	72123	64856
465 049	**NW**	F	*SE*	SG	64807	72124	72125	64857
465 050	**NW**	F	*SE*	SG	64808	72126	72127	64858

Class 465/1. Built by ABB. As Class 465/0 but with detail differences.

465 151	**NW**	F	*SE*	SG	65800	72900	72901	65847
465 152	**NW**	F	*SE*	SG	65801	72902	72903	65848
465 153	**NW**	F	*SE*	SG	65802	72904	72905	65849
465 154	**NW**	F	*SE*	SG	65803	72906	72907	65850
465 155	**NW**	F	*SE*	SG	65804	72908	72909	65851
465 156	**NW**	F	*SE*	SG	65805	72910	72911	65852
465 157	**NW**	F	*SE*	SG	65806	72912	72913	65853
465 158	**NW**	F	*SE*	SG	65807	72914	72915	65854
465 159	**NW**	F	*SE*	SG	65808	72916	72917	65855
465 160	**NW**	F	*SE*	SG	65809	72918	72919	65856

465 161	**NW**	F	*SE*	SG	65810	72920	72921	65857
465 162	**NW**	F	*SE*	SG	65811	72922	72923	65858
465 163	**NW**	F	*SE*	SG	65812	72924	72925	65859
465 164	**NW**	F	*SE*	SG	65813	72926	72927	65860
465 165	**NW**	F	*SE*	SG	65814	72928	72929	65861
465 166	**NW**	F	*SE*	SG	65815	72930	72931	65862
465 167	**NW**	F	*SE*	SG	65816	72932	72933	65863
465 168	**NW**	F	*SE*	SG	65817	72934	72935	65864
465 169	**NW**	F	*SE*	SG	65818	72936	72937	65865
465 170	**NW**	F	*SE*	SG	65819	72938	72939	65866
465 171	**NW**	F	*SE*	SG	65820	72940	72941	65867
465 172	**NW**	F	*SE*	SG	65821	72942	72943	65868
465 173	**NW**	F	*SE*	SG	65822	72944	72945	65869
465 174	**NW**	F	*SE*	SG	65823	72946	72947	65870
465 175	**NW**	F	*SE*	SG	65824	72948	72949	65871
465 176	**NW**	F	*SE*	SG	65825	72950	72951	65872
465 177	**NW**	F	*SE*	SG	65826	72952	72953	65873
465 178	**NW**	F	*SE*	SG	65827	72954	72955	65874
465 179	**NW**	F	*SE*	SG	65828	72956	72957	65875
465 180	**NW**	F	*SE*	SG	65829	72958	72959	65876
465 181	**NW**	F	*SE*	SG	65830	72960	72961	65877
465 182	**NW**	F	*SE*	SG	65831	72962	72963	65878
465 183	**NW**	F	*SE*	SG	65832	72964	72965	65879
465 184	**NW**	F	*SE*	SG	65833	72966	72967	65880
465 185	**NW**	F	*SE*	SG	65834	72968	72969	65881
465 186	**NW**	F	*SE*	SG	65835	72970	72971	65882
465 187	**NW**	F	*SE*	SG	65836	72972	72973	65883
465 188	**NW**	F	*SE*	SG	65837	72974	72975	65884
465 189	**NW**	F	*SE*	SG	65838	72976	72977	65885
465 190	**NW**	F	*SE*	SG	65839	72978	72979	65886
465 191	**NW**	F	*SE*	SG	65840	72980	72981	65887
465 192	**NW**	F	*SE*	SG	65841	72982	72983	65888
465 193	**NW**	F	*SE*	SG	65842	72984	72985	65889
465 194	**NW**	F	*SE*	SG	65843	72986	72987	65890
465 195	**NW**	F	*SE*	SG	65844	72988	72989	65891
465 196	**NW**	F	*SE*	SG	65845	72990	72991	65892
465 197	**NW**	F	*SE*	SG	65846	72992	72993	65893

Class 465/2. Built by Metro-Cammell.

465 201	**NW**	F	*SE*	SG	65700	72719	72720	65750
465 202	**NW**	F	*SE*	SG	65701	72721	72722	65751
465 203	**NW**	F	*SE*	SG	65702	72723	72724	65752
465 204	**NW**	F	*SE*	SG	65703	72725	72726	65753
465 205	**NW**	F	*SE*	SG	65704	72727	72728	65754
465 206	**NW**	F	*SE*	SG	65705	72729	72730	65755
465 207	**NW**	F	*SE*	SG	65706	72731	72732	65756
465 208	**NW**	F	*SE*	SG	65707	72733	72734	65757
465 209	**NW**	Γ	*SE*	SG	65708	72735	72736	65758
465 210	**NW**	F	*SE*	SG	65709	72737	72738	65759
465 211	**NW**	F	*SE*	SG	65710	72739	72740	65760
465 212	**NW**	F	*SE*	SG	65711	72741	72742	65761

465 213	**NW**	F	*SE*	SG	65712	72743	72744	65762
465 214	**NW**	F	*SE*	SG	65713	72745	72746	65763
465 215	**NW**	F	*SE*	SG	65714	72747	72748	65764
465 216	**NW**	F	*SE*	SG	65715	72749	72750	65765
465 217	**NW**	F	*SE*	SG	65716	72751	72752	65766
465 218	**NW**	F	*SE*	SG	65717	72753	72754	65767
465 219	**NW**	F	*SE*	SG	65718	72755	72756	65768
465 220	**NW**	F	*SE*	SG	65719	72757	72758	65769
465 221	**NW**	F	*SE*	SG	65720	72759	72760	65770
465 222	**NW**	F	*SE*	SG	65721	72761	72762	65771
465 223	**NW**	F	*SE*	SG	65722	72763	72764	65772
465 224	**NW**	F	*SE*	SG	65723	72765	72766	65773
465 225	**NW**	F	*SE*	SG	65724	72767	72768	65774
465 226	**NW**	F	*SE*	SG	65725	72769	72770	65775
465 227	**NW**	F	*SE*	SG	65726	72771	72772	65776
465 228	**NW**	F	*SE*	SG	65727	72773	72774	65777
465 229	**NW**	F	*SE*	SG	65728	72775	72776	65778
465 230	**NW**	F	*SE*	SG	65729	72777	72778	65779
465 231	**NW**	F	*SE*	SG	65730	72779	72780	65780
465 232	**NW**	F	*SE*	SG	65731	72781	72782	65781
465 233	**NW**	F	*SE*	SG	65732	72783	72784	65782
465 234	**NW**	F	*SE*	SG	65733	72785	72786	65783
465 235	**NW**	F	*SE*	SG	65734	72787	72788	65784
465 236	**NW**	F	*SE*	SG	65735	72789	72790	65785
465 237	**NW**	F	*SE*	SG	65736	72791	72792	65786
465 238	**NW**	F	*SE*	SG	65737	72793	72794	65787
465 239	**NW**	F	*SE*	SG	65738	72795	72796	65788
465 240	**NW**	F	*SE*	SG	65739	72797	72798	65789
465 241	**NW**	F	*SE*	SG	65740	72799	72800	65790
465 242	**NW**	F	*SE*	SG	65741	72801	72802	65791
465 243	**NW**	F	*SE*	SG	65742	72803	72804	65792
465 244	**NW**	F	*SE*	SG	65743	72805	72806	65793
465 245	**NW**	F	*SE*	SG	65744	72807	72808	65794
465 246	**NW**	F	*SE*	SG	65745	72809	72810	65795
465 247	**NW**	F	*SE*	SG	65746	72811	72812	65796
465 248	**NW**	F	*SE*	SG	65747	72813	72814	65797
465 249	**NW**	F	*SE*	SG	65748	72815	72816	65798
465 250	**NW**	F	*SE*	SG	65749	72817	72818	65799

CLASS 466   NETWORKER

DMSO–DTSO. New units with Aluminium bodies. Sliding doors. Disc, rheo-static and regenerative brakes.

Electrical Equipment: Networker.
Bogies: P3 (motor car) and T3 (trailer).
Gangways: Within set.
Traction Motors: Four GEC Alsthom G354CX three-phase induction motors per car driven by four GTO inverters.
Dimensions: 20.89 x 2.82 m.

Maximum Speed: 75 m.p.h.

DMSO. Dia. EA271. Lot No. 31128 Metro-Cammell 1992–3. –/86. 39.2 t.
DTSO. Dia. EE279. Lot No. 31129 Metro-Cammell 1991–2. –/82. 33.2 t.

466 001	**NW**	F	*SE*	SG	64860	78312
466 002	**NW**	F	*SE*	SG	64861	78313
466 003	**NW**	F	*SE*	SG	64862	78314
466 004	**NW**	F	*SE*	SG	64863	78315
466 005	**NW**	F	*SE*	SG	64864	78316
466 006	**NW**	F	*SE*	SG	64865	78317
466 007	**NW**	F	*SE*	SG	64866	78318
466 008	**NW**	F	*SE*	SG	64867	78319
466 009	**NW**	F	*SE*	SG	64868	78320
466 010	**NW**	F	*SE*	SG	64869	78321
466 011	**NW**	F	*SE*	SG	64870	78322
466 012	**NW**	F	*SE*	SG	64871	78323
466 013	**NW**	F	*SE*	SG	64872	78324
466 014	**NW**	F	*SE*	SG	64873	78325
466 015	**NW**	F	*SE*	SG	64874	78326
466 016	**NW**	F	*SE*	SG	64875	78327
466 017	**NW**	F	*SE*	SG	64876	78328
466 018	**NW**	F	*SE*	SG	64877	78329
466 019	**NW**	F	*SE*	SG	64878	78330
466 020	**NW**	F	*SE*	SG	64879	78331
466 021	**NW**	F	*SE*	SG	64880	78332
466 022	**NW**	F	*SE*	SG	64881	78333
466 023	**NW**	F	*SE*	SG	64882	78334
466 024	**NW**	F	*SE*	SG	64883	78335
466 025	**NW**	F	*SE*	SG	64884	78336
466 026	**NW**	F	*SE*	SG	64885	78337
466 027	**NW**	F	*SE*	SG	64886	78338
466 028	**NW**	F	*SE*	SG	64887	78339
466 029	**NW**	F	*SE*	SG	64888	78340
466 030	**NW**	F	*SE*	SG	64889	78341
466 031	**NW**	F	*SE*	SG	64890	78342
466 032	**NW**	F	*SE*	SG	64891	78343
466 033	**NW**	F	*SE*	SG	64892	78344
466 034	**NW**	F	*SE*	SG	64893	78345
466 035	**NW**	F	*SE*	SG	64894	78346
466 036	**NW**	F	*SE*	SG	64895	78347
466 037	**NW**	F	*SE*	SG	64896	78348
466 038	**NW**	F	*SE*	SG	64897	78349
466 039	**NW**	F	*SE*	SG	64898	78350
466 040	**NW**	F	*SE*	SG	64899	78351
466 041	**NW**	F	*SE*	SG	64900	78352
466 042	**NW**	F	*SE*	SG	64901	78353
466 043	**NW**	F	*SE*	SG	64902	78354

CLASS 483   'NEW' ISLE OF WIGHT STOCK

DMBSO(A)–DMBSO(B). Tube stock built 1938 onwards for LTE. Converted 1989–90 for Isle of Wight Line. Sliding doors. End doors. Former London Underground numbers are shown in parentheses.

System: 660 V d.c. third rail.
Gangways: Non-gangwayed.
Traction Motors: Two of 130 kW.
Dimensions: 15.95 x 2.69 m.
Maximum Speed: 45 m.p.h.

DMSO (A). Lot No. 31071. Dia. EA265. –/42. 27.5 t.
DMSO (B). Lot No. 31072. Dia. EA266. –/42. 27.5 t.

483 001	**NW**	A		RY	121	(10184)	222	(11142)
483 002	**NW**	A	IL	RY	122	(10221)	225	(11116)
483 003	**NW**	A		RY	123	(10116)	221	(11221)
483 004	**NW**	A		RY	124	(10205)	224	(11205)
483 005	**NW**	A		RY	125	(10142)	223	(11184)
483 006	**NW**	A		RY	126	(10297)	226	(11297)
483 007	**NW**	A	IL	RY	127	(10291)	227	(11291)
483 008	**NW**	A	IL	RY	128	(10255)	228	(11255)
483 009	**NW**	A	IL	RY	129	(10289)	229	(11229)

CLASS 507

BDMSO–TSO–DMSO. Tightlock couplers. Sliding doors. Disc and rheostatic brakes.
System: 750 V d.c. third rail.
Bogies: BX1.
Gangways: Gangwayed within unit. End doors.
Traction Motors: Four GEC G310AZ of 82.125 kW per car.
Dimensions: 19.80 x 2.82 m (outer cars), 19.92 x 2.82 m (inner cars).
Maximum Speed: 75 m.p.h.

BDMSO. Dia. EI202. Lot No. 30906 York 1978–80. –/74 (–/68*). 37.06 t.
TSO. Dia. EH205. Lot No. 30907 York 1978–80. –/82 (–/86*). 25.60 t.
DMSO. Dia. EA201. Lot No. 30908 York 1978–80. –/74 (–/68*). 35.62 t.

507 001		**MT**	A	ME	BD	64367 71342 64405
507 002		**MT**	A	ME	BD	64368 71343 64406
507 003		**MT**	A	ME	BD	64369 71344 64407
507 004		**MT**	A	ME	BD	64370 71345 64408
507 005		**MT**	A	ME	BD	64371 71346 64409
507 006	*	**MT**	A	ME	BD	64372 71347 64410
507 007		**MT**	A	ME	BD	64373 71348 64411
507 008		**MT**	A	ME	BD	64374 71349 64412
507 009		**MT**	A	ME	BD	64375 71350 64413
507 010		**MT**	A	ME	BD	64376 71351 64414
507 011		**MT**	A	ME	BD	64377 71352 64415
507 012		**MT**	A	ME	BD	64378 71353 64416

507 013		**MT**	A	*ME*	BD	64379 71354 64417
507 014		**MT**	A	*ME*	BD	64380 71355 64418
507 015		**MT**	A	*ME*	BD	64381 71356 64419
507 016		**MT**	A	*ME*	BD	64382 71357 64420
507 017	*	**MT**	A	*ME*	BD	64383 71358 64421
507 018		**MT**	A	*ME*	BD	64384 71359 64422
507 019		**MT**	A	*ME*	BD	64385 71360 64423
507 020		**MT**	A	*ME*	BD	64386 71361 64424
507 021		**MT**	A	*ME*	BD	64387 71362 64425
507 023		**MT**	A	*ME*	BD	64389 71364 64427
507 024	*	**MT**	A	*ME*	BD	64390 71365 64428
507 025		**MT**	A	*ME*	BD	64391 71366 64429
507 026		**MT**	A	*ME*	BD	64392 71367 64430
507 027		**MT**	A	*ME*	BD	64393 71368 64431
507 028		**MT**	A	*ME*	BD	64394 71369 64432
507 029		**MT**	A	*ME*	BD	64395 71370 64433
507 030		**MT**	A	*ME*	BD	64396 71371 64434
507 031		**MT**	A	*ME*	BD	64397 71372 64435
507 032		**MT**	A	*ME*	BD	64398 71373 64436
507 033		**MT**	A	*ME*	BD	64399 71374 64437

CLASS 508

DMSO–TSO–BDMSO. Tightlock couplers. Sliding doors. Disc and rheostatic brakes. Originally built as 4-car units and numbered 508 001–043. One trailer removed and used for Class 455/7 on transfer from the former Southern Region of BR. Certain of these units are being transferred to Connex South Eastern.
System: 750 V d.c. third rail.
Bogies: BX1.
Gangways: Gangwayed within unit. End doors.
Traction Motors: Four GEC G310AZ of 82.125 kW per car.
Dimensions: 19.80 x 2.82 m (outer cars), 19.92 x 2.82 m (inner cars).
Maximum Speed: 75 m.p.h.

DMSO. Dia. EA208. Lot No. 30979 York 1979–80. –/74. 36.15 t.
TSO. Dia. EH218. Lot No. 30980 York 1979–80. –/82. 26.72 t.
BDMSO. Dia. EI203. Lot No. 30981 York 1979–80. –/74. 36.61 t.

508 101		A		ZG	64649 71483 64692
508 102	**MT**	A	*ME*	Kirkdale (S)	64650 71484 64693
508 103	**MT**	A	*ME*	BD	64651 71485 64694
508 104	**MT**	A	*ME*	BD	64652 71486 64695
508 105		A		ZG	64653 71487 64696
508 106		A		KI	64654 71488 64697
508 107		A		KI	64655 71489 64698
508 108		A	*ME*	West Kirby (S)	64656 71490 64699
508 109		A		KI	64657 71491 64700
508 110	**MT**	A	*ME*	DD	64658 71492 64701
508 111	**MT**	A	*ME*	BD	64659 71493 64702
508 112	**MT**	A	*ME*	BD	64660 71494 64703
508 113		A		KI	64661 71495 64704

508 114	**MT**	A	*ME*	BD	64662	71496	64705
508 115	**MT**	A	*ME*	BD	64663	71497	64706
508 116		A		KI	64664	71498	64707
508 117	**MT**	A	*ME*	BD	64665	71499	64708
508 118	**MT**	A	*ME*	BD	64666	71500	64709
508 119		A		KI	64667	71501	64710
508 120	**MT**	A	*ME*	BD	64668	71502	64711
508 121		A		ZG	64669	71503	64712
508 122	**MT**	A	*ME*	West Kirby (S)	64670	71504	64713
508 123	**MT**	A	*ME*	West Kirby (S)	64671	71505	64714
508 124	**MT**	A	*ME*	BD	64672	71506	64715
508 125	**MT**	A	*ME*	BD	64673	71507	64716
508 126	**MT**	A	*ME*	BD	64674	71508	64717
508 127	**MT**	A	*ME*	BD	64675	71509	64718
508 128	**MT**	A	*ME*	Kirkdale (S)	64676	71510	64719
508 129		A		ZG	64677	71511	64720
508 130	**MT**	A	*ME*	BD	64678	71512	64721
508 131	**MT**	A	*ME*	BD	64679	71513	64722
508 132		A		ZG	64680	71514	64723
508 133		A		ZG	64681	71515	64724
508 134	**MT**	A	*ME*	BD	64682	71516	64725
508 135	**MT**	A	*ME*	Southport CS (S)	64683	71517	64726
508 136	**MT**	A	*ME*	BD	64684	71518	64727
508 137	**MT**	A	*ME*	BD	64685	71519	64728
508 138	**MT**	A	*ME*	BD	64686	71520	64729
508 139	**MT**	A	*ME*	BD	64687	71521	64730
508 140	**MT**	A	*ME*	BD	64688	71522	64731
508 141	**MT**	A	*ME*	BD	64689	71523	64732
508 142	**MT**	A	*ME*	Kirkdale (S)	64690	71524	64733
508 143	**MT**	A	*ME*	BD	64691	71525	64734

4.3. EUROSTAR SETS (CLASS 373)

Eurostar sets work services through the Channel Tunnel between London and Paris and Brussels. They are based on the French TGV design concept, and the individual cars are numbered like French TGVs. Each train consists of two 9-coach sets back-to-back with a power car at the outer end. Regional sets for operation north of London consist of two 7-coach half-sets. All sets are articulated with an extra motor bogie on the coach next to the power car. Coaches are numbered R1–R9 (and in traffic R10–R18 in the second set). Coaches R18–R10 are identical to R1–R9.

Note: The pairs of sets are also known by designations as follows:
F/FN Assembled in France,   UK/UN Assembled in the UK.
Systems: 25 kV a.c. overhead, 3000 V d.c. overhead and 750 V d.c. third rail.
* Also fitted for 1500 V d.c. operation.

BR Sets:

3001	F15	LC	*ES*	PI	3730010	3730011	3730012	3730013
3002	F15	LC	*ES*	PI	3730020	3730021	3730022	3730023
3003	UK3	LC	*ES*	PI	3730030	3730031	3730032	3730033
3004	UK3	LC	*ES*	PI	3730040	3730041	3730042	3730043
3005	UK4	LC	*ES*	PI	3730050	3730051	3730052	3730053
3006	UK4	LC	*ES*	PI	3730060	3730061	3730062	3730063
3007	UK5	LC	*ES*	PI	3730070	3730071	3730072	3730073
3008	UK5	LC	*ES*	PI	3730080	3730081	3730082	3730083
3009	UK8	LC	*ES*	PI	3730090	3730091	3730092	3730093
3010	UK8	LC	*ES*	PI	3730100	3730101	3730102	3730103
3011	UK9	LC	*ES*	PI	3730110	3730111	3730112	3730113
3012	UK9	LC	*ES*	PI	3730120	3730121	3730122	3730123
3013	UK10	LC	*ES*	PI	3730130	3730131	3730132	3730133
3014	UK10	LC	*ES*	PI	3730140	3730141	3730142	3730143
3015	UK11	LC	*ES*	PI	3730150	3730151	3730152	3730153
3016	UK11	LC	*ES*	PI	3730160	3730161	3730162	3730163
3017	UK12	LC	*ES*	PI	3730170	3730171	3730172	3730173
3018	UK12	LC	*ES*	PI	3730180	3730181	3730182	3730183
3019	UK14	LC	*ES*	PI	3730190	3730191	3730192	3730193
3020	UK14	LC	*ES*	PI	3730200	3730201	3730202	3730203
3021	UK15	LC	*ES*	PI	3730210	3730211	3730212	3730213
3022	UK15	LC	*ES*	PI	3730220	3730221	3730222	3730223
3999		LC	*ES*	PI	3739990	Spare power car.		

SNCB/NMBS Sets:

3101	UK1	CB	*ES*	FF	3731010	3731011	3731012	3731013
3102	UK1	CB	*ES*	FF	3731020	3731021	3731022	3731023
3103	UK2	CB	*ES*	FF	3731030	3731031	3731032	3731033
3104	UK2	CB	*ES*	FF	3731040	3731041	3731042	3731043
3105	UK6	CB	*ES*	FF	3731050	3731051	3731052	3731053
3106	UK6	CB	*ES*	FF	3731060	3731061	3731062	3731063
3107	UK7	CB	*ES*	FF	3731070	3731071	3731072	3731073
3108	UK7	CB	*ES*	FF	3731080	3731081	3731082	3731083

Built: 1992–3 by GEC Alsthom at various works.
Wheel Arrangement: Bo–Bo + Bo–2–2–2–2–2–2–2–2–2.
Length: 22.15 + 21.845 + (7 x 18.70) + 21.845 m.
Max. Speed: 300 km/h (187.5 m.p.h.).
Livery: White with dark blue window band roof and yellow bodysides.

Car	Type	Lot No.	Accommodation
M	DM	31118	
R1	MSOL	31119	–/48 + 3 tip-up 2T.
R2	TSOL	31120	–/58 + 4 tip-up 1T + train manager's compartment.
R3	TSOL	31121	–/58 + 4 tip-up 2T.
R4	TSOL	31122	–/58 + 4 tip-up 1T + public telephone.
R5	TSOL	31123	–/58 + 4 tip-up 2T.
R6	RB	31124	Kitchen/bar.
R7	TFOL	31125	39/– + 3 tip-up + 1 settee 1T.
R8	TFOL	31126	39/– + 3 tip-up + 1 settee 1T + public telephone.
R9	TBFOL	31127	25/– + 1 tip-up + 1 settee 1DT + staff compartment.

3730014	3730015	3730016	3730017	3730018	3730019
3730024	3730025	3730026	3730027	3730028	3730029
3730034	3730035	3730036	3730037	3730038	3730039
3730044	3730045	3730046	3730047	3730048	3730049
3730054	3730055	3730056	3730057	3730058	3730059
3730064	3730065	3730066	3730067	3730068	3730069
3730074	3730075	3730076	3730077	3730078	3730079
3730084	3730085	3730086	3730087	3730088	3730089
3730094	3730095	3730096	3730097	3730098	3730099
3730104	3730105	3730106	3730107	3730108	3730109
3730114	3730115	3730116	3730117	3730118	3730119
3730124	3730125	3730126	3730127	3730128	3730129
3730134	3730135	3730136	3730137	3730138	3730139
3730144	3730145	3730146	3730147	3730148	3730149
3730154	3730155	3730156	3730157	3730158	3730159
3730164	3730165	3730166	3730167	3730168	3730169
3730174	3730175	3730176	3730177	3730178	3730179
3730184	3730185	3730186	3730187	3730188	3730189
3730194	3730195	3730196	3730197	3730198	3730199
3730204	3730205	3730206	3730207	3730208	3730209
3730214	3730215	3730216	3730217	3730218	3730219
3730224	3730225	3730226	3730227	3730228	3730229

3731014	3731015	3731016	3731017	3731018	3731019
3731024	3731025	3731026	3731027	3731028	3731029
3731034	3731035	3731036	3731037	3731038	3731039
3731044	3731045	3731046	3731047	3731048	3731049
3731054	3731055	3731056	3731057	3731058	3731059
3731064	3731065	3731066	3731067	3731068	3731069
3731074	3731075	3731076	3731077	3731078	3731079
3731084	3731085	3731086	3731087	3731088	3731089

SNCF Sets:

3201	F16	CF	*ES*	LY	3732010	3732011	3732012	3732013
3202	F16	CF	*ES*	LY	3732020	3732021	3732022	3732023
3203*	F1	CF	*ES*	LY	3732030	3732031	3732032	3732033
3204*	F1	CF	*ES*	LY	3732040	3732041	3732042	3732043
3205	F2	CF	*ES*	LY	3732050	3732051	3732052	3732053
3206	F2	CF	*ES*	LY	3732060	3732061	3732062	3732063
3207*	F3	CF	*ES*	LY	3732070	3732071	3732072	3732073
3208*	F3	CF	*ES*	LY	3732080	3732081	3732082	3732083
3209	F4	CF	*ES*	LY	3732090	3732091	3732092	3732093
3210	F4	CF	*ES*	LY	3732100	3732101	3732102	3732103
3211	F5	CF	*ES*	LY	3732110	3732111	3732112	3732113
3212	F5	CF	*ES*	LY	3732120	3732121	3732122	3732123
3213	F6	CF	*ES*	LY	3732130	3732131	3732132	3732133
3214	F6	CF	*ES*	LY	3732140	3732141	3732142	3732143
3215*	F7	CF	*ES*	LY	3732150	3732151	3732152	3732153
3216*	F7	CF	*ES*	LY	3732160	3732161	3732162	3732163
3217	F8	CF	*ES*	LY	3732170	3732171	3732172	3732173
3218	F8	CF	*ES*	LY	3732180	3732181	3732182	3732183
3219	F9	CF	*ES*	LY	3732190	3732191	3732192	3732193
3220	F9	CF	*ES*	LY	3732200	3732201	3732202	3732203
3221	F10	CF	*ES*	LY	3732210	3732211	3732212	3732213
3222	F10	CF	*ES*	LY	3732220	3732221	3732222	3732223
3223	F11	CF	*ES*	LY	3732230	3732231	3732232	3732233
3224	F11	CF	*ES*	LY	3732240	3732241	3732242	3732243
3225*	F12	CF	*ES*	LY	3732250	3732251	3732252	3732253
3226*	F12	CF	*ES*	LY	3732260	3732261	3732262	3732263
3227*	F13	CF	*ES*	LY	3732270	3732271	3732272	3732273
3228*	F13	CF	*ES*	LY	3732280	3732281	3732282	3732283
3229	F14	CF	*ES*	LY	3732290	3732291	3732292	3732293
3230	F14	CF	*ES*	LY	3732300	3732301	3732302	3732303
3231	UK13	CF	*ES*	LY	3732310	3732311	3732312	3732313
3232	UK13	CF	*ES*	LY	3732320	3732321	3732322	3732323

'Regional Eurostar' Sets for services from the North of England & Scotland: These are 7 coach sets consisting of PC + R1/3/2/5/6/7/9 only.

3301	FN1	LC	*ES*	PI	3733010	3733011	3733013	3733012
3302	FN1	LC	*ES*	PI	3733020	3733021	3733023	3733022
3303	FN2	LC	*ES*	PI	3733030	3733031	3733033	3733032
3304	FN2	LC	*ES*	PI	3733040	3733041	3733043	3733042
3305	UN1	LC	*ES*	PI	3733050	3733051	3733053	3733052
3306	UN1	LC	*ES*	PI	3733060	3733061	3733063	3733062
3307	UN2	LC	*ES*	PI	3733070	3733071	3733073	3733072
3308	UN2	LC	*ES*	PI	3733080	3733081	3733083	3733082
3309	UN3	LC	*ES*	PI	3733090	3733091	3733093	3733092
3310	UN3	LC	*ES*	PI	3733100	3733101	3733103	3733102
3311	UN4	LC	*ES*	PI	3733110	3733111	3733113	3733112
3312	UN4	LC	*ES*	PI	3733120	3733121	3733123	3733122
3313	UN5	LC	*ES*	PI	3733130	3733131	3733133	3733132
3314	UN5	LC	*ES*	PI	3733140	3733141	3733143	3733142

3732014 3732015 3732016 3732017 3732018 3732019
3732024 3732025 3732026 3732027 3732028 3732029
3732034 3732035 3732036 3732037 3732038 3732039
3732044 3732045 3732046 3732047 3732048 3732049
3732054 3732055 3732056 3732057 3732058 3732059
3732064 3732065 3732066 3732067 3732068 3732069
3732074 3732075 3732076 3732077 3732078 3732079
3732084 3732085 3732086 3732087 3732088 3732089
3732094 3732095 3732096 3732097 3732098 3732099
3732104 3732105 3732106 3732107 3732108 3732109
3732114 3732115 3732116 3732117 3732118 3732119
3732124 3732125 3732126 3732127 3732128 3732129
3732134 3732135 3732136 3732137 3732138 3732139
3732144 3732145 3732146 3732147 3732148 3732149
3732154 3732155 3732156 3732157 3732158 3732159
3732164 3732165 3732166 3732167 3732168 3732169
3732174 3732175 3732176 3732177 3732178 3732179
3732184 3732185 3732186 3732187 3732188 3732189
3732194 3732195 3732196 3732197 3732198 3732199
3732204 3732205 3732206 3732207 3732208 3732209
3732214 3732215 3732216 3732217 3732218 3732219
3732224 3732225 3732226 3732227 3732228 3732229
3732234 3732235 3732236 3732237 3732238 3732239
3732244 3732245 3732246 3732247 3732248 3732249
3732254 3732255 3732256 3732257 3732258 3732259
3732264 3732265 3732266 3732267 3732268 3732269
3732274 3732275 3732276 3732277 3732278 3732279
3732284 3732285 3732286 3732287 3732288 3732289
3732294 3732295 3732296 3732297 3732298 3732299
3732304 3732305 3732306 3732307 3732308 3732309
3732314 3732315 3732316 3732317 3732318 3732319
3732324 3732325 3732326 3732327 3732328 3732329

3733015 3733016 3733017 3733019
3733025 3733026 3733027 3733029
3733035 3733036 3733037 3733039
3733045 3733046 3733047 3733049
3733055 3733056 3733057 3733059
3733065 3733066 3733067 3733069
3733075 3733076 3733077 3733079
3733085 3733086 3733087 3733089
3733095 3733096 3733097 3733099
3733105 3733106 3733107 3733109
3733115 3733116 3733117 3733119
3733125 3733126 3733127 3733129
3733135 3733136 3733137 3733139
3733145 3733146 3733147 3733149

HIGH SPEED IN EUROPE

by David Haydock

Today's Railways editor David Haydock examines the development of European high speed railways over the past 20 years, and looks at the trains and new lines which will be making international travel easier in the next 20. First published in 1995, the book includes the following:

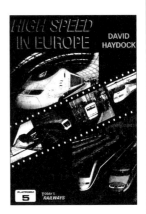

★ The Channel Tunnel Rail Link
★ Eurostar: A Truly International Train
★ Development of the TGV
★ Les Lignes à Grand Vitesse
★ TGV Duplex & Thalys Trainsets
★ The German ICE & Swedish X 2000
★ Pendolino: Tilting Through Italy
★ High Speed Lines in Belgium, Netherlands, Germany, Sweden, Italy and Spain.

High Speed in Europe also includes an index of European High Speed Line Statistics, and details of European High Speed Train Numbering. 80 pages including 38 in full colour. Thread Sewn. £9.95.

Available from the Platform 5 Mail Order Department. To place an order, please follow the instructions on page 383 of this book.

HIGH SPEED IN JAPAN

by Peter Semmens

When Japan's legendary 'Bullet Trains' entered service in 1964, a new era in high speed rail travel had begun. Brand new trains operating over brand new railway lines cut journey times in half and opened the world's eyes to the possibilities of high speed rail travel. Never before had a new railway provoked such interest, with many new ideas now copied by engineers the world over.

HIGH SPEED IN JAPAN

SHINKANSEN – THE WORLD'S
BUSIEST HIGH-SPEED RAILWAY

Peter Semmens

With photographs by Mikio Miura

HIGH SPEED IN JAPAN tells the story of Japan's high speed rail network from the earliest beginnings to the present day. Using unique first hand knowledge, author Peter Semmens guides the reader through the development of the network using a clear and explanatory narative. Civil engineering challenges, day to day operating procedures, political aspects and the high speed trains themselves are all examined. A useful glossary and a chronology of events are also included.

The combination of highest quality photographic material and the exclusive knowledge of Peter Semmens make this the definitive work on the subject. It contains 122 colour illustrations, 13 black & white illustrations, 23 diagrams, 53 tables and 3 maps. A4 size. 112 pages. Hardback. £16.95.

Available from the Platform 5 Mail Order Department. To place an order, please follow the instructions on page 383 of this book.

4.4. SERVICE EMUs

INDIVIDUAL VEHICLES

–		SO	*TE*	ZA	977335	(76277)	MTA Pool generator coach for DB999550.
–	**RT**	RT	*SA*	SU	977364	(10400)	Works with DEMU 930 301.
930 078	**RT**	RT	*SA*	HE	977578	(77101)	Works with Class 313/317.
930 079		RT	*SA*	SU	977579	(77109)	Works with Class 319.
–	**SO**	SO	*TE*	ZA	999602	(62483)	Works with DMU 977391/2. Ultrasonic test coach.

COMPLETE UNITS

Note: Some service units do not carry '93x' numbers.

Class 930/0. Sandite & De-icing Units. 750 V d.c. SR design.

930 001	**RT**	RT	*SA*	FR	975604	(10939)	975605	(10940)
930 002	**RT**	RT	*SA*	RE	975896	(11387)	975897	(11388)
930 003	**RT**	RT	*SA*	SU	975594	(12658)	975595	(10904)
930 004	**RT**	RT	*SA*	WD	975586	(10907)	975587	(10908)
930 005	**RT**	RT	*SA*	WD	975588	(10981)	975589	(10982)
930 006	**RT**	RT	*SA*	WD	975590	(10833)	975591	(10834)
930 007	**RT**	RT	*SA*	GI	975592	(10933)	975593	(12659)
930 008	**N**	RT	*SA*	GI	975596	(10844)	975597	(10987)
930 009	**RT**	RT	*SA*	SU	975598	(10989)	975599	(10990)
930 010		RT	*SA*	BI	975600	(10988)	975601	(10843)
930 011	**RT**	RT	*SA*	SU	975602	(10991)	975603	(10992)

Class 930/1. Tractor Unit. 750 V d.c. BR design.

930 101	**RT**	RT	*TE*	WD	977609	(65414)	977207	(61658)

Class 930/1. Sandite Unit. 750 V d.c. SR design.

930 102		RT	*SA*	FR	977533	(14273)	977534	(14384)

Class 930/2. Sandite & De-icing Units. 750 V d.c. BR design.

930 201	**RT**	RT	*SA*	BM	977566	(65312)	977567	(65314)
930 202	**RT**	RT	*SA*	FR	977804	(65336)	977805	(65357)
930 203	**RT**	RT	*SA*	RE	977864	(65341)	977865	(65355)
930 204	**RT**	RT	*SA*	RE	977874	(65302)	977875	(65304)
930 205	**RT**	RT	*SA*	RE	977871	(65353)	977872	(65367)
930 206	**RT**	RT	*SA*	WD	977924	(65382)	977925	(65379)

Classes 930 & 931. Route Learning Units. 750 V d.c. BR design.

930 082	**CX**	CX	*CR*	SU	977861	(61044)	977862	(70039)
					977863	(61038)		
931 001	**N**	RT	*CR*	SL	977857	(65346)	977856	(77531)
931 002	**N**	LC	*CR*	RE	977917	(65331)	977918	(77516)

Class 932. Development Units. 750 V d.c. (ex Class 411/5).

932 545	**ST**	P	*RS*	ZD	61359	70330	70287	61358
932 620	**PL**	P	*RS*	ZM	61948	70653		61949

Class 936/0. Sandite Unit. Merseyrail 750 V d.c. (ex Class 501).

936 003	**MD**	RT	*SA*	BD	977349	(61183)	977350	(75183)

Class 936/1. Sandite Units. 25 kV a.c. (ex Class 311).

936 103	**RT**	RT	*SA*	GW	977844	(76414)	977845	(62174)
					977846	(76433)		
936 104	**RT**	RT	*SA*	GW	977847	(76415)	977848	(62175)
					977849	(76434)		

Class 937. Sandite Units. 25 kV a.c. (ex Class 302, 305† or 308§).

937 908	†	RT	*SA*	IL	977741	(75469)	977742	(61436)
					977743	(75521)		
937 990	§	RT	*SA*	EM	977876	(75905)	977877	(61901)
					977878	(75938)		
937 991	§	RT	*SA*	IL	977926	(75900)	977927	(61896)
					977928	(75933)		
937 998		RT	*SA*	IL	977604	(75077)	977605	(61062)
					977606	(75070)		

Class 316. Test Unit. Ex Class 307.

316 997		SO	*TE*	EH	977708	(75118)	977709	(61018)
					977710	(75018)		

Service numbers not carried.

CLASS 931 (Formerly 419)   1957 type MLV

DMLV. Built 1959–61. Dual braked. These units are now officially in service stock, but they retain their capital stock side numbers.

Electrical Equipment: 1957-type.
Bogies: Mk 3B.
Gangways: Non-gangwayed.
Traction Motors: Two EE507 of 185 kW.
Dimensions: 19.64 x 2.82 m.
Maximum Speed: 90 mph.

68001. DMLV. Lot No. 30458 Ashford/Eastleigh. 1959. 45.5 t.
68003–10. DMLV. Lot No. 30623 Ashford/Eastleigh. 1960–61. 45.5 t.

931 090	(9010)	**J**	P	BM	68010
931 091	(9001)	**N**	P	BM	68001
931 092	(9002)	**N**	P	BM	68002
931 093	(9003)	**B**	P	BM	68003
931 094	(9004)	**N**	P	BM	68004
931 095	(9005)	**N**	P	BM	68005
931 097	(9007)	**N**	P	BM	68007
931 098	(9008)	**N**	P	BM	68008
931 099	(9009)	**J**	P	BM	68009

4.5. EMUs AWAITING DISPOSAL

The following withdrawn EMUs are awaiting disposal with the last known storage location shown.

FORMER CAPITAL STOCK UNITS

304 003	Crewe Brook Sidings	75047	61047	75647		
304 008	Crewe Brook Sidings	75052	61052	75652		
304 021	Crewe Brook Sidings	75685	61633	75665		
304 024	Crewe Brook Sidings	75688	61636	75668		
305 403	LG	75506	61473	75558		
306 017	IL	65217	65417	65617		
4308	Long Marston	61275	75395			
4311	Long Marston	61287	75407			
4732	Long Marston	12795	10239	12354	12796	
5001	Long Marston	14001	15207	15101	14002	
5176	Long Marston	14352	15396	15354	14351	
6213	Long Marston	65327	77512			
6308	Long Marston	14564	16108			
6309	Long Marston	14562	16106			
6402	Long Marston	65362	77547			
7001	ZG	67300	67401	67301		

FORMER SERVICE STOCK VEHICLES

975032	(75165)	SH
977296	(65319)	ZG
977345	(61180)	BD
977347	(61178)	BD
977639	(75548)	Southall ECD
977640	(61463)	Southall ECD
977641	(75214)	Southall ECD
977763	(70871)	ZG
977764	(70866)	ZG

LOOSE CARS

61433	LG	75003	Kineton
70003	OM	75015	Kineton
70008	OM	75019	Kineton
70010	OM	75020	OM
70612	Crewe Brook Sidings	75023	Kineton
70621	Crewe Brook Sidings	75025	OM
70622	Crewe Brook Sidings	75026	OM
70631	Crewe Brook Sidings	75030	OM
70640	Crewe Brook Sidings	75773	Yoker
75002	OM		

5.1. NON-PASSENGER-CARRYING COACHING STOCK

The notes shown for locomotive-hauled passenger stock generally apply also to non-passenger-carrying coaching stock (often abbreviated to NPCCS).

TOPS CODES

TOPS codes for NPCCS are made up as follows:

(1) Two letters denoting the type of the vehicle:

NA	Propelling control vehicle.
NB	High security brake van (100 m.p.h.).
NC	Gangwayed brake van modified for newspaper conveyance (100 m.p.h.).
ND	Gangwayed brake van (90 m.p.h.).
NE	Gangwayed brake van (100 m.p.h.).
NF	Gangwayed brake van with guard's safety equipment removed.
NG	Motorail loading wagon.
NH	Gangwayed brake van (110 m.p.h.).
NJ	General utility van (90 m.p.h.).
NK	High Security general utility van (100 m.p.h.).
NL	Newspaper van.
NN	Courier vehicle.
NO	General utility van (100 m.p.h. e.t.h. wired).
NP	General utility van for Post Office use or Motorail van (110 m.p.h.).
NR	BAA Container van (100 m.p.h.).
NS	Post office sorting van.
NT	Post office stowage van.
NU	Brake post office stowage van.
NX	Motorail van (100 m.p.h.).
NY	Exhibition van.
NZ	Driving brake van (also known as driving van trailer).
YR	Ferry van (special SR version of NJ with two pairs of side doors instead of three).

(2) A third letter denoting the brake type:

A	Air braked
V	Vacuum braked
X	Dual braked

OPERATOR CODES

The normal operator codes are given in brackets after the TOPS codes. These are as follows:

BG	Gangwayed brake van.
BPOT	Brake post office stowage van.
DLV	Driving brake van (also known as driving van trailer – DVT).
GUV	General utility van.
POS	Post office sorting van.
POT	Post office stowage van.

AK51 (RK) KITCHEN CAR

Dia. AK503. Mark 1. Gas cooking. Converted 1989 from RBR. Fluorescent lighting. ETH 2X. Commonwealth bogies.

Lot No. 30628 Pressed Steel 1960–61. 39 t.

Note: Kitchen cars have traditionally been numbered in the NPCCS series, but have passenger coach diagram numbers!

80041	(1690)	x	**CC**	RS	*SS*	BN

NN COURIER VEHICLE

Dia. NN504. Converted 1986–7 from Mark 1 BSKs. One compartment retained for courier use. Roller shutter doors. ETH 2.

80207. Lot No. 30721 Wolverton 1963. Commonwealth bogies. 37 t.
80211–7/23–5. Lot No. 30699 Wolverton 1962. Commonwealth bogies. 37 t.
80220. Lot No. 30573 Gloucester 1960. B4 bogies. 33 t.

Non-Standard Livery: 80211 is purple.

80207	(35466)	x	**PC**	VS	*SS*	SL
80211	(35296)	x	**0**	RS	*SS*	BN
80212	(35307)	` x	**RM**	E		OM
80213	(35316)	x	**CH**	O		CP
80214	(35323)	x	**RY**	CC		Ferme Park
80216	(35295)	x	**RM**	E		OM
80217	(35299)	x	**M**	O	*SU*	GT
80220	(35276)	x	**G**	MH	*SU*	RL
80223	(35331)	x	**RY**	CC		DY
80225	(35327)	x	**BG**	O		SZ

Name: 80207 is branded 'BAGGAGE CAR No.11'.

NP POST OFFICE GUV

Dia. NP502. Converted 1991–93 from newspaper vans. Mark 1. Short frames (57'). Originally converted from GUV. Fluorescent lighting, toilets and gangways fitted. Load 14 t. ETH 3X. B5 bogies.

Lot No. 30922 Wolverton or Doncaster 1977–8. 31 t.

80250	(86838, 94008)		**RM**	E	BK
80251	(86467, 94017)	x	**RM**	E	OM
80252	(86718, 94022)		**RM**	E	OM
80253	(86170, 94018)		**RM**	E	OM
80254	(86082, 94012)	x	**RM**	E	OM
80255	(86098, 94019)	x	**RM**	E	OM
80256	(86408, 94013)	x	**RM**	E	OM
80257	(86221, 94023)	x	**RM**	E	OM
80258	(86651, 94002)		**RM**	E	OM
80259	(86845, 94005)	x	**RM**	E	OM

NS (POS)   POST OFFICE SORTING VAN

Used in travelling post office (TPO) trains. Mark 1. Various diagrams.

The following lots have BR Mark 1 bogies except * B5 bogies. (subtract 2 t from weight).

80301–80305. Lot No. 30486 Wolverton 1959. Dia. NS501. Originally built with nets for collecting mail bags in motion. Equipment now removed. ETH 3X. 36 t.
80306–80308. Lot No. 30487 Wolverton 1959. Dia. NS502. ETH 3. 36 t.
80309–80314. Lot No. 30661 Wolverton 1961. Dia. NS501. ETH 3. 37 t.
80315–80316. Lot No. 30662 Wolverton 1961. Dia. NS501. ETH 3X. 36 t.

80301	v	**RM**	E		BK	80310	v	**RM** E	OM
80303	x*	**RM**	E		OM	80312	v	**RM** E	OM
80305	x*	**RM**	E		OM	80313	v	**RM** E	ZH
80306	v	**RM**	E		OM	80314	x*	**RM** E	OM
80308	x*	**RM**	E		OM	80315	v	**RM** E	OM
80309	x*	**RM**	E		OM	80316	x*	**RM** E	OM

The following lots are pressure ventilated and have B5 bogies.

80319–80327. Dia. NS504. Lot No. 30778 York 1968–9. ETH 4. 35 t.
80328–80338. Dia. NS505. Lot No. 30779 York 1968–9. ETH 4. 35 t.
80339–80355. Dia. NS506. Lot No. 30780 York 1968–9. ETH 4. 35 t.

80319	**RM** E	*EW*	EN		80338	**RM** E		BK
80320	**RM** E	*EW*	NC		80339	**RM** E	*EW*	EN
80321	**RM** E	*EW*	BK		80340	**RM** E	*EW*	BK
80322	**RM** E	*EW*	EN		80341	**RM** E	*EW*	EN
80323	**RM** E	*EW*	EN		80342	**RM** E	*EW*	BK
80324	**RM** E	*EW*	EN		80343	**RM** E	*EW*	BK
80325	**RM** E	*EW*	EN		80344	**RM** E	*EW*	BK
80326	**RM** E	*EW*	NC		80345	**RM** E	*EW*	NC
80327	**RM** E	*EW*	BK		80346	**RM** E	*EW*	EN
80328	**RM** E	*EW*	EN		80347	**RM** E	*EW*	NC
80329	**RM** E		OM		80348	**RM** E	*EW*	BZ
80330	**RM** E		OM		80349	**RM** E	*EW*	EN
80331	**RM** E	*EW*	EN		80350	**RM** E	*EW*	BK
80332	**RM** E	*EW*	EN		80351	**RM** E	*EW*	EN
80333	**RM** E	*EW*	EN		80352	**RM** E	*EW*	BK
80334	**RM** E	*EW*	BK		80353	**RM** E	*EW*	EN
80335	x	**RM** E		BK	80354	**RM** E	*EW*	BK
80336	**RM** E		BK		80355	**RM** E	*EW*	EN
80337	**RM** E	*EW*	NC					

Names:

80320	The Borders Mail	80327	George James

80356–80380. Lot No. 30839 York 1972–3. Dia. NS501. Pressure ventilated. Fluorescent lighting. B5 bogies. ETH 4X. 37 t.

80356	**RM** E	*EW*	BK		80358	**RM** E	*EW*	EN
80357	**RM** E	*EW*	BK		80359	**RM** E	*EW*	BK

80360	**RM** E	*EW*	EN		80371	**RM** E	*EW*	BZ
80361	**RM** E	*EW*	EN		80372	**RM** E	*EW*	EN
80362	**RM** E	*EW*	EN		80373	**RM** E	*EW*	EN
80363	**RM** E	*EW*	BK		80374	**RM** E	*EW*	EN
80364	**RM** E	*EW*	BK		80375	**RM** E	*EW*	BK
80365	**RM** E	*EW*	NC		80376	**RM** E	*EW*	BK
80366	**RM** E	*EW*	EN		80377	**RM** E	*EW*	NC
80367	**RM** E	*EW*	BK		80378	**RM** E	*EW*	EN
80368	**RM** E	*EW*	BK		80379	**RM** E	*EW*	EN
80369	**RM** E	*EW*	EN		80380	**RM** E	*EW*	BZ
80370	**RM** E	*EW*	BK					

Names:

80360	Derek Carter		80380	Ernie Gosling
80367	M.G. Berry			

80381–80395. Lot No. 30900 Wolverton 1977. Dia NS531. Converted from SK. Pressure ventilated. Fluorescent lighting. B5 bogies. ETH 4X, 38 t.

80381	(25112)	**RM** E	*EW*	NC		80389	(25103)	**RM** E		ZG
80382	(25109)	**RM** E	*EW*	EN		80390	(25047)	**RM** E	*EW*	EN
80383	(25033)	**RM** E	*EW*	EN		80392	(25082)	**RM** E	*EW*	EN
80384	(25078)	**RM** E	*EW*	EN		80393	(25118)	**RM** E	*EW*	EN
80385	(25083)	**RM** E	*EW*	EN		80394	(25156)	**RM** E	*EW*	NC
80386	(25099)	**RM** E	*EW*	EN		80395	(25056)	**RM** E	*EW*	EN
80387	(25045)	**RM** E	*EW*	EN						

NT (POT)   POST OFFICE STOWAGE VAN

Mark 1. Open vans used for stowage of mail bags in conjunction with POS. Various diagrams.

Lot No. 30488 Wolverton 1959. Dia. NT502. Originally built with nets for collecting mail bags in motion. Equipment now removed. B5 bogies. ETH 3. 35 t.

80400	**RM** E	*EW*	BK		80402	**RM** E	*EW*	BK
80401	**RM** E	*EW*	NC					

The following eight vehicles were converted at York from BSK to lot 30143 (80403) and 30229 (80404–80414). No new lot number was issued. Dia. NT503. B5 bogies. 35 t. (* Dia. NT501 BR2 bogies 38 t. ETH 3 (3X*).

80403	(34361)	**RM** E	*EW*	BZ		80411	(35003)	*	**RM** E	*EW*	BZ
80404	(35014)	**RM** E	*EW*	BZ		80412	(35002)	*	**RM** E	*EW*	NC
80405	(35009)	**RM** E	*EW*	BZ		80413	(35004)	*	**RM** E	*EW*	EN
80406	(35022)	**RM** E	*EW*	EN		80414	(35005)	*	**RM** E	*EW*	BK

Lot No. 30781 York 1968. Dia. NT505. Pressure ventilated. B5 bogies. ETH 4. 34 t.

80415	**RM** E	*EW*	EN		80421	**RM** E	*EW*	EN
80416	**RM** E	*EW*	EN		80422	**RM** E	*EW*	EN
80417	**RM** E	*EW*	EN		80423	**RM** E	*EW*	EN
80419	**RM** E	*EW*	BK		80424	**RM** E	*EW*	EN
80420	**RM** E	*EW*	EN					

Lot No. 30840 York 1973. Dia. NT504. Pressure ventilated. fluorescent lighting. B5 bogies. ETH 4X. 35 t.

80425	**RM** E	*EW*	BK	80428	**RM** E	*EW*	EN
80426	**RM** E	*EW*	EN	80429	**RM** E	*EW*	BK
80427	**RM** E	*EW*	EN	80430	**RM** E	*EW*	EN

Lot No. 30901 Wolverton 1977. converted from SK. Dia. NT521. Pressure ventilated. Fluorescent lighting. B5 bogies. ETH 4X. 35 t.

80431	(25104)	**RM** E	*EW*	EN	80436	(25077)	**RM** E	*EW*	EN
80432	(25071)	**RM** E	*EW*	EN	80437	(25068)	**RM** E	*EW*	EN
80433	(25150)	**RM** E	*EW*	BK	80438	(25139)	**RM** E	*EW*	BK
80434	(25119)	**RM** E	*EW*	EN	80439	(25127)	**RM** E	*EW*	BK
80435	(25117)	**RM** E	*EW*	EN					

NU (BPOT)   BRAKE POST OFFICE STOWAGE VAN

As NT but with brake. Mark 1.

Lot No. 30782 York 1968. Dia. NU502. Pressure ventilated. B5 bogies. ETH 4. 36 t.

80456	**RM** E	*EW*	EN	80458	**RM** E	*EW*	EN
80457	**RM** E	*EW*	EN				

NZ (DLV)   DRIVING BRAKE VAN (110 m.p.h.)

Dia. NZ501. Mark 3B. Air conditioned. T4 bogies. dg. Cab to shore communication. ETH 5X.

Lot No. 31042 Derby 1988. 45.18 t.

82101	**V**	P	*VW*	OY	82122		P	*VW*	PC
82102		P	*VW*	OY	82123		P	*VW*	PC
82103		P	*VW*	OY	82124		P	*VW*	PC
82104		P	*VW*	PC	82125		P	*VW*	PC
82105	**V**	P	*VW*	PC	82126	**V**	P	*VW*	MA
82106		P	*VW*	OY	82127	**V**	P	*VW*	OY
82107		P	*VW*	MA	82128		P	*VW*	OY
82108		P	*VW*	PC	82129		P	*VW*	OY
82109		P	*VW*	PC	82130		P	*VW*	PC
82110	**V**	P	*VW*	MA	82131		P	*VW*	MA
82111		P	*VW*	PC	82132		P	*VW*	OY
82112		P	*VW*	PC	82133	**V**	P	*VW*	OY
82113		P	*VW*	OY	82134	**V**	P	*VW*	MA
82114		P	*VW*	PC	82135		P	*VW*	MA
82115		P	*VW*	PC	82136		P	*VW*	OY
82116		P	*VW*	PC	82137		P	*VW*	PC
82117	**V**	P	*VW*	MA	82138		P	*VW*	PC
82118		P	*VW*	OY	82139		P	*VW*	PC
82119		P	*VW*	PC	82140		P	*VW*	PC
82120		P	*VW*	PC	82141		P	*VW*	PC
82121	**V**	P	*VW*	PC	82142		P	*VW*	MA

82143		P	VW	OY	82148		P	VW	OY
82144		P	VW	OY	82149	V	P	VW	MA
82145		P	VW	OY	82150		P	VW	PC
82146	V	P	VW	MA	82151		P	VW	OY
82147	V	P	VW	MA	82152	V	P	VW	MA

Names:

82115	Liverpool John Moores University
82120	Liverpool Chamber of Commerce
82121	Carlisle Cathedral
82132	West Midlands
82134	Sir Henry Doulton 1820–1897
82135	Spirit of Cumbria
82147	The Red Devils
82148	International Spring Fair

NZ (DLV) DRIVING BRAKE VAN (140 m.p.h.)

Dia. NZ502. Mark 4. Air conditioned. Swiss-built (SIG) bogies. dg. Cab to shore communication. ETH 6X.

Lot No. 31043 Metro-Cammell 1988. 45.18 t.

82200	**GN**	F	GN	BN	82216	**GN**	F	GN	BN
82201	**GN**	F	GN	BN	82217	**GN**	F	GN	BN
82202	**GN**	F	GN	BN	82218	**GN**	F	GN	BN
82203	**GN**	F	GN	BN	82219	**GN**	F	GN	BN
82204	**GN**	F	GN	BN	82220		F	GN	BN
82205	**GN**	F	GN	BN	82221	**GN**	F	GN	BN
82206	**GN**	F	GN	BN	82222	**GN**	F	GN	BN
82207	**GN**	F	GN	BN	82223	**GN**	F	GN	BN
82208	**GN**	F	GN	BN	82224	**GN**	F	GN	BN
82209	**GN**	F	GN	BN	82225	**GN**	F	GN	BN
82210	**GN**	F	GN	BN	82226	**GN**	F	GN	BN
82211	**GN**	F	GN	BN	82227	**GN**	F	GN	BN
82212	**GN**	F	GN	BN	82228	**GN**	F	GN	BN
82213	**GN**	F	GN	BN	82229	**GN**	F	GN	BN
82214	**GN**	F	GN	BN	82230	**GN**	F	GN	BN
82215	**GN**	F	GN	BN	82231	**GN**	F	GN	BN

ND (BG) GANGWAYED BRAKE VAN (90 m.p.h.)

Mark 1. ND501. Short frames (57'). Load 10t. All vehicles were built with BR1 bogies. ETH 1. Vehicles numbered 81xxx had 3000 added to the original numbers to avoid confusion with Class 81 locomotives. The full lot number list is listed here for reference purposes with renumbered vehicles. No unmodified vehicles remain in service.

80525. Lot No. 30009 Derby 1952–3. 31 t.
80621. Lot No. 30046 York 1954. 31.5 t.
80700–80703. Lot No. 30136 Metro-Cammell 1955. 31.5 t.
80731–80791. Lot No. 30140 BRCW 1955–6. 31.5 t.

80805–80848. Lot No. 30144 Cravens 1955. 31.5 t.
80855–80962. Lot No. 30162 Pressed Steel 1956–7. 32 t.
80971–81014. Lot No. 30173 York 1956. 31.5 t.
81019–81051. Lot No. 30224 Cravens 1956. 31.5 t.
81055–81175. Lot No. 30228 Metro-Cammell 1957–8. 31.5 t.
81182–81188. Lot No. 30234 Cravens 1956–7. 31.5 t.
81205–81265. Lot No. 30163 Pressed Steel 1957. 31.5 t.
81266–81309. Lot No. 30323 Pressed Steel 1957. 32 t.
81313–81497. Lot No. 30400 Pressed Steel 1957–8. 32 t.
81498–81568. Lot No. 30484 Pressed Steel 1958. 32 t.
81590. Lot No. 30715 Gloucester 1962. 31 t.
81604–81606. Lot No. 30716 Gloucester 1962. 31 t.

NB   HIGH SECURITY BRAKE VAN

Mark 1. NB501. High security brake van. Converted at WB from ND 1985. Gangways removed. B4 bogies. Now used for movement of materials between EWS maintenance depots.

Lot No. 30400 Pressed Steel 1957–8. 30.5 t.

84382	(81382, 80460)	x	**RX**	E		Cambridge
84387	(81387, 80461)	x	**B**	E	EW	BK
84477	(81477, 80463)	x	**B**	E	EW	BK

NJ (GUV)   GENERAL UTILITY VAN

Mark 1. NJ501. Short frames. Load 14 t. Screw couplings. All vehicles were built with BR2 bogies. ETH 0 or 0X*. These vehicles had 7000 added to the original numbers to avoid confusion with Class 86 locomotives. The full lot number list is listed here for reference purposes with renumbered vehicles. No unmodified vehicles remain in service.

86081–86499. Lot No. 30417 Pressed Steel 1958–9. 30 t.
86508–86518. Lot No. 30343 York 1957. 30 t.
86521–86651. Lot No. 30403 York/Glasgow 1958–60. 30 t.
86656–86834. Lot No. 30565 Pressed Steel 1959. 30 t.
86836–86980. Lot No. 30616 Pressed Steel 1959–60. 30 t.

NE/NH (BG)   100/110 m.p.h. GANGWAYED BRAKE VAN

As ND but rebogied with B4 bogies suitable for 100 m.p.h.–NE (110 m.p.h. with special maintenance–NH). ETH 1 (1X* and NHA). For lot numbers refer to original number series. Deduct 1.5t from weights. All NHA are *pg.

92100	(81391)	to		CC		Ferme Park
92111	(81432)	NHA		F		CP
92112	(81440)	x	**RY**	E		BK
92114	(81443)	NHA		F		LT
92116	(81450)	to	**0**	RS	SS	BN
92121	(81457)	*	**RX**	E		CF
92122	(81459)	x*to	**RY**	E		KM
92125	(81470)	to		A		OY

92146	(81498)	NHA		F		LT
92159	(81534)	NHA		F	SR	IS
92174	(81567)	NHA		F	SR	IS
92175	(81568)	pg		F	GW	LA
92193	(81604)	pg		E	EW	EN
92194	(81606)	to		F	GW	LA
92211	(81267)	*	R	E		KM
92229	(80902)	*	R	E		KM
92234	(81336, 84336)	*	RX	E		DY
92238	(81563, 84563)		RY	E		DY
92243	(81489, 84489)	*	R	E		KM
92252	(80959)	x*	RY	E		Crewe South Yard
92258	(81346, 84346)		RY	E		KM
92259	(81313, 84313)	x	RY	E		BK
92261	(80988)	x*	RY	E		KM
92265	(80945)	x	RY	E		KM
92267	(81404, 84404)	x		E		Crewe South Yard
92271	(80962)	x*	R	E		OM

92193 is kept at Preston station.

NE (BG) 100 m.p.h. GANGWAYED BRAKE VAN

As ND but rebogied with Commonwealth bogies suitable for 100 m.p.h. ETH 1 (1X*). For lot numbers refer to original number series. Add 1.5 t to weights to allow for the increased weight of the Commonwealth bogies.

92302	(81501, 84501)		RX	E	KM
92303	(81427, 84427)		RX	E	DY
92306	(81217, 84217)	*	RY	E	KM
92307	(80805)	*		E	KM
92309	(81043, 84043)	x*	RX	E	KM
92311	(81453, 84453)	x	RY	E	Crewe South Yard
92312	(81548, 84548)		RX	E	KM
92314	(80777)	x*	RY	E	Crewe South Yard
92316	(80980)	x*	RY	E	KM
92319	(81055, 84055)	*	RY	E	KM
92321	(81566, 84566)		RY	E	BN
92323	(80832)	*	R	E	KM
92324	(81087, 84087)		RY	E	KM
92325	(80791)		RY	E	KM
92328	(80999)	x*	RY	E	KM
92329	(81001, 84001)	*	RY	E	KM
92330	(80995)	x*	RY	E	KM
92332	(80845)	*	RX	E	KM
92333	(80982)	*	RY	E	KM
92334	(80983)	x*	R	E	BK
92337	(81140, 84140)	*	RX	E	KM
92340	(81059, 84059)	*	RY	E	KM
92341	(81316, 84316)	x	RY	E	KM
92343	(81505, 84505)	x	R	E	KM
92344	(81154, 84154)	*	RY	E	KM

92345	(81083, 84083)	x*	**RY**	E		KM
92346	(81091, 84091)		**RY**	E		KM
92347	(81326, 84326)		**RX**	E		DY
92348	(81075, 84075)	x*	**R**	E		KM
92350	(81049, 84049)	*	**RY**	E		DY
92353	(81323, 84323)		**R**	E		KM
92355	(81517, 84517)	x	**RX**	E		DY
92356	(81535, 84535)	x		E		KM
92357	(81136, 84136)		**RX**	E		KM
92362	(81188, 84188)	x	**RY**	E		KM
92363	(81294, 84294)	x	**RY**	E		Crewe South Yard
92364	(81030, 84030)	x*	**R**	E		KM
92365	(81122, 84122)		**RX**	E		KM
92366	(81551, 84551)		**RX**	E		KM
92369	(80960)	x*		E	*EW*	EN
92370	(81324, 84324)		**RX**	E		KM
92377	(80928)	*	**RX**	E		DY
92379	(80914)	*	**RX**	E		KM
92380	(81247, 84247)	*	**R**	E		KM
92381	(81476, 84476)		**RX**	E		KM
92382	(81561, 84561)		**RX**	E		DY
92384	(80893)		**RY**	E		Crewe South Yard
92385	(81261, 84261)	x*	**RY**	E		KM
92387	(81380, 84380)	x		E		BK
92389	(81026, 84026)	*	**RY**	E		KM
92390	(80834)	*		E		KM
92392	(80861)	*	**RY**	E		KM
92395	(81274, 84274)			E		KM
92398	(80859)	x*	**RY**	E		KM
92399	(80781)	x*	**RY**	E		BK
92400	(81211, 84211)	*		E		Crewe South Yard
92401	(81280, 84280)	x	**RX**	E		KM
92402	(81099, 84099)	*	**RY**	E		KM
92403	(81273, 84273)	x	**RY**	E		OM
92404	(81051, 84051)	x		E		KM
92409	(81370, 84370)	x		E		OM
92410	(81469, 84469)	x		E		Crewe South Yard
92411	(81252, 84252)	x	**RY**	E		BK
92412	(81354, 84354)	*	**RY**	E		Crewe South Yard
92413	(81472, 84472)	x	**RY**	E		Crewe South Yard
92414	(81458, 84458)	x		E		OM
92415	(81388, 84388)		**RX**	E		KM
92416	(81250, 84250)	*	**RY**	E		KM
92417	(80885)	*	**RX**	E		KM
92418	(81512, 84512)	*	**RX**	E		OM

92369 is kept at Doncaster station for use when the lifts are out of order.

NF (BG) 100/110 m.p.h. GANGWAYED BRAKE VAN

As NE but with emergency equipment removed. For details and lot numbers refer to original number series. 92503–92750 have B4 bogies whilst 92804–92897 have Commonwealth bogies.

b (Dia. NB501). High security brake van. Converted at WB from ND 1985. Gangways removed. Now used for movement of materials between EWS maintenance depots.

92503	(80864, 92903)	x	RY	E		Carlisle Currock
92505	(80876, 92905)	x		E		OM
92509	(80897, 92909)		RX	E		KM
92510	(80900, 92910)		RX	E		KM
92513	(80916, 92913)	x	RX	E		KM
92518	(80941, 92918)		RX	E	EW	EN
92521	(80956, 92921)	x		E		OM
92530	(81461, 84461)	xb	RX	E	EW	BK
92542	(81207, 92942)		RX	E		KM
92547	(81216, 92947)		RX	E		BK
92550	(81220, 92950)		RX	E		BK
92555	(81225, 92955)		RX	E		BK
92558	(81228, 92958)		RX	E		BK
92562	(81232, 92962)		RX	E	EW	EN
92566	(81238, 92966)			E		BK
92568	(81244, 92968)		RX	E		OM
92576	(81257, 92976)		RX	E		OM
92577	(81258, 92977)		RX	E		BK
92582	(81265, 92982)		RY	E	EW	EN
92607	(81410, 92107)		RX	E		OM
92649	(81509, 92149)	x		E		KM
92709	(80873, 92209)	x	RX	E		OM
92714	(81504, 92214)		RX	E		KM
92716	(81376, 92216)	x	RY	E		BK
92718	(81314, 92218)		RY	E		BK
92720	(80924, 92220)	x	RY	E		OM
92722	(80887, 92222)		RX	E	EW	EN
92725	(80891, 92225)		RX	E		OM
92728	(80921, 92228)		RX	E	EW	EN
92740	(80703, 92240)	x	RY	E		BK
92748	(80935, 92248)		RX	E		OM
92750	(81235, 92250)		RX	E	EW	EN
92804	(81339, 92304)	x	RX	E		KM
92805	(81590, 92305)	x	RX	E		KM
92808	(80784, 92308)	x	RX	E		BK
92810	(81105, 92310)		RX	E		KM
92815	(80848, 92315)	*	RX	E		KM
92817	(80836, 92317)	x	RX	E		KM
92822	(80771, 92322)	x	RX	E		KM
92827	(80842, 92327)	x	RX	E		KM
92831	(81365, 92331)	x	RX	E		KM

92842	(81397, 92342)	x	**RY**	E		KM
92852	(81182, 92352)	*	**RX**	E		DY
92854	(81353, 92354)	x	**RX**	E		KM
92858	(81393, 92358)	x	**RX**	E		KM
92859	(81275, 92359)	*	**RX**	E		DY
92860	(81431, 92360)		**RX**	E	*EW*	EN
92861	(81463, 92361)		**R**	E		KM
92867	(81293, 92367)	x	**RX**	E		KM
92872	(81362, 92372)	x	**RY**	E		MA
92873	(81528, 92373)		**RX**	E		KM
92876	(81374, 92376)	*	**RX**	E		KM
92883	(81429, 92383)	*	**RX**	E		OM
92886	(80843, 92386)	x	**RX**	E		KM
92888	(80868, 92388)	*	**RX**	E		BK
92893	(80701, 92393)	x	**RX**	E		BK
92894	(81322, 92394)		**RX**	E		BK
92897	(80700, 92397)	x*	**RY**	E		KM

NE/NH (BG)   100/110 m.p.h. GANGWAYED BRAKE VAN

Renumbered from 920xx series by adding 900 to number to avoid conflict
with Class 92 locos. Class continued from 92271.

92901	(80855, 92001)	NHA		F	*SR*	IS
92904	(80867, 92004)	*pg	**G**	VS	*SS*	SL
92907	(80880, 92007)	*pg	**RX**	E		KM
92908	(80895, 92008)	NHA		F	*SR*	IS
92912	(80910, 92012)	*pg		F		LM
92916	(80930, 92016)	x*pg	**RY**	E		KM
92917	(80940, 92017)	*to	**RX**	E		KM
92922	(80958, 92022)	x*pg	**RX**	E		BK
92923	(80971, 92023)	*pg		F		LT
92926	(81060, 92026)	NHA		F		ZB
92927	(81061, 92027)	NHA		F		LT
92928	(81064, 92028)	NHA		F		LT
92929	(81077, 92029)	NHA		F		LM
92931	(81102, 92031)	NHA		F	*SR*	IS
92932	(81117, 92032)	NHA		F		ZD
92933	(81123, 92033)	NHA		F		LM
92934	(81142, 92034)	NHA		F		LT
92935	(81150, 92035)	*pg		F	*SR*	IS
92936	(81158, 92036)	NHA		F	*SR*	IS
92937	(81165, 92037)	NHA		F		PC
92938	(81173, 92038)	NHA		F	*SR*	IS
92939	(81175, 92039)	NHA		F		LM
92940	(81186, 92040)	pg		F	*GW*	LA
92946	(81214, 92046)	NHA		F	*SR*	IS
92948	(81218, 92048)	NHA		F	*SR*	IS
92961	(81231, 92061)			F		LT
92986	(81282, 92086)	to		F		CP
92988	(81284, 92088)	to		F		LT

| 92991 | (81308, 92091) | to | F | | LT |
| 92998 | (81381, 92098) | NHA | F | | LT |

NL   NEWSPAPER VAN

Dia. NL501. Mark 1. Short frames (57'). Converted from NJ (GUV). Fluorescent lighting, toilets and gangways fitted. Load 14 t. Not now used for news traffic. ETH 3X. B5 bogies.

Lot No. 30922 Wolverton or Doncaster 1977–8. 31 t.

94003	(86281, 93999)	x	RX	E		OM
94004	(86156, 85504)		RY	E		OM
94006	(86202, 85506)		RX	E		OM
94007	(86572, 85507)		B	E		OM
94009	(86144, 85509)		RY	E		OM
94010	(86151, 85510)	x	RX	E		OM
94011	(86437, 85511)		RX	E		OM
94015	(86484, 85515)	x	B	E		BK
94016	(86317, 85516)	x	B	E		OM
94020	(86220, 85520)	x	RY	E		OM
94021	(86204, 85521)	x	B	E		OM
94024	(86106, 85524)		B	E		OM
94025	(86377, 85525)		RY	E		OM
94026	(86703, 85526)	x	RY	E		OM
94027	(86732, 85527)		R	E		Ferme Park
94028	(86733, 85528)	x	RX	E		OM
94029	(86740, 85529)	x	RY	E		OM
94030	(86746, 85530)	x	B	E		OM
94031	(86747, 85531)	x	B	E		BK
94032	(86730, 85532)		RX	E		OM
94033	(86731, 85533)	x	RY	E		BK

NKA   HIGH SECURITY GENERAL UTILITY VAN

Dia. NK501. These vehicles are GUVs further modified with new floors, three roller shutter doors per side and the end doors removed. For lot Nos. see original number series. Add 2 t to weight. ETH 0X. Commonwealth bogies.

94100	(86668, 95100)	RX	E	EW	EN
94101	(86142, 95101)	RX	E	EW	BK
94102	(86762, 95102)	RX	E	EW	BK
94103	(86956, 95103)	RX	E	EW	BK
94104	(86942, 95104)	RX	E	EW	EN
94106	(86353, 95106)	RX	E	EW	BK
94107	(86576, 95107)	RX	E	EW	EN
94108	(86600, 95108)	RX	E	EW	BK
94110	(86393, 95110)	RX	E	EW	BK
94111	(86578, 95111)	RX	E	EW	EN
94112	(86673, 95112)	RX	E	EW	EN
94113	(86235, 95113)	RX	E	EW	BK
94114	(86081, 95114)	RX	E	EW	BK

94116	(86426, 95116)	**RX**	E	*EW*	BK
94117	(86534, 95117)	**RX**	E	*EW*	BK
94118	(86675, 95118)	**RX**	E	*EW*	EN
94119	(86167, 95119)	**RX**	E	*EW*	EN
94121	(86518, 95121)	**RX**	E	*EW*	BK
94123	(86376, 95123)	**RX**	E	*EW*	BK
94126	(86692, 95126)	**RX**	E	*EW*	EN
94132	(86607, 95132)	**RX**	E	*EW*	EN
94133	(86604, 95133)	**RX**	E	*EW*	BK
94137	(86610, 95137)	**RX**	E	*EW*	EN
94138	(86212, 95138)	**RX**	E	*EW*	EN
94140	(86571, 95140)	**RX**	E	*EW*	BK
94146	(86648, 95146)	**RX**	E	*EW*	BK
94147	(86091, 95147)	**RX**	E	*EW*	BK
94148	(86416, 95148)	**RX**	E	*EW*	EN
94150	(86560, 95150)	**RX**	E	*EW*	BK
94153	(86798, 95153)	**RX**	E	*EW*	EN
94155	(86820, 95155)	**RX**	E	*EW*	EN
94157	(86523, 95157)	**RX**	E	*EW*	EN
94160	(86581, 95160)	**RX**	E	*EW*	BK
94164	(86104, 95164)	**RX**	E	*EW*	EN
94166	(86112, 95166)	**RX**	E	*EW*	BK
94168	(86914, 95168)	**RX**	E	*EW*	BK
94170	(86395, 95170)	**RX**	E	*EW*	BK
94172	(86429, 95172)	**RX**	E	*EW*	EN
94174	(86852, 95174)	**RX**	E	*EW*	EN
94175	(86521, 95175)	**RX**	E	*EW*	BK
94176	(86210, 95176)	**RX**	E	*EW*	EN
94177	(86411, 95177)	**RX**	E	*EW*	BK
94180	(86362, 95141)	**RX**	E	*EW*	EN
94182	(86710, 95182)	**RX**	E	*EW*	BK
94190	(86624, 95350)	**RX**	E	*EW*	EN
94191	(86596, 95351)	**RX**	E	*EW*	BK
94192	(86727, 95352)	**RX**	E	*EW*	EN
94193	(86514, 95353)	**RX**	E	*EW*	EN
94195	(86375, 95355)	**RX**	E	*EW*	BK
94196	(86478, 95356)	**RX**	E	*EW*	BK
94197	(86508, 95357)	**RX**	E	*EW*	BK
94198	(86195, 95358)	**RX**	E	*EW*	BK
94199	(86854, 95359)	**RX**	E	*EW*	BK
94200	(86207, 95360)	**RX**	E	*EW*	BK
94202	(86563, 95362)	**RX**	E	*EW*	EN
94203	(86345, 95363)	**RX**	E	*EW*	BK
94204	(86715, 95364)	**RX**	E	*EW*	BK
94205	(86857, 95365)	**RX**	E	*EW*	BK
94207	(86529, 95367)	**RX**	E	*EW*	EN
94208	(86656, 95368)	**RX**	E	*EW*	EN
94209	(86390, 95369)	**RX**	E	*EW*	BK
94211	(86713, 95371)	**RX**	E	*EW*	EN
94212	(86728, 95372)	**RX**	E	*EW*	EN
94213	(86258, 95373)	**RX**	E	*EW*	EN

94214	(86367, 95374)	**RX**	E	*EW*	BK
94215	(86862, 94077)	**RX**	E	*EW*	BK
94216	(86711, 93711)	**RX**	E	*EW*	EN
94217	(86131, 93131)	**RX**	E	*EW*	BK
94218	(86541, 93541)	**RX**	E	*EW*	BK
94221	(86905, 93905)	**RX**	E	*EW*	BK
94222	(86474, 93474)	**RX**	E	*EW*	EN
94223	(86660, 93660)	**RX**	E	*EW*	BK
94224	(86273, 93273)	**RX**	E	*EW*	BK
94225	(86849, 93849)	**RX**	E	*EW*	BK
94226	(86525, 93525)	**RX**	E	*EW*	BK
94227	(86585, 93585)	**RX**	E	*EW*	BK
94228	(86511, 93511)	**RX**	E	*EW*	BK
94229	(86720, 93720)	**RX**	E	*EW*	BK

NAA   PROPELLING CONTROL VEHICLE

Dia. NA508. Mark 1. Class 307 driving trailers converted for use in propelling parcels trains out of termini. Fitted with roller shutter doors. Equipment fitted for communication between cab of PCV and locomotive. B5 bogies. ETH 2X.

Lot No. 30206 Ashford/Eastleigh 1954–6. Converted at RTC Derby 1993 (94300–1), Hunslet-Barclay, Kilmarnock 1994–6 (remainder).

94300	(75114)	**RX**	E	*EW*	BK	94322	(75111)	**RX**	E	*EW*	BK
94301	(75102)	**RX**	E	*EW*	BK	94323	(75110)	**RX**	E	*EW*	EN
94302	(75124)	**RX**	E	*EW*	BK	94324	(75103)	**RX**	E	*EW*	EN
94303	(75131)	**RX**	E	*EW*	EN	94325	(75113)	**RX**	E	*EW*	EN
94304	(75107)	**RX**	E	*EW*	EN	94326	(75123)	**RX**	E	*EW*	BK
94305	(75104)	**RX**	E	*EW*	EN	94327	(75116)	**RX**	E	*EW*	EN
94306	(75112)	**RX**	E	*EW*	BK	94331	(75022)	**RX**	E	*EW*	BK
94307	(75127)	**RX**	E	*EW*	EN	94332	(75011)	**RX**	E	*EW*	EN
94308	(75125)	**RX**	E	*EW*	BK	94333	(75016)	**RX**	E	*EW*	BK
94309	(75130)	**RX**	E	*EW*	EN	94334	(75017)	**RX**	E	*EW*	EN
94310	(75119)	**RX**	E	*EW*	EN	94335	(75032)	**RX**	E	*EW*	BK
94311	(75105)	**RX**	E	*EW*	BK	94336	(75031)	**RX**	E	*EW*	EN
94312	(75126)	**RX**	E	*EW*	BK	94337	(75029)	**RX**	E	*EW*	EN
94313	(75129)	**RX**	E	*EW*	BK	94338	(75008)	**RX**	E	*EW*	EN
94314	(75109)	**RX**	E	*EW*	BK	94339	(75024)	**RX**	E	*EW*	BK
94315	(75132)	**RX**	E	*EW*	EN	94340	(75012)	**RX**	E	*EW*	BK
94316	(75108)	**RX**	E	*EW*	EN	94341	(75007)	**RX**	E	*EW*	EN
94317	(75117)	**RX**	E	*EW*	EN	94342	(75005)	**RX**	E	*EW*	BK
94318	(75115)	**RX**	E	*EW*	EN	94343	(75027)	**RX**	E	*EW*	BK
94319	(75128)	**RX**	E	*EW*	EN	94344	(75014)	**RX**	E	*EW*	BK
94320	(75120)	**RX**	E	*EW*	EN	94345	(75004)	**RX**	E	*EW*	EN
94321	(75122)	**RX**	E	*EW*	EN						

NBA   HIGH SECURITY BRAKE VAN

Dia. NB501. These vehicles are NEs further modified with sealed gangways, new floors, built-in tail lights and roller shutter doors. For lot Nos. see original number series. 31.4 t. ETH 1X. B4 bogies.

94400	(81224, 92954)	**RX**	E	*EW*	EN
94401	(81277, 92224)	**RX**	E	*EW*	EN
94403	(81479, 92629)	**RX**	E	*EW*	BK
94404	(81486, 92135)	**RX**	E	*EW*	BK
94405	(80890, 92233)	**RX**	E	*EW*	EN
94406	(81226, 92956)	**RX**	E	*EW*	EN
94407	(81223, 92553)	**RX**	E	*EW*	BK
94408	(81264, 92981)	**RX**	E	*EW*	BK
94409	(81511, 92249)	**RX**	E	*EW*	EN
94410	(81205, 92941)	**RX**	E	*EW*	EN
94411	(81378, 92997)	**RX**	E	*EW*	EN
94412	(81210, 92945)	**RX**	E	*EW*	EN
94413	(80909, 92236)	**RX**	E	*EW*	BK
94414	(81377, 92996)	**RX**	E	*EW*	EN
94415	(81309, 92992)	**RX**	E	*EW*	EN
94416	(80929, 92746)	**RX**	E	*EW*	BK
94418	(81248, 92244)	**RX**	E	*EW*	BK
94419	(80858, 92902)	**RX**	E	*EW*	BK
94420	(81325, 92263)	**RX**	E	*EW*	EN
94421	(81230, 92960)	**RX**	E	*EW*	EN
94422	(81516, 92651)	**RX**	E	*EW*	BK
94423	(80923, 92914)	**RX**	E	*EW*	BK
94424	(81400, 92103)	**RX**	E	*EW*	EN
94425	(80937, 92212)	**RX**	E	*EW*	BK
94426	(81283, 92987)	**RX**	E	*EW*	BK
94427	(80894, 92754)	**RX**	E	*EW*	BK
94428	(81550, 92166)	**RX**	E	*EW*	BK
94429	(80870, 92232)	**RX**	E	*EW*	NC
94430	(80908, 94235)	**RX**	E	*EW*	BZ
94431	(81401, 92604)	**RX**	E	*EW*	BK
94432	(81383, 92999)	**RX**	E	*EW*	EN
94433	(81495, 92643)	**RX**	E	*EW*	EN
94434	(81268, 92584)	**RX**	E	*EW*	BK
94435	(81485, 92134)	**RX**	E	*EW*	EN
94436	(81237, 92565)	**RX**	E	*EW*	BK
94437	(81403, 92208)	**RX**	E	*EW*	BK
94438	(81425, 92251)	**RX**	E	*EW*	BK
94439	(81480, 92130)	**RX**	E	*EW*	BK
94440	(81497, 92645)	**RX**	E	*EW*	BK
94441	(81492, 92140)	**RX**	E	*EW*	BK
94442	(80932, 92723)	**RX**	E	*EW*	BZ
94443	(81473, 92127)	**RX**	E	*EW*	BK
94444	(81484, 92133)	**RX**	E	*EW*	BK
94445	(81444, 92615)	**RX**	E	*EW*	EN
94446	(80857, 92242)	**RX**	E	*EW*	NC
94447	(81515, 92266)	**RX**	E	*EW*	EN
94448	(81541, 92664)	**RX**	E	*EW*	BK
94449	(81536, 92747)	**RX**	F	*EW*	EN
94450	(80927, 92915)	**RX**	E	*EW*	BK
94451	(80955, 92257)	**RX**	E	*EW*	EN

94452	(81394, 92602)	**RX**	E	*EW*	BK
94453	(81170, 92239)	**RX**	E	*EW*	EN
94454	(81465, 92124)	**RX**	E	*EW*	EN
94455	(81239, 92264)	**RX**	E	*EW*	BK
94456	(80879, 92226)	**RX**	E	*EW*	BK
94457	(81454, 92119)	**RX**	E	*EW*	BK
94458	(81255, 92974)	**RX**	E	*EW*	BK
94459	(81490, 92138)	**RX**	E	*EW*	BK
94460	(81266, 92983)	**RX**	E	*EW*	EN
94461	(81487, 92136)	**RX**	E	*EW*	EN
94462	(81289, 92270)	**RX**	E	*EW*	EN
94463	(81375, 92995)	**RX**	E	*EW*	EN
94464	(81240, 92262)	**RX**	E	*EW*	EN
94465	(81481, 92131)	**RX**	E	*EW*	BK
94466	(81236, 92964)	**RX**	E	*EW*	NC
94467	(81245, 92969)	**RX**	E	*EW*	BK
94468	(81259, 92978)	**RX**	E	*EW*	BK
94469	(81260, 92979)	**RX**	E	*EW*	BK
94470	(81442, 92113)	**RX**	E	*EW*	BK
94471	(81518, 92152)	**RX**	E	*EW*	BK
94472	(81526, 92975)	**RX**	E	*EW*	BK
94473	(81262, 92272)	**RX**	E	*EW*	EN
94474	(81452, 92618)	**RX**	E	*EW*	EN
94475	(81208, 92943)	**RX**	E	*EW*	EN
94476	(81209, 92944)	**RX**	E	*EW*	BK
94477	(81494, 92642)	**RX**	E	*EW*	BK
94478	(81488, 92637)	**RX**	E	*EW*	EN
94479	(81482, 92132)	**RX**	E	*EW*	BK
94480	(81411, 92608)	**RX**	E	*EW*	EN
94481	(81493, 92641)	**RX**	E	*EW*	BK
94482	(81491, 92639)	**RX**	E	*EW*	EN
94483	(81500, 92647)	**RX**	E	*EW*	NC
94484	(81426, 92110)	**RX**	E	*EW*	BK
94485	(81496, 92644)	**RX**	E	*EW*	EN
94486	(81254, 92973)	**RX**	E	*EW*	EN
94487	(81413, 92609)	**RX**	E	*EW*	BK
94488	(81405, 92105)	**RX**	E	*EW*	BK
94489	(81423, 92230)	**RX**	E	*EW*	BK
94490	(81409, 92606)	**RX**	E	*EW*	EN
94491	(80936, 92753)	**RX**	E	*EW*	BK
94492	(80888, 92721)	**RX**	E	*EW*	BK
94493	(80944, 92919)	**RX**	E	*EW*	BK
94494	(81451, 92617)	**RX**	E	*EW*	BK
94495	(80871, 92755)	**RX**	E	*EW*	BK
94496	(81514, 92650)	**RX**	E	*EW*	BK
94497	(80877, 92717)	**RX**	E	*EW*	BK
94498	(,)				
94499	(,)				
94500	(,)				

NO (GUV) GENERAL UTILITY VAN (100 MPH ETH WIRED)

Dia. NO513. Commonwealth bogies except where shown otherwise. For lot Nos. see original number series. Add 2 t to weight (Subtract 1 t for B4). ETH 0X.

95105	(86126, 93126)		**RX**	E	KM
95109	(86269, 93269)	x	**B**	E	KM
95120	(86468, 93468)	x	**RY**	E	KM
95124	(86836, 93836)	x	**R**	E	KM
95125	(86143, 93143)	x	**B**	E	KM
95128	(86764, 93764)	x	**RY**	E	Crewe South Yard
95129	(86347, 93347)	x	**RY**	E	Crewe South Yard
95131	(86860, 93860)		**RX**	E	OM
95135	(86249, 93249)	x	**RY**	E	KM
95136	(86396, 93396)	x	**RX**	E	OM
95142	(86844, 93844)	x	**RX**	E	BK
95144	(86165, 93165)	x	**RY**	E	OM
95145	(86293, 93293)	x	**RX**	E	KM
95151	(86606, 93606)	x	**RX**	E	OM
95152	(86969, 93969)	x	**RY**	E	KM
95156	(86160, 93160)	x	**RX**	E	OM
95165	(86262, 93262)	x	**RX**	E	KM
95167	(86255, 93255)		**RX**	E	OM
95169	(86277, 93277)		**RX**	E	BK
95171	(86110, 93110)	x	**RX**	E	OM
95173	(86842, 94076)	x	**RX**	E	BK
95181	(86971, 95361)	x	**B**	E	BK
95190	(86643, 95393)	B4	**RY**	E	OM
95191	(86278, 95391)	x B4	**B**	E	OM
95192	(86495, 95392)	x B4	**R**	E	BK
95194	(86192, 93192)	x B4	**RX**	E	OM
95195	(86539, 93539)	x B4	**RX**	E	OM
95196	(86775, 93775)	x B4	**RX**	E	OM
95197	(86590, 93590)	x B4	**RX**	E	OM
95198	(86134, 93134)	x B4	**RX**	E	OM
95199	(86141, 93141)	x B4	**RX**	E	OM

NCX   NEWSPAPER VAN (100 m.p.h.)

Dia. NC501. BGs modified to carry newspapers. ETH 3 (3X*). Commonwealth bogies. For lot Nos. refer to original number series. Add 2 t to weight. Not now used for news traffic.

95200	(81019, 84019)	x*	**RY**	E	*EW*	BK
95201	(80875)	x	**RX**	E		KM
95204	(80947)	x*	**RX**	E		OM
95209	(81047, 84047)	x	**RX**	E		BK
95210	(80731)	x	**RX**	E		OM
95211	(80949)	x	**RY**	E		KM
95217	(81385, 84385)	x	**B**	E	*EW*	BK

95223	(80933)	x*	**RY**	E		NC
95227	(81292, 95310)	x	**RX**	E		KM
95228	(81014, 95332)	x	**RX**	E		NC
95229	(81341, 95329)	x	**RX**	E		OM
95230	(80525, 95321)	x	**RX**	E		DY

95200 and 95217 are used for movement of materials between EWS maintenance depots.

NOV   GENERAL UTILITY VAN (100 m.p.h. ETH WIRED)

Dia. NO513. ETH 0X. Commonwealth bogies. For lot Nos. refer to original number series. Add 2 t to weight.

| 95366 | (86251, 93251) | v | **B** | E | *EW* | BK |

Kept at Gloucester station.

NRX   BAA CONTAINER VAN (100 m.p.h.)

Dia. NR503. Modified for carriage of British Airports Authority containers with roller shutter doors and roller floors and gangways removed. ETH 3. Commonwealth bogies. For lot Nos. see original number series. Add 2 t to weight. Now used for movement of materials between EWS maintenance depots.

| 95400 | (80621, 95203) | x | **RX** | E | *EW* | BK |
| 95410 | (80826, 95213) | x | **RX** | E | *EW* | BK |

NOA   SUPER GENERAL UTILITY VAN

Dia. NO502. ETH 0X. Commonwealth bogies. For lot Nos. see original number series. Add 2 t to weight. These vehicles are GUVs further modified with new floors, two roller shutter doors per side, middle doors sealed and end doors removed.

95715	(86174, 95115)	**R**	E	*EW*	EN
95727	(86323, 95127)	**R**	E	*EW*	EN
95734	(86462, 95134)	**RX**	E	*EW*	EN
95739	(86172, 95139)	**R**	E	*EW*	EN
95743	(86485, 95143)	**RX**	E	*EW*	EN
95749	(86265, 95149)	**R**	E	*EW*	EN
95754	(86897, 95154)	**R**	E	*EW*	EN
95758	(86499, 95158)	**RX**	E	*EW*	EN
95759	(86084, 95159)	**RX**	E	*EW*	EN
95761	(86205, 95161)	**RX**	E	*EW*	EN
95762	(86122, 95162)	**RX**	E	*EW*	EN
95763	(86407, 95163)	**RX**	E	*EW*	EN

NX (GUV)   MOTORAIL VAN (100 m.p.h.)

Mark 1. Dia. NX501. For details and lot numbers see original number series. ETH 0 (0X*). 100 m.p.h.

| 96100 | (86734, 93734) | *B5 | F | | KI |

96101	(86741, 93741)	*B5		F	KI
96110	(86738, 93738)	*C		F	KI
96111	(86742, 93742)	*C		F	KI
96112	(86750, 93750)	*C		F	LT
96130	(86736, 93736)	*C		F	KI
96131	(86737, 93737)	*C		F	KI
96132	(86754, 93754)	*C		F	LT
96133	(86685, 93685)	C		F	LT
96134	(86691, 93691)	C		F	LT
96135	(86755, 93755)	C		F	LM
96136	(86735, 93735)	C		F	LT
96137	(86748, 93748)	C	B	F	ZN
96138	(86749, 93749)	C		F	LT
96139	(86751, 93751)	C		F	LM
96141	(86753, 93753)	C	B	F	LT
96150	(86097, 93097)	*B5		F	KI
96155	(86334, 93334)	*B5		F	KI
96156	(86337, 93337)	*B5		F	KI
96157	(86344, 93344)	*B5		F	KI
96162	(86647, 93647)	*C		F	LT
96163	(86646, 93646)	*C		F	KI
96164	(86880, 93880)	*C		F	LT
96165	(86784, 93784)	*C		F	KI
96166	(86834, 93834)	*C		F	KI
96167	(86756, 93756)	*C		F	KI
96168	(86978, 93978)	*C		F	LT
96170	(86159, 93159)	x*C		F	KI
96171	(86326, 93326)	x*C		F	LT
96172	(86363, 93363)	x*C		F	KI
96173	(86440, 93440)	x*C		F	KI
96174	(86453, 93453)	x*C		F	LT
96175	(86628, 93628)	x*C		F	KI
96176	(86641, 93641)	x*C		F	KI
96177	(86980, 93980)	*C		F	KI
96178	(86782, 93782)	*C		F	KI
96179	(86910, 93910)	*C		F	LT
96181	(86875, 93875)	*C		F	LT
96182	(86944, 93944)	*C		F	ZH
96185	(86083, 93083)	x*C		F	LT
96186	(86087, 93087)	x*C		F	LT
96187	(86168, 93168)	x*C		F	LT
96188	(86320, 93320)	x*C		F	KI
96189	(86447, 93447)	x*C		F	LT
96190	(86448, 93448)	x*C		F	LT
96191	(86665, 93665)	x*C		F	KI
96192	(86669, 93669)	x*C		F	KI
96193	(86874, 93874)	x*C		F	LT
96194	(86949, 93949)	x*C		F	LT
96195	(86958, 93958)	x*C		F	LT

NP (GUV)   MOTORAIL VAN (110 m.p.h.)

Mark 1. Dia. NP503. Vehicles modified with concertina end doors. For details and lot numbers see original number series. B5 Bogies. ETH 0X*.

96210	(86355, 96159)	F	LT
96212	(86443, 96161)	F	LT
96213	(86324, 96152)	F	KI
96215	(86351, 96158)	F	KI
96216	(86385, 96160)	F	KI
96217	(86327, 96153)	F	KI
96218	(86286, 96151)	F	LT

AX5G   NIGHTSTAR GENERATOR VAN

Dia. AX502. Generator vans for European Night Services trains. Operate sandwiched between two Class 37 locomotives. Converted from Mark 3A sleeping cars. Gangways removed. Two Cummins diesel generator groups providing a 1500 V train supply. Hydraulic parking brake. 61-way ENS interface jumpers. BT10 bogies.

Lot No. 30960 Derby 1981–3.   t.

96371	(10545, 6371)	E	LC	*ES*	PI
96372	(10564, 6372)	E	LC	*ES*	PI
96373	(10568, 6373)	E	LC	*ES*	PI
96374	(10585, 6374)	E	LC	*ES*	PI
96375	(10587, 6375)	E	LC	*ES*	PI

NG   MOTORAIL LOADING WAGON

Dia. NG503. These vehicles have been converted andrenumbered from weltrol wagons and are used for side loading purposes.

Built Swindon 1960. wagon Lot No. 3102 (3192*).

96450	(B900920)		F	KI
96451	(B900912)		F	KI
96452	(B900917)		F	LT
96453	(B900926)	*	F	LT
96454	(B900938)	*	F	ZH

NY   EXHIBITION VAN

Various interiors. Converted from various vehicle types. Electric heating from shore supply. In some cases new lot numbers were issued for conversions, but not always.

Lot 30842 Swindon 1972–3. Dia. NY503. Converted from BSK to Lot No. 30156 Wolverton 1955.

Non-Standard Livery: Varies according to job being undertaken.

Mk4 denotes a Southern Region Mark 4 EMU trailer bogie.

99621	(34697)	x	BR1**0**	E	OM	Exhibition Coach.
99625	(34693)	x	Mk4**0**	E	OM	Generator Van.

Converted Salisbury 1981 from RB to Lot No. 30636 Pressed Steel 1962. Dia NY523/4 respectively.

99645	(1765)	v	C	**0**	E	Ferme Park	Club Car.
99646	(1766)	v	C	**0**	E	Ferme Park	Club Car.

Converted Railway Age, Crewe 1996 from TSO to Lot No. 30822 Derby 1971.

99662	(5689)		B4	**0**	RS	SS	BN

Converted Railway Age, Crewe 1996 from SO to Lot No. 30821 Derby 1971. Originally FO.

99663	(3194, 6223)		B4	**0**	RS	SS	BN
99664	(3189, 6231)		B4	**0**	RS	SS	BN

Converted Railway Age, Crewe 1996 from TSO to Lot No. 30837 Derby 1972.

99665	(5755)		B4	**0**	RS	SS	BN

Converted Railway Age, Crewe 1996 from FO to Lot No. 30843 Derby 1972–3.

99666	(3250)		B4	**0**	RS	SS	BN

YR   FERRY VAN

Dia. YR025. This vehicle was built to a wagon lot although the design closely resembles that of NJ except it only has two sets of doors per side. Short Frames. Load 14 t. Commonwealth bogies.

Built Eastleigh 1958. Wagon Lot. No. 2849. 30 t.

889202		PC	VS	SS	SL

Name: 889202 is branded 'BAGGAGE CAR No.8'.

5.2. NPCCS AWAITING DISPOSAL

This list contains the last known locations of non-passenger-carrying coaching stock awaiting disposal. The definition of which vehicles are "awaiting disposal" is somewhat vague, but generally speaking these are vehicles of types not now in normal service or vehicles which have been damaged by fire, vandalism or collision.

80735	Perth Holding Sidings
80865	Hornsey Sand Terminal
80977	LL
84197	Shrewsbury Road Sidings, Sheffield
84361	Cambridge Station Yard
84364	Doncaster West Yard
84519	Crewe Coal Sidings
92067	Doncaster West Yard
92172	NC
92198	Doncaster West Yard
92199	Doncaster West Yard
92378	Cambridge Coldham Lane Sidings
93149	OY
93180	Derby South Dock Siding
93234	Hayes & Harlington
93259	LL
93358	Mossend Yard
93446	Crewe South Yard
93457	Cricklewood Rubbish Terminal
93482	Bedford Civil Engineers Sidings
93542	Hayes & Harlington
93579	Derby Etches Park T&RSMD
93723	Bletchley T&RSMD
93930	Crewe South Yard
93952	Willesden Brent Sidings
93979	Willesden Brent Sidings
96250	Oxford Hinksey Yard
96256	Oxford Hinksey Yard
96260	Oxford Hinksey Yard
96265	Oxford Hinksey Yard
99648	Eastleigh Locomotive Holdings Sidings

STEAM DAYS ON BR 1:
THE MIDLAND LINE IN SHEFFIELD

by Peter Fox

This book is a pictorial record of train services on the
Midland Line in Sheffield during the BR steam era.
The book concentrates on the wide variety of motive
power in evidence at that time and the trains which
they operated. Whilst primarily a pictorial account,
The Midland Line in Sheffield also includes some use-

ful reference information including a complete allocation history for
Millhouses and Grimesthorpe Motive Power Depots from 1948 to closure. 60
pages £4.95.

RAILS ALONG THE SEA WALL

by Peter Kay

The picturesque Exeter-Newton Abbot line is captured
in photographs through the years, showing the build-
ing of the line, the broad gauge era, and British Rail
operation. The development of every station on the
line is described in some detail, as are the many
lineside features that make this line so popular. Many

colour and black & white photograph of an important historical nature are
included, particularly from the BR steam era. 60 pages. £4.95.

Both available from the Platform 5 Mail Order Department. To place an
order, please follow the instructions on page 383 of this book.

6.1. LIVERY CODES

AC	ARC (yellow and grey with grey lettering and cast numberplates).
B	Plain BR blue.
BG	BR blue & grey.
BR	Revised BR blue (blue with yellow cabs, grey roof, large numbers and full height BR logo).
BS	BR blue with red solebar stripe.
C	Civil Engineers (grey and yellow with black cab doors and window surrounds).
CC	BR carmine & cream ('Blood & Custard').
CE	Centro (WMPTE) (grey/light blue/white/green/white).
CH	BR/GWR chocolate & cream.
CI	Chiltern Railways (grey and blue with red stripe).
CN	Connex (yellow and white with Network SouthEast blue retained on upper bodyside).
CS	Central Services (grey and red).
CT	Civil Engineers livery with Transrail lettering and markings (large white 'T' on a blue circle with a red outline underlined with red stripes).
CW	Connex (white with blue solebar stripe).
CX	Connex (yellow and white with blue solebar stripe).
D	Departmental (plain grey with black cab doors and window surrounds).
DR	Direct Rail Services (dark blue with light blue roof and green lettering).
E	English, Welsh & Scottish Railway Company (maroon with large maroon EW&S or EWS lettering and number on a broad gold band between cabs).
EP	European Passenger Services (two-tone grey with dark blue roof).
F	New Railfreight (two-tone grey with black cab doors and window surrounds. Some locos still retain Trainload Coal, Construction, Metals, Petroleum or Railfreight Distribution markings).
FD	Forward Trust Rail (blue & white).
FG	Fragonset Railways (black with a silver roof and a broad red stripe between the cabs).
FH	New Railfreight Livery with Loadhaul lettering.
FL	Freightliner (as F with black Freightliner lettering and red markings (diagonal stripes behind right hand cab door)).
FM	New Railfreight Livery with Mainline markings.
FO	Old Railfreight (grey sides, yellow cabs and full height BR logo).
FR	Old Railfreight Revised (as FO but with a red solebar stripe and a slightly smaller BR logo).
FT	New Railfreight Livery with Transrail lettering and markings (large white 'T' on a blue circle with a red outline underlined with red stripes).
FY	Foster-Yeoman (blue/silver/blue livery with white lettering and cast numberplates).
G	BR, GWR or Southern Region green.
GE	Great Eastern Railway (grey and blue with green stripe).
GM	New Greater Manchester PTE (dark grey/red/white/light grey).

GN	Great North Eastern Railway (dark blue with an orange bodyside stripe and gold or silver GNER lettering).
GW	Great Western Trains (green and ivory with Great Western Trains logo and lettering).
GX	Gatwick Express (white and dark grey with claret stripe and Gatwick Express lettering and motif).
H	LNER Tourist green and cream ('Highland Heritage').
HB	Hunslet-Barclay (two-tone grey with red solebars and black lettering).
HE	Heathrow Express (silver with Heathrow Express lettering).
I	InterCity (white and dark grey with red stripe and Intercity lettering and swallow motif).
IO	Old InterCity (light grey and dark grey with red stripe, yellow lower cab sides and BR logo).
J	Jaffacake (grey/dark brown with orange stripe).
LH	Loadhaul (black with orange cabsides and Loadhaul lettering).
LS	LTS Rail (grey/white/green/white/blue/white).
M	*Locomotives.* BR maroon.
M	*Coaching Stock & Multiple Units.* Mainline (as IO but without the yellow lower cabsides and BR logo).
MD	Merseyrail Departmental (dark grey and yellow with Merseyrail logo).
ML	Mainline Freight (blue with silver body stripe and Mainline logo and lettering).
MM	Midland Mainline (grey and green with three orange bodyside stripes and Midland Mainline logo and lettering).
MT	Merseytravel (yellow/blue/white/yellow).
N	Network SouthEast (grey/white/red/white/blue/white).
NP	National Power (grey/red/white/blue with white and red lettering and cast numberplates).
NR	Network SouthEast livery with the red stripe repainted light blue.
NT	North Western Trains (blue with gold cantrail stripe and star).
NW	Network SouthEast (white/red/white/blue/white).
O	Other livery (non-standard - refer to text).
P	Provincial Services (grey/light blue/white/dark blue).
PC	Pullman Car Company (umber and cream).
PL	Porterbrook Leasing (purple at one end and white at the other. The livery represents an enlarged portion of the Porterbrook logo).
PR	Provincial Services railbus variant (dark blue/white/light blue).
R	*Locomotives.* Parcels (Post Office red and dark grey).
R	*Coaching Stock & Multiple Units.* Plain red.
RA	Railfilms (dark blue & cream).
RD	New Railfreight Distribution (light grey and dark grey with dark blue roof, black Railfreight Distribution lettering and RfD markings (red diamonds on a yellow background)).
RE	Regional Railways Express (buff/light grey/dark grey/light grey/buff with dark blue white and light blue stripes).
RM	Royal Mail (red with two yellow stripes above solebar and 'Royal Mail' insignia).
RN	North West Regional Railways (as RR but with light blue stripe replaced with green stripe).
RP	Royal Train (purple with red lining and light grey roof and bogies).

RR	Regional Railways (grey/light blue/white/dark blue/white).
RT	Railtrack (orange/white/orange with Railtrack lettering).
RX	Rail express systems (Post Office red with Res blue & black markings).
RY	Red with yellow stripes above solebar and BR logo.
S	Strathclyde PTE (orange and black).
SL	Silverlink (green, white & mauve).
SO	Serco (grey with broad diagonal red stripe and Serco lettering).
ST	Stagecoach (white/orange/red/blue).
T	Tyne & Wear PTE (yellow blue & white).
TL	Old Thameslink (grey with orange with Thameslink logo and lettering).
TN	New Thameslink (grey, yellow and navy blue with Thameslink lettering).
V	Virgin Trains (red with black ends extending into bodysides and three white lower bodysides stripes or red with black inner ends and large full height Virgin logo on Class 43 power cars and DVTs).
W	Waterman Railways (maroon with cream stripes).
WV	Waterman Railways VIP (West Coast Joint Stock lined purple lake).
Y	West Yorkshire PTE (red with cream band and Metrotrain logo and lettering).

6.2. OWNER & OPERATION CODES

This book now uses a (generally) logical system of codes instead of the gobbledygook codes of the BR Rolling Stock Library (RSL). We have decided to do this since RSL information is not officially available to the general public these days and a system of coding which is fairly obvious to the reader is preferred. For passenger train operating companies these are generally based on those used by Railtrack in the Great Britain passenger timetable, but there are a few changes for clarity or to reflect changes since the timetable was printed.

OWNER CODES

A	Angel Trains Contracts Ltd.
AC	ARC (Southern) Limited
AD	ADtranz
AR	Anglia Railways Train Services
B	British Airports Authority
CA	Cardiff Railway Company
CB	SNCB/NMBS (Belgian Railways)
CC	The Carriage & Traction Company Ltd.
CF	SNCF (French Railways)
CT	Central Trains
CX	Connex Leasing Company
D	Direct Rail Services
E	English Welsh & Scottish Railway Company Ltd.
F	Forward Trust Rail Ltd. (formerly Eversholt Holdings)
FF	The Fifty Fund
FG	Fragonset Railways
FL	Freightliners Ltd.
FS	Flying Scotsman Railways Ltd.
FY	Foster-Yeoman Ltd.
GN	Great North Eastern Railway
GS	Great Scottish & Western Railway Co. Ltd.
GW	Great Western Trains
H	Hunslet-Barclay Ltd.
HD	Hastings Diesels Ltd.
I	ICI Chemicals & Polymers Ltd.
IL	Island Line
LC	London & Continental Eurostar (UK) Ltd.
LN	London & North Western Railway Co. Ltd.
ME	Merseyrail Electrics
MH	Mid-Hants Railway plc
ML	Midland Mainline
MR	Midland & Northern Railroad Co. Ltd.
NP	National Power plc
NT	9000 Locomotives Ltd.
O	Other Owners (refer to the list at the end of this section)
P	Porterbrook Leasing Co. Ltd.

R	Royal Mail
RA	Railfilms Ltd.
RC	Railcare Ltd.
RE	Resco Railways Ltd. (owners of the 'Queen of Scots' set)
RF	RFS (E) Ltd.
RO	Rolltrack Trains Ltd.
RR	Rail Rider Tours Ltd.
RS	Rail Charter Services Ltd.
RT	Railtrack plc
RV	Riviera Trains Ltd. (operators of the 'Riviera Limited' set)
SC	Connex South Central
SL	Sea Containers Rail Services Ltd.
SO	Serco Railtest Ltd.
SP	Scottish Railway Preservation Society
SR	ScotRail Railways
SW	South West Trains
VS	Venice Simplon Orient Express Ltd.
VT	Virgin Trains
W	West Coast Railway Co. Ltd.
WT	Wessex Traincare Ltd.

Other Owners

17007	Merchant Navy Locomotive Preservation Society Ltd.
17019	North Eastern Locomotive Preservation Group
21096	A4 Locomotive Society Ltd.
35333	6024 Preservation Society Ltd.
35449	75014 Locomotive Operators Group
35457	Ian Storey Engineering Ltd.
35468	National Railway Museum
35486	Severn Valley Railway
80213	Princess Elizabeth Locomotive Society
80217	75014 Locomotive Operators Group
80225	Merchant Navy Locomotive Preservation Society Ltd.

OPERATION CODES

AD	ADtranz
AR	Anglia Railways Train Services
AT	Angel Trains Contracts Ltd.
CA	Cardiff Railway Company
CH	Chiltern Railway Company
CR	Crew Training
CT	Central Trains
DR	Direct Rail Services
ES	Eurostar (UK)
EW	English Welsh & Scottish Railway Company Ltd.
FL	Freightliners Ltd
GE	Great Eastern Railway
GN	Great North Eastern Railway
GW	Great Western Trains

GX	Gatwick Express
HB	Hunslet-Barclay Ltd.
HE	Heathrow Express
IC	ICI Chemicals & Polymers Ltd.
IL	Island Line
JF	Jarvis Facilities plc
LS	LTS Rail
MD	Mendip Rail Ltd.
ME	Merseyrail Electrics
ML	Midland Mainline
NE	Regional Railways North East
NP	National Power plc
NW	North Western Trains
PL	Porterbrook Leasing Co. Ltd.
RC	Railcare Ltd.
RS	Research
RT	Royal Train
SA	Sandite spraying or Sandite spraying/De-icing
SC	Connex South Central
SE	Connex South Eastern
SL	Silverlink
SO	Serco Railtest Ltd.
SR	ScotRail Railways
SS	Used normally on special or charter services
ST	Stores vehicle
SU	Support coach
SW	South West Trains
TE	Test Train
TL	Thameslink Rail
TT	Thames Trains
VW	Virgin West Coast
VX	Virgin Cross Country
WN	West Anglia Great Northern Railway
WT	Wessex Traincare Ltd.
WW	Wales & West Passenger Trains

6.3. DEPOT, LOCATION & WORKS CODES

DEPOT & LOCATION CODES

AB	Aberdeen Guild Street Yard *
AL	Aylesbury TMD
AN	Allerton T&RSMD (Liverpool)
AY	Ayr T&RSMD
BD	Birkenhead North T&RSMD
BH	Billingham Works (ICI Chemical & Polymers Ltd.) *
BI	Brighton T&RSMD
BK	Bristol Barton Hill T&RSMD
BL	Brush Traction Ltd., Loughborough *
BM	Bournemouth T&RSMD
BN	Bounds Green T&RSMD (London)
BO	Bo'Ness (West Lothian) (Bo'Ness & Kinneil Railway)
BQ	Bury (Greater Manchester) (East Lancashire Railway)
BS	Bescot TMD (Walsall)
BY	Bletchley T&RSMD
BZ	St Blazey TMD
CB	Crewe Brook Sidings *
CD	Crewe Diesel TMD
CE	Crewe International EMD
CF	Cardiff Canton T&RSMD
CJ	Clapham Yard CSD (London)
CK	Corkerhill DMUD (Glasgow)
CL	Carlisle Upperby CWMD (closed)
CN	Carnforth *
CO	Cranmore (Somerset) (East Somerset Railway)
CP	Crewe Carriage Shed (London & North Western Railway Co. Ltd.)
CQ	Crewe, The Railway Age (London & North Western Railway Co. Ltd.)
CS	Carnforth (West Coast Railway Co. Ltd.)
CU	Coquelles (France) (Eurotunnel) *
DI	Didcot Railway Centre (Great Western Society)
DR	Doncaster T&RSMD
DY	Derby Etches Park T&RSMD
EC	Craigentinny T&RSMD (Edinburgh)
EH	Eastleigh T&RSMD
EM	East Ham EMUD (London)
EN	Euston Downside CARMD (London)
EX	Exeter LIP
FB	Ferrybridge (National Power plc)
FF	Bruxelles Forest/Brussel Vorst (Belgium) (SNCB/NMBS)
FH	Frodingham Yard (Scunthorpe) *
FR	Fratton EMUD (Portsmouth)
FW	Fort William LIP

GI	Gillingham EMUD
GT	Grosmont (North Yorkshire) (North Yorkshire Moors Railway)
GW	Glasgow Shields TMD
HA	Haymarket TMD (Edinburgh)
HE	Hornsey TMD (London)
HG	Hither Green TMD
HM	Healey Mills FP
HT	Heaton T&RSMD (Newcastle)
IL	Ilford T&RSMD (London)
IM	Immingham TMD (Lincolnshire)
IS	Inverness T&RSMD
KI	MoD B.A.D. Kineton *
KM	Carlisle New Yard *
KR	Kidderminster (Severn Valley Railway)
KY	Knottingley TMD
LA	Laira T&RSMD (Plymouth)
LE	Landore T&RSMD (Swansea)
LG	Longsight TMD (E) (Manchester)
LL	Liverpool Edge Hill CARMD
LM	MoD Long Marston (Warwickshire) *
LO	Longsight TMD (D) (Manchester)
LT	MoD C.A.D. Longtown (Cumbria) *
LY	Le Landy (Paris) (SNCF)
MA	Manchester Longsight CARMD
MD	Merehead (Foster Yeoman Ltd.)
MG	Margam TMD
MH	Millerhill LIP
ML	Motherwell TMD
NC	Norwich Crown Point T&RSMD
NH	Newton Heath T&RSMD (Manchester)
NL	Neville Hill T&RSMD (Leeds)
OC	Old Oak Common TMD (London)
OH	Old Oak Common Heathrow Express EMUD (London)
OM	Old Oak Common CARMD (London)
OO	Old Oak Common HSTD (London)
OY	Oxley CARMD (Wolverhampton)
PB	Peterborough LIP
PC	Polmadie CARMD (Glasgow)
PH	Perth New Yard *
PI	North Pole International T&RSMD (London)
PM	St. Phillips Marsh T&RSMD (Bristol)
PZ	Penzance T&RSMD
RE	Ramsgate T&RSMD
RG	Reading T&RSMD
RL	Ropley (Mid-Hants Railway)
RY	Ryde T&RSMD (Isle of Wight)
SA	Salisbury TMD
SD	Sellafield (Direct Rail Services)
SE	St Leonards Railway Engineering Co.
SF	Stratford TMD (London)
SG	Slade Green T&RSMD

SH	Strawberry Hill EMUD (London)
SK	Swanwick Junction (Derbyshire) (Midland Railway Centre)
SL	Stewarts Lane T&RSMD (London)
SP	Springs Branch TMD (Wigan) (closed)
SU	Selhurst T&RSMD (London)
SY	Saltley LIP (Birmingham)
SZ	Southall (Greater London) (Flying Scotsman Railways Ltd.)
TE	Thornaby TMD
TI	Tinsley TMD (Sheffield)
TM	Tyseley Railway Museum (Birmingham)
TO	Toton TMD (Nottinghamshire)
TS	Tyseley TMD (Birmingham)
TY	Tyne Yard LIP
WA	Warrington Arpley LIP
WB	Wembley InterCity CARMD (London)
WD	East Wimbledon EMUD (London)
WH	Whatley (ARC (Southern) Limited)
WN	Willesden TMD (London)
YM	National Railway Museum (York)

WORKS CODES

ZA	Railway Technical Centre (Derby)
ZB	RFS (E) Ltd., Doncaster
ZC	ADtranz Crewe Works
ZD	ADtranz Derby Carriage Works
ZF	ADtranz Doncaster Works
ZG	Wessex Traincare Ltd., Eastleigh Works
ZH	Railcare Ltd., Springburn Works (Glasgow)
ZI	ADtranz Ilford Works
ZK	Hunslet-Barclay Ltd., Kilmarnock Works
ZM	GEC-Alsthom Metro-Cammell Ltd., Washwood Heath, Birmingham
ZN	Railcare Ltd., Wolverton Works
ZP	Bombardier Prorail, Horbury Junction Works, West Yorkshire
ZT	ADtranz Trafford Park Wheel Works (Manchester) *

* unofficial code

6.4. DEPOT TYPE CODES

CARMD	Carriage Maintenance Depot
CS	Carriage Sidings
CSD	Carriage Servicing Depot
CWMD	Carriage and Wagon Maintenance Depot
DMUD	DMU Maintenance Depot
EMD	Electric Maintenance Depot
EMUD	EMU Maintenance Depot
FP	Fuelling Point
HSTD	High Speed Train Maintenance Depot
LIP	Locomotive Inspection Point
TMD	Traction Maintenance Depot
TMD (D)	Traction Maintenance Depot (Diesel)
TMD (E)	Traction Maintenance Depot (Electric)
T&RSMD	Traction and Rolling Stock Maintenance Depot
WRD	Wagon Repair Depot

6.5. GENERAL ABBREVIATIONS

BR	British Railways
EWS	English Welsh & Scottish Railway Company
GWR	Great Western Railway
LNER	London & North Eastern Railway
LMS	London Midland & Scottish Railway
SR	Southern Railway
NMBS	Nationale Maatschappij Belgische Spoorwegen *
SNCB	Société Nationale des Chemins de Fer Belges *
SNCF	Société Nationale des Chemins de Fer Francais +
DEMU	Diesel Electric Multiple Unit
DHMU	Diesel Hydraulic Multiple Unit
DMMU	Diesel Mechanical Multiple Unit
DMU	Diesel Multiple Unit (general term)
EMU	Electric Multiple Unit
a.c.	alternating current
BSI	Bergische Stahl Industrie
d.c.	direct current
ETH	electric train heat
GTO	gate turn-off
hp	horsepower
IGBT	insulated gate bi-polar transistor
km/h	kilometres per hour
kN	kilonewtons
kV	kilovolts
kW	kilowatts
lbf	pounds force
m	metres
mm	millimetres
m.p.h.	miles per hour
pa	public address
RA	route availability
RCH	Railway Clearing House
rpm	revolutions per minute
T	toilets
TD	toilets (suitable for disabled passengers)
TDM	time-division multiplex
TOPS	Total Operations Processing System
t	tons
UIC	Union Internationale des Chemins de fer
V	volts
(S)	stored servicable
(U)	stored unservicable

* Belgian Railways in Dutch and French respectively
+ French Railways

6.6. BUILDERS

These are shown in class headings where the following abbreviations are used:

Alexander	Walter Alexander Ltd., Falkirk
ABB Derby	ABB Transportation Ltd., Derby Carriage Works
ABB York	ABB Transportation Ltd., York Works
ADtranz Derby	ADtranz Ltd., Derby Carriage Works
AEI	Associated Electrical Industries Ltd.
Barclay	Andrew Barclay Ltd., Kilmarnock
BRCW	The Birmingham Railway Carriage & Wagon Co. Ltd.
Brush	Brush Traction Ltd., Loughborough
BTH	The British Thomson Houston Co. Ltd.
Cravens	Cravens Ltd., Sheffield
CP	Crompton-Parkinson Ltd.
EE	The English Electric Company Ltd.
GEC	The General Electric Company Ltd. (Now GEC Alsthom).
Gloucester	The Gloucester Railway Carriage and Wagon Co. Ltd.
Hunslet	Hunslet Transportation Projects Ltd.
Leyland Bus	Leyland Bus Ltd., Workington
Metro	The Metropolitan Railway Carriage and Wagon Co. Ltd., Birmingham
Metro-Cammell	The Metropolitan Cammell Railway Carriage and Wagon Co. Ltd., Birmingham
Midland	The Midland Railway Carriage and Wagon Co. Ltd,. Oldbury, Worcs.
MV	The Metropolitan-Vickers Co. Ltd.
Pressed Steel	Pressed Steel Ltd., Swindon

This list generally excludes BR/BREL workshops which are denoted in the text by their town/city. Where a dual BR works builder is shown (e.g. Ashford/Eastleigh) the first named built the underframe and the last named built the body and assembled the vehicle. For second generation vehicles, the first name is that of the main contractor with the second name baing the underframe and final assembly sub-contractor.

7. UK LIGHT RAIL SYSTEMS & METROS

7.1. BLACKPOOL & FLEETWOOD TRAMWAY

System: 660 V d.c. overhead.
Depot: Rigby Road.
Livery: Cream and green unless indicated. (many in advertising livery).

Note: Numbers in brackets are pre-1968 numbers.

ONE-MAN CARS

Rebuilt 1972–76 from English Electric railcoaches built 1934–5. Radio fitted.
13 converted (1–13).
Seats: 48.
Traction Motors: Two EE305 of 40 kW.

Note: First numbers in brackets are post 1968 numbers prior to conversion.

5	(609, 221)(U)	11	(615, 268)

OPEN BOAT CARS

Built 1934–5 by English Electric. 12 built (225–236).
Seats: 56.
Traction Motors: Two EE327 of 30 kW.

† Yellow and black livery.
§ Red and white livery.
b Blue & yellow livery.

600 *	(225)	604 §	(230)	606 b	(235)
602 †	(227)	605	(233)	607	(236)

REPLICA VANGUARD

Built 1987 on underframe of one man car No. 7 (619,282).
Seats: .
Traction Motors: Two EE327 of 30 kW.

619

BRUSH RAILCOACHES

Built 1937 by Brush. 20 built (284–303).
Seats: 48.
Traction Motors: Two EE305 of 40 kW. (EE327 of 30 kW*).

621	(284)	626	(289)	631	(294)	634	(297)
622 *	(285)	627	(290)	632	(295)	636	(299)
623	(286)	630	(293)	633	(296)	637	(300)
625	(288)						

CENTENARY CLASS

Built 1984–7. Body by East Lancs. Coachbuilders, Blackburn. One man operated. Radio fitted.
Seats: 52.
Traction Motors: Two EE305 of 40 kW.

* Rebuilt from GEC car 651.

641	643	645	647
642	644	646	648 *

CORONATION CLASS

Built 1953 by Charles Roberts & Co. Resilient wheels. 25 built (304–328).
Seats: 56.
Traction motors: Four Crompton-Parkinson 92 of 34 kW.

660   (324)

PROGRESS TWIN CARS

Motor cars (671–677) rebuilt 1958–60 from English Electric railcoaches.
Seats: 53.
Traction Motors: Two EE305 of 40 kW.

Driving trailers (681–687) built 1960 by Metro-Cammell.
Seats: 53.

671+681 (281+T1)	674+684 (284+T4)	676+686 (286+T6)
672+682 (282+T2)	675+685 (285+T5)	677+687 (287+T7)
673+683 (283+T1)		

SINGLE CARS

Rebuilt 1958–60 from English Electric railcoaches. Originally ran with trailers.
Seats: 48.
Traction Motors: Two EE305 of 40 kW.

678 (278)	679 (279)	680 (280)

"BALLOON" DOUBLE DECKERS

Built 1934–5 by English Electric. 700–712 were originally built with open tops, and 706 has now reverted to that condition and is named 'PRINCESS ALICE'.
Seats: 94.
Traction Motors: Two EE305 of 40 kW.

* Converted to ice cream tram seating 64 with an ice cream sales area in one of the lower saloons.
§ Red and white livery.

700	(237)	709	(246)	718	(255)
701 §	(238)	710	(247)	719	(256)
702	(239)	711	(248)	720	(257)
703	(240)	712	(249)	721	(258)
704	(241)	713	(250)	722	(259)
706	(243)	715	(252)	723	(260)
707	(244)	716	(253)	724	(261)
708	(245)	717	(254)	726	(263)

ILLUMINATED CARS

732	(168)	Rocket	Seats: 47
733	(209)	Western Train loco. & tender	Seats: 35
734	(174)	Western Train coach	Seats: 60
735	(222)	Hovertram	Seats: 99
736	(170)	HMS Blackpool	Seats: 71

WORKS CARS

259	(748, 624)	PW gang towing car.
260	(751, 628, 291)	Crane car and rail carrier.
749	(S)	Tower wagon trailer.
750		Cable drum trailer.
752	(2, 1)	Rail grinder and snowplough.
754		New works car (unnumbered).

JUBILEE CLASS DOUBLE DECKERS

Rebuilt 1979/82 from Balloon cars. Standard bus ends, thyristor control and stairs at each end. 761 has one door per side whereas 762 has two. Radio fitted.
Seats: 100.
Traction Motors: Two EE305 of 40 kW.

761   (725, 262) | 762   (714, 251)

PRESERVED CARS

Blackpool & Fleetwood 40	Box car. Bogie single decker built 1914
Bolton 66	Bogie double-decker built 1901

LIGHT RAIL REVIEW

Light Rail Review is the widely acclaimed series from Platform 5 Publishing examining the development of light rail transit projects worldwide. Each volume is comprised of topical articles from a team of specialist transport writers, all highly regarded in light rail circles. Much use is made of illustrations and diagrams, the majority of which are reproduced in colour. All volumes also in-

cludes a round up of developments in the UK and Ireland at the time of publication and a full world list of LRT systems. The series builds into an important reference work which is widely used by enthusiasts and transport professionals alike. A4 size, softback with thread sewn binding.

Light Rail Review 4 (1992) ... £7.50 | Light Rail Review 6 (1994) ... £7.50
Light Rail Review 5 (1993) ... £7.50 | Light Rail Review 7 (1996) ... £8.95

SPECIAL NOTE: Volume 8 in the series, which covers the very latest light rail developments will be published in April 1998, price £9.50.

Also available from the Platform 5 Mail Order Department. To place an order, please follow the instructions on page 383 of this book.

7.2. DOCKLANDS LIGHT RAILWAY

This is a light rail line running in London's East End from Bank, Tower Gateway and Stratford to Island Gardens and Beckton. It is being extended to Lewisham. Originally owned by London Transport, it is now owned by the London Docklands Development Corporation.

System: 750 V d.c. third rail (bottom contact).
Depots: Poplar, Beckton.
Livery: Blue/red/blue with red doors unless indicated.

## CLASS B90							B–2–B

Built 1991–2 by BN Construction, Bruges, Belgium. (now Bombardier BN). 28.80 x 2.65 m. Sliding doors. End doors for staff use. Chopper control. Scharfenberg Couplers. These units are to be converted for Seltrack signalling.
Weight: 36 t.
Seats: 66 + 4 tip-up.
Traction Motors: Two Brush of 140 kW.
Max. Speed: 80 km/h.
Electric Brake: Rheostatic.

22	28	34	40
23	29	35	41
24	30	36	42
25	31	37	43
26	32	38	44
27	33	39	

CLASS B92 B–2–B

Built 1992–5 by BN Construction, Bruges, Belgium. (now Bombardier BN). 28.80 x 2.65 m. Sliding doors. End doors for staff use. Chopper control. Scharfenberg Couplers. Fitted with Seltrack signalling.

Weight: 36 t.
Seats: 66 + 4 tip-up.
Traction Motors: Two Brush of 140 kW.
Max. Speed: 80 km/h.
Electric Brake: Rheostatic.

† Blue & white livery.

45	†	57	69	81		
46		58	70	82		
47		59	71	83		
48		60	72	84		
49		61	73	85		
50		62	74	86		
51		63	75	87		
52		64	76	88		
53		65	77	89		
54		66	78	90		
55		67	79	91		
56		68	80			

7.3. GREATER MANCHESTER METROLINK

This light rail system runs from Bury to Altrincham through the streets of Manchester, with a spur to Piccadilly. An extension is being built to Salford Quay and Eccles.

System: 750 V d.c. overhead.
Depot: Queens Road.
Livery: Dark grey/white.

SIX-AXLE ARTICULATED CARS   Bo–2–Bo

Built 1991–2 by Firema, Italy. Power operated sliding doors. Chopper control. Scharfenberg Couplers.
Weight: 45 t.
Seats: 84.
Dimensions: 29.00 x 2.65 m.
Traction Motors: Four GEC of 130 kW.
Braking: Rheostatic, regenerative, disc and emergency track brakes.

1001	
1002	
1003	
1004	THE ROBERT OWEN
1005	
1006	
1007	
1008	MANCHESTER AIRPORT
1009	
1010	MANCHESTER CHAMPION
1011	
1012	KERRY
1013	THE FUSILIER
1014	THE CITY OF DRAMA
1015	SPARKY
1016	
1017	
1018	
1019	
1020	THE DAVID GRAHAM CBE
1021	THE GREATER MANCHESTER RADIO
1022	THE GRAHAM ASHWORTH
1023	
1024	THE JOHN GREENWOOD
1025	
1026	THE POWER

SPECIAL PURPOSE VEHICLE

Built 1991 by RFS Industries, Kilnhurst and Brown Root. Used for shunting and track maintenance. Includes a crane.

Unnumbered.

7.4. MIDLAND METRO

This light rail system is under construction and will operate between Birmingham Snow Hill and Central Wolverhampton. It is due to open on 3rd August 1998.

System: 750 V d.c. overhead.
Depot: Wednesbury.
Livery: Grey/blue/red with yellow doors.

SIX-AXLE ARTICULATED CARS   Bo–2–Bo

Built 1998 by Ansaldo Trasporti, Italy. Power operated sliding plug doors. IGBT control. Scharfenberg Couplers.
Weight: 35.6 t.
Seats: 58.
Dimensions: 24.00 x 2.65 m.
Traction Motors: Four.
Max. Speed: 75 km/h (47 m.p.h.).
Braking: Rheostatic, regenerative, disc and emergency track brakes.

Note: Actual fleet numbers not definite at this stage.

01	05	09	13
02	06	10	14
03	07	11	15
04	08	12	

7.5. SOUTH YORKSHIRE SUPERTRAM

This light rail system has three lines, to Halfway in the south east of Sheffield with a spur from Gleadless Townend to Herdings, to Middlewood in the north west with a spur from Hillsborough to Malin Bridge and to Meadowhall Interchange in the north east adjacent to the large shopping complex. Because of the severe gradients in Sheffield (up to 1 in 10), all axles are powered on these vehicles. The vehicles are owned by South Yorkshire Light Rail Ltd., a subsidiary of South Yorkshire Passenger Transport Executive, but are mortgaged to Lloyd's Bank, whilst the operating company, South Yorkshire Supertram Ltd. has been leased to Stagecoach Holdings Ltd. for 27 years.

System: 750 V d.c. overhead.
Depot: Nunnery.
Livery: Blue/grey.

EIGHT-AXLE ARTICULATED UNITS   B–B–B–B

Built 1993–4 by Duewag, Düsseldorf, Germany.
Weight: 52 t.
Seats: 88.
Dimensions: 34.75 x 2.65 m.
Traction Motors: Four monomotors.
Braking: Rheostatic, regenerative, disc and emergency track brakes.

01	08	14	20
02	09	15	21
03	10	16	22
04	11	17	23
05	12	18	24
06	13	19	25
07			

FOUR WHEELED WORKS CAR   B

Built 1968 by Reichsbahn Ausbesserungswerke Schöneweide, Berlin, East Germany as single-ended passenger car with electrical equipment by LEW Henningsdorf. Converted 1980 to double-ended works car. Delivered to Sheffield on 7th November 1996. To be converted to a rail grinder.
Weight: .
Dimensions: .
Traction Motors: Two.

721 039-4   (5104, 217 303-7)

TRAM TO SUPERTRAM

by Peter Fox, Paul Jackson & Roger Benton

The official publication concerning the Sheffield Supertram system traces the history of Sheffield's old street tramway and tells the story behind the development of the new modern tramway network. It is lavishly illustrated in colour throughout, and includes maps of both old & new systems. The book is divided into three main sections:

1. SHEFFIELD'S FORMER TRAM SYSTEM. From horse-drawn trams to early electric trams and later the cream and blue double-deckers familiar to Sheffield's residents. The development of the early system, 20th century progress and post war decline. Sheffield's tramcars in preservation.

2. SUPERTRAM. The largest section of the book covers why the light rail option was chosen; Funding and implementation of the pioneering scheme; Construction — track, power supply, overhead line equipment, tramstops, Nunnery depot; The new trams; Operations and ticketing. Each of the three lines to Meadowhall, Halfway and Middlewood is described in detail.

3. SHEFFIELD TRAMWAY PICTORIAL. A combination of high quality photographs from both tramway eras.

48 pages. A4 size. 75 colour illustrations. £4.95.

Available from the Platform 5 Mail Order Department. To place an order, please follow the instructions on page 383 of this book.

7.6. STRATHCLYDE PTE UNDERGROUND

This circular 4' gauge underground line in Glasgow is generally referred to as the "Subway".

System: 750 V d.c. third rail.
Depot: Broomloan.
Livery: Orange.

SINGLE CARS   Bo–Bo

Built 1978–9 by Metro-Cammell. Power-operated sliding doors. 12.58 x 2.34 m.
Seats: 36.
Traction Motors: Two GEC G312AZ of 35.6 kW.

101	110	118	126
102	111	119	127
103	112	120	128
104	113	121	129
105	114	122	130
106	115	123	131
107	116	124	132
108	117	125	133
109			

INTERMEDIATE TRAILERS   2–2

Built 1992 by Hunslet TPL. Power-operated sliding doors. 12.58 x 2.34 m.
Seats: 40.

201	203	205	207
202	204	206	208

7.7. TYNE & WEAR METRO

System: 1500 V d.c. overhead.
Depot: South Gosforth.
Livery: Yellow/white unless indicated.

BATTERY/OVERHEAD ELECTRIC LOCOS

Built 1989–80 by Hunslet, Leeds. BSI couplers.
Traction Motors: Hunslet-Greenbat T9-4P.
Weight: 26 t.

| BL1 | BL2 | BL3 |

SIX-AXLE ARTICULATED UNITS   B–2–B

Built 1976, 1978–81 by Metro-Cammell. 27.80 x 2.65m. BSI couplers.
Weight: 39 t.
Seats: 84 (68 r – refurbished units, 70 p – Prototype refurbished unit).
Traction Motors: Two 187 kW monomotor bogies.
Max. Speed: 80 km/h.

Liveries:

A   Advertising livery.
B   Blue.
G   Green.
R   Red.

4001			4019	r	**R**	4037			4055	r	**R**	4073		
4002			4020	r	**R**	4038			4056		**A**	4074	r	**R**
4003	r	**R**	4021	r	**R**	4039	r	**A**	4057			4075	r	**B**
4004	r	**G**	4022			4040			4058			4076		
4005	r	**R**	4023			4041			4059			4077	r	**R**
4006			4024			4042			4060			4078	r	**R**
4007	r	**R**	4025			4043	r	**R**	4061	r	**G**	4079		
4008	r	**B**	4026	r	**A**	4044	r	**R**	4062			4080	r	**R**
4009			4027	r	**R**	4045	r	**A**	4063			4081		
4010	r	**R**	4028			4046	r	**R**	4064	r	**R**	4082	r	**G**
4011			4029			4047			4065	r	**R**	4083	r	**A**
4012	r	**A**	4030	r	**R**	4048	r	**B**	4066	r	**B**	4084		
4013			4031			4049	r	**A**	4067			4085	r	**B**
4014			4032			4050			4068	r	**R**	4086	r	**B**
4015			4033	r	**B**	4051	r	**R**	4069			4087	p	**A**
4016	r	**B**	4034	r	**R**	4052			4070	r	**R**	4088	r	**R**
4017	r	**R**	4035	r	**B**	4053			4071			4089	r	**R**
4018			4036	r	**G**	4054			4072			4090	r	**R**

Names:

| 4041 | HARRY COWANS | | 4065 | Catherine Cookson |

NEW FREE SERVICE FROM PLATFORM 5 - THE EASY WAY TO KEEP UP TO DATE

Following comments from a number of our readers who have experienced difficulties obtaining Platform 5 books from shops, we have launched a new information service to keep readers informed of new publications.

Our aim is to ensure that all readers are aware of new books which may be of interest to them and to make ordering those books as simple as possible.

Any readers taking advantage of the new service will automatically be sent details of new Platform 5 books shortly before they are published, with the option to order new books on publication.

However, as the range of books we publish is quite diverse and some titles may not be of interest to everyone, we are asking readers to let us know which subject areas they are interested in. This will enable us to send only appropriate information and not 'junk mail'.

To take advantage of this new service, simply complete the form overleaf (or a photocopy, or use a plain piece of paper) and return it to us. It really is as simple as that!

Readers taking advantage of this service should be aware that there is no obligation to purchase any books whatsoever, and that any reader wishing to remove their name from our circulation list may do so at any time.

Readers can also keep up to date with new publications by following our advertisements in every issue of Today's Railways magazine.

PLATFORM 5 PUBLISHING

INFORMATION SERVICE REGISTRATION FORM

NAME: ....

ADDRESS: ....

....

....

.... POSTCODE: ....

DAYTIME TELEPHONE NUMBER: ....

I would like to be kept informed of new books available from Platform 5 Publishing. I am interested in the following subject areas:

SUBJECT	✓ (tick as appropriate)
BRITISH RAILWAYS POCKET BOOKS/ LOCOMOTIVES & COACHING STOCK	
OVERSEAS RAILWAYS	
LIGHT RAIL TRANSIT	
PRESERVATION	
NOSTALGIA	
MAPS & TRACK DIAGRAMS	
ROAD TRANSPORT	

SIGNATURE: ....

RETURN YOUR COMPLETED FORM TO: Platform 5 Information Service, 3 Wyvern House, Sark Road, SHEFFIELD, S2 4HG, GREAT BRITAIN.

PLATFORM 5 PUBLISHING LTD
MAIL ORDER

New Titles	Price
British Railways Locomotives & Coaching Stock 1998	10.50
BR Pocket Book No.1: Locomotives, Spring 1998	2.60
BR Pocket Book No.2: Coaching Stock 1998	2.60
BR Pocket Book No.3: DMUs & Light Rail Systems 1998	2.60
BR Pocket Book No.4: Electric Multiple Units 1998	2.60
Railway Track Diagrams 2: England East (Quail)	7.95
British Railways Locomotives - The First 12 Years (SCTP)	18.95
Railways Around Lake Luzern (Bairstow)	9.95
The Railways of Tunisia (Simms)	8.50
The 1998 Cowie Bus Handbook (British Bus)	15.00
The 1998 Stagecoach Bus Handbook (British Bus)	15.00
The South West Bus Handbook (British Bus)	12.50
Fire Brigade Handbook Special Appliances Vol. 2 (British Bus)	12.50
London Tilbury & Southend Railway Part 2 (Kay)	9.95
Midland Railway System Maps Vol. 2: Leicester–Leeds (Kay)	9.95
Midland Railway System Maps Vol. 4: Birmingham–Bristol (Kay)	7.95
Show Me The Way To Go Home (Medloc) ..... [Story of Medloc]	5.95
The History of the Gotthard Railway (Swiss Book Service)	44.95
Steam on 4 Continents Part 4: China (Wardale)	19.95
Johnson's Atlas & Gazetteer of the Railways of Ireland (Midland)	19.99
The Londonderry & Lough Swilly Railway (Midland)	8.99
The Cavan & Leitrim Railway (Midland)	8.99
London Coach Handbook (Capital)	15.00
London Bus Handbook (Capital)	17.50
Truckin' Round Scotland (Arthur Southern)	10.95

Modern British Railways Titles

Preserved Locomotives of British Railways 9th ed.	7.95
Preserved Coaching Stock Part 1: BR Design Stock	7.95

Preserved Coaching Stock Part 2: Pre-Nationalisation Stock 8.95
Diesel & Electric Loco Register 3rd edition 7.95
Valley Lines - The People's Railway 9.95
Air Braked Series Wagon Fleet (SCTP) 7.95
Departmental Coaching Stock 5th edition (SCTP) 6.95
On-Track Plant on British Railways 5th edition (SCTP) 7.95
Engineers Series Wagon Fleet 970000-999999 (SCTP) 6.95
British Rail Wagon Fleet - B-Prefix Series (SCTP) 6.95
British Rail Internal Users (SCTP) 7.95
British Rail Depot Directory (Metro) 6.95
Private Owner Wagons Vol. 1 (Metro) 7.95
Miles & Chains Volume 2 - London Midland (Milepost) 1.95
Miles & Chains Volume 3 - Scottish (Milepost) 1.95
Miles & Chains Volume 5 - Southern (Milepost) 1.95

Metro Systems

The Twopenny Tube (Capital) .... [History of the Central Line] 5.95
Circles Under the Clyde (Capital) .... [Glasgow Subway] 15.95
The 1938 Tube Stock (Capital) 9.95
London Underground Rolling Stock (Capital) 9.95
Underground Official Handbook (Capital) 7.95
Docklands Light Rail Official Handbook (Capital) 7.95
Paris Metro Handbook (Capital) 7.95
World Metro Systems 2nd ed. (Capital) 10.95
The Berlin S-Bahn (Capital) 7.50
The Berlin U-Bahn (Capital) 7.50
The Victoria Line (Rose) 3.45
The Bakerloo Line (Rose) 3.95

Light Rail Transit and Trams

Tram to Supertram .... [Sheffield Trams] 4.95
Light Rail Review 3 .... 7.50
Light Rail Review 4 .... 7.50
Light Rail Review 5 .... 7.50
Light Rail Review 6 .... 7.50

Light Rail Review 7 .... 8.95
Manx Electric 8.95
Light Rail in Europe (Capital) 9.95
London Tramways (Capital) 19.95
Tramway & Light Railway Atlas Germany 1996 (LRTA) 10.45
The Tramways of Portugal (LRTA) 9.05
Tramtracks & Trolleybooms (Headstock) ... [Chesterfield Trams] 6.95

Postcards
Sheffield Supertram - crosses Sheffield Canal 0.30
Manchester Metrolink 0.30

Overseas Railways
High Speed in Europe 9.95
High Speed in Japan 16.95
European Handbook No. 1: Benelux Railways 3rd ed. 10.50
European Handbook No. 3: Austrian Railways 3rd ed. 10.50
European Handbook No. 5: Swiss Railways 2nd ed. 13.50
European Handbook No. 6: Italian Railways 1st ed. 13.50
European Handbook No. 7: Irish Railways 1st ed. 9.95
The Swiss Railway Saga (AS Verlag) 49.95
Bahnpanorama Schweiz/Swiss Rail Review (Fachpresse Goldach) 27.50
Railways in the Austrian Tirol (Bairstow) 8.95
The Railways of Greece (Simms) 8.10
The Railways of Corsica (Simms) 5.10
Irish Railways In Colour: From Steam to Diesel 1955-1967 (Midland) ... 16.99
Irish Railways In Colour: A Second Glance 1947-1970 (Midland) 19.99
Irish Narrow Gauge - Pictorial History Part 1 (Midland) 15.99
Irish Narrow Gauge - Pictorial History Part 2 (Midland) 15.99
Locomotives & Railcars of Bord Na Mona (Midland) 4.99
The County Donegal Railway (Midland) 7.99
Midland & Great Western Railway of Ireland (Midland) 18.99

Atlases, Maps and Track Diagrams
Railway Track Diagrams No. 1: Scotland & Isle of Man (Quail) .. 6.50

British Railway Track Diagrams No. 4: Midland - 1990 Reprint (Quail) 6.95
Railway Track Diagrams No. 6: Ireland (Quail) 5.50
London Transport Railway Track Map (Quail) 1.75
Harzer Schmalspurbahnen Track Diagram (Quail) 0.60
Czech Republic & Slovakia Railway Map (Quail) 1.70
Berlin Track Map (Quail) 2.20
Moscow Railway Map (Quail) 2.20
Portugal Railway Map (Quail) 2.00
Greece Railway Map (Quail) 1.20
Poland Railway Map (Quail) 2.00
Albanian Railway Guide (Quail) 1.20
Adelaide Track Maps (Quail) 1.20
Estonia Railway Map (Quail) 1.20
Latvia & Lithuania Railway Map (Quail) 2.00
Korea Railway Map (Quail) 2.00
European Railway Atlas: France, Benelux (Ian Allan) 10.99
Track Diagram - South Yorkshire Supertram (HRT Rail Sales) 1.50
Track Diagram - Blackpool & Fleetwood (HRT Rail Sales) 1.00
Track Diagram - Tyne & Wear (HRT Rail Sales) 2.00

Rambling
Rambles by Rail 2 - Liskeard–Looe 1.95
Rambles by Rail 4 - The New Forest 1.95
Buxton Spa Line Rail Rambles 1.20

Historical Railway Titles
6203 'Princess Margaret Rose' 19.95
Steam Days on BR 1 - The Midland Line in Sheffield 4.95
Rails along the Sea Wall .... [Dawlish–Teignmouth Pictorial] 4.95
The Rolling Rivers 6.95
British Baltic Tanks 6.95
Rails in the Isle of Wight (Midland) 16.99
LNWR Branch Lines of West Leics & East Warwicks (Milepost) .. 7.95
Register of Closed Railways 1948-1991 (Milepost) 5.95
Metropolitan Steam Locomotives (Capital) 9.95

Private Owner Wagons Volume 1 (Headstock) 9.95
Private Owner Wagons Volume 2 (Headstock) 9.95
Private Owner Wagons Volume 3 (Headstock) 9.95
Private Owner Wagons Volume 4 (Headstock) 9.95
London Tilbury & Southend Railway Part 1 (Kay) 9.95
Signalling Atlas & Signal Box Directory Great Britain & Ireland (Kay) .. 9.95
The Midland's Settle & Carlisle Distance Diagrams (CRA) 3.50
Midland Railway System Maps Vol. 3: Leicester–London (Kay) ... 8.95
Bradshaw's Guide 1850 (Kay) 7.95
Power Railway Signalling Part 1A (Kay) 12.50
Mechanical Railway Signalling Part 1 (Kay) 9.50
Locomotive Management - Cleaning, Driving, Maintenance Part 2 (Kay) . 8.95
Railway Signal Engineering - Mechanical (Kay) 12.50
Railway Carriages & Wagons (Kay) 8.95

Bargain Books *

Today's Railways Review of the Year Volume 1 (was 11.95) 6.95
Today's Railways Review of the Year Volume 3 (was 13.95) 6.95
The Handbook of British Railways Steam Motive Power Depots
    Volume 2 - Central England, East Anglia & Wales (was 8.95).. 3.95
    Volume 4 - Northern England & Scotland (was 9.95)............ 3.95
North West Rails in Colour (was 8.50) 3.95
The Battle for the Settle & Carlisle (was 6.95) 2.95
The Fifty 50s in Colour (was 5.95) 2.95
Steam Alive (Friends of the NRM) 2.95
* Postage on these titles must be based on original book price.

Buses and Fire Engines

The First RTs (Capital) 19.95
National Express Handbook (Capital) 7.95
Buses In Britain 2 (Capital) 19.95
London Trolleybus Routes (Capital) 18.95
Greater Manchester Buses (Capital) 19.95
London Buses Before the War (Capital) 19.95
Routemaster Volume 1 (Capital) 19.95

Last Years of the General (Capital) 16.95
South East Bus Handbook (Capital) 12.50
Routemaster Handbook (Capital) 10.95
London's Utility Buses (Capital) 19.95
London's Wartime Gas Buses (Capital) 5.95
The 1997 Firstbus Bus Handbook (British Bus) 12.50
The Fire Brigade Handbook 2nd edition (British Bus) 9.95
Fire Brigade Handbook Special Appliances Vol. 1 (British Bus) .. 12.50
Yorkshire Bus Handbook (British Bus) 12.50
Ireland & Islands Bus Handbook (British Bus) 9.95
The Scottish Bus Handbook (British Bus) 12.50
The South Midlands Bus Handbook (British Bus) 9.95
The East Midlands Bus Handbook (British Bus) 9.95
The South Wales Bus Handbook (British Bus) 9.95
The North & West Wales Bus Handbook (British Bus) 9.95
The North & West Midlands Bus Handbook (British Bus) 9.95
The Lancashire, Cumbria & Manchester Bus Handbook (British Bus) ... 9.95
The Merseyside & Cheshire Bus Handbook (British Bus) 9.95
The North East Bus Handbook (British Bus) 9.95
The Eastern Bus Handbook (British Bus) 9.95
The Leyland Lynx Bus Handbook (British Bus) 8.95
The Model Bus Handbook (British Bus) 9.95
The Toy & Model Bus Handbook 1 - Early Diecast Models (British Bus) .. 9.95
The Hong Kong Bus Handbook 1997 (British Bus) 9.95
Bus Review 11 (Bus Enthusiast) 6.95
Bus Review 12 (Bus Enthusiast) 7.50
Holidays By Coach (Bus Enthusiast) 11.95

Locomotives & Coaching Stock Book Back Numbers

1986 .... 3.30	1992 .... 7.00
1987 .... 3.30	1993 .... 7.25
1988 .... 3.95	1994 .... 7.50
1989 .... 4.95	1995 .... 8.50
1990 .... 5.95	1996 .... 8.95
1991 .... 6.60	1997 .... 9.95

HOW TO ORDER

All these and many other publications are available from our Mail Order Department. Please send your sterling cheque (drawn on a UK bank), money order, Eurocheque, or British Postal Order payable to 'Platform 5 Publishing Ltd' to:

Platform 5 Mail Order Department (LCS)
3 Wyvern House, Sark Road
SHEFFIELD, S2 4HG, ENGLAND

Tel: (+44) 0114 255 2625
Fax: (+44) 0114 255 2471

Postage & Packing

For postage & packing please add: 10% UK (2nd Class); 20% Europe (Airmail); 30% Rest of World (Airfreight); 50% Rest of World (Airmail). If p&p works out at less than 40p, then please send 40p, this is the minimum post & packing accepted.

* Postage on reduced price titles must be based on original price.

NOTE. When ordering publications in conjunction with a **Today's Railways** subscription offer please add on post and packing **before** deducting the voucher. Vouchers may **not** be combined.

Credit card orders can be accepted by post, 'phone or fax:
from UK - Tel 0114 255 2625, or Fax 0114 255 2471;
from Overseas - Tel +44 114 255 2625, or Fax: +44 114 255 2471.
Please state type of card, card holders name and address, card number and the expiry date.

Please note that we cannot accept foreign currency cheques.

Details correct as at 31st March 1998. Prices are not guaranteed, we reserve the right to alter details without further notification. Please allow 28 days for delivery in the UK.

FOR ORDER FORM SEE OVER

Quantity	Title	Price	Total
		SUB-TOTAL	
	Postage & Packing (see previous page for details)		
		TOTAL REMITTANCE	

Name: ....

Address: ....

....

.... Postcode: ....

Daytime Telephone Number: ....

Payment (Delete as appropriate):
I enclose my cheque/postal order for £ .... made
payable to: 'PLATFORM 5 PUBLISHING LTD'.
Please debit my Visa/Delta/Mastercard/Access/Eurocard

Card No.: .... Expiry Date: ....

Signature: .... Date: ....